# Communications in Computer and Information Science 1923

**Rationale**

The CCIS series is devoted to the publication of proceedings of computer science conferences. Its aim is to efficiently disseminate original research results in informatics in printed and electronic form. While the focus is on publication of peer-reviewed full papers presenting mature work, inclusion of reviewed short papers reporting on work in progress is welcome, too. Besides globally relevant meetings with internationally representative program committees guaranteeing a strict peer-reviewing and paper selection process, conferences run by societies or of high regional or national relevance are also considered for publication.

**Topics**

The topical scope of CCIS spans the entire spectrum of informatics ranging from foundational topics in the theory of computing to information and communications science and technology and a broad variety of interdisciplinary application fields.

**Information for Volume Editors and Authors**

Publication in CCIS is free of charge. No royalties are paid, however, we offer registered conference participants temporary free access to the online version of the conference proceedings on SpringerLink (http://link.springer.com) by means of an http referrer from the conference website and/or a number of complimentary printed copies, as specified in the official acceptance email of the event.

CCIS proceedings can be published in time for distribution at conferences or as post-proceedings, and delivered in the form of printed books and/or electronically as USBs and/or e-content licenses for accessing proceedings at SpringerLink. Furthermore, CCIS proceedings are included in the CCIS electronic book series hosted in the SpringerLink digital library at http://link.springer.com/bookseries/7899. Conferences publishing in CCIS are allowed to use Online Conference Service (OCS) for managing the whole proceedings lifecycle (from submission and reviewing to preparing for publication) free of charge.

**Publication process**

The language of publication is exclusively English. Authors publishing in CCIS have to sign the Springer CCIS copyright transfer form, however, they are free to use their material published in CCIS for substantially changed, more elaborate subsequent publications elsewhere. For the preparation of the camera-ready papers/files, authors have to strictly adhere to the Springer CCIS Authors' Instructions and are strongly encouraged to use the CCIS LaTeX style files or templates.

**Abstracting/Indexing**

CCIS is abstracted/indexed in DBLP, Google Scholar, EI-Compendex, Mathematical Reviews, SCImago, Scopus. CCIS volumes are also submitted for the inclusion in ISI Proceedings.

**How to start**

To start the evaluation of your proposal for inclusion in the CCIS series, please send an e-mail to ccis@springer.com.

Haofen Wang · Xianpei Han · Ming Liu ·
Gong Cheng · Yongbin Liu · Ningyu Zhang
Editors

# Knowledge Graph and Semantic Computing

## Knowledge Graph Empowers Artificial General Intelligence

8th China Conference, CCKS 2023
Shenyang, China, August 24–27, 2023
Revised Selected Papers

*Editors*
Haofen Wang
Tongji University
Shanghai, China

Xianpei Han
Chinese Academy of Sciences
Beijing, China

Ming Liu
Harbin Institute of Technology
Harbin, China

Gong Cheng
Nanjing University
Nanjing, China

Yongbin Liu
University of South China
Hengyang, China

Ningyu Zhang
Zhejiang University
Hangzhou, China

ISSN 1865-0929 ISSN 1865-0937 (electronic)
Communications in Computer and Information Science
ISBN 978-981-99-7223-4 ISBN 978-981-99-7224-1 (eBook)
https://doi.org/10.1007/978-981-99-7224-1



This Springer imprint is published by the registered company Springer Nature Singapore Pte Ltd.
The registered company address is: 152 Beach Road, #21-01/04 Gateway East, Singapore 189721, Singapore

Paper in this product is recyclable.

# Preface

This volume contains the papers presented at CCKS 2023: the China Conference on Knowledge Graph and Semantic Computing held during August 24–27, 2023, in Shenyang.

CCKS is organized by the Technical Committee on Language and Knowledge Computing of the Chinese Information Processing Society, and has previously been held in Beijing (2016), Chengdu (2017), Tianjin (2018), Hangzhou (2019), Nanchang (2020), Guangzhou (2021), and Qinhuangdao (2022). CCKS is the merger of two previously held relevant forums, i.e., the Chinese Knowledge Graph Symposium (CKGS) and the Chinese Semantic Web and Web Science Conference (CSWS). CKGS was previously held in Beijing (2013), Nanjing (2014), and Yichang (2015). CSWS was first held in Beijing in 2006 and has been the main forum for research on Semantic (Web) technologies in China for a decade. Since 2016, CCKS has brought together researchers from both forums and covers wider fields, including the knowledge graph, the Semantic Web, linked data, natural language processing, knowledge representation, graph databases, information retrieval and knowledge-aware machine learning. It aims to become the top forum on knowledge graph and semantic technologies for Chinese researchers and practitioners from academia, industry, and government.

The theme of this year is *Knowledge Graph Empowers Artificial General Intelligence*. Enclosing this theme, the conference scheduled various activities, including keynotes, academic forums, industrial forums, tutorials, paper presentations, evaluations, etc. The conference invited Qinghua Zheng (Professor, President of Tongji University, National Outstanding Youth, Chang Jiang Scholar), Jirong Wen (the dean of the School of Information at Renmin University of China and the executive dean of Gaoling School of Artificial Intelligence), and Denny Vrandečić (the founder of Wikidata, co-creator of Semantic MediaWiki, and former elected member of the Wikimedia Foundation Board of Trustees) to present the latest progress and development trends in Knowledge Engineering, Large Language Models, and free knowledge graphs, respectively. The conference also invited industrial practitioners to share their experiences and promote industry-university-research cooperation.

As for peer-reviewed papers, 106 submissions were received in the following six areas,

- Knowledge Representation and Knowledge Graph Reasoning
- Knowledge Acquisition and Knowledge Base Construction
- Knowledge Integration and Knowledge Graph Management
- Natural Language Understanding and Semantic Computing
- Knowledge Graph Applications
- Knowledge Graph Open Resources

During the reviewing process, each submission was adopting a double-blind peer review process to at least three Program Committee members. The committee decided to accept 43 full papers (20 papers in English). The CCIS volume contains revised versions of 20 English full papers.

Additionally, the evaluation track this year set up 4 topics and 7 tasks, which attracted more than 2771 teams to participate, forming a highly influential event. Besides the bonuses and issued certificates for the top three teams in each task, the committee also encouraged them to submit evaluation papers. After peer review by experienced researchers and task organizers, 8 papers were accepted (after revision) for inclusion in this volume of proceedings.

The hard work and close collaboration of a number of people have contributed to the success of this conference. We would like to thank the Organizing Committee and Program Committee members for their support, and the authors and participants who are the primary reason for the success of this conference. We also thank Springer for their trust and for publishing the proceedings of CCKS 2023.

Finally, we appreciate the sponsorships from Global Tone Communication Technology as platinum sponsor, Baidu, Ant Group, Meituan, 360, Du Xiaoman as gold sponsors, Top AI, Vesoft, Yunfu Technology, Zhipu AI, NiuTrans, Qutke, and HashData as the silver sponsors.

August 2023

Haofen Wang
Xianpei Han
Ming Liu
Gong Cheng
Yongbin Liu
Ningyu Zhang

# Organization

## General Chairs

| | |
|---|---|
| Haofen Wang | Tongji University, China |
| Xianpei Han | Institute of Software, Chinese Academy of Sciences, China |

## Program Committee Chairs

| | |
|---|---|
| Ming Liu | Harbin Institute of Technology, China |
| Gong Cheng | Nanjing University, China |

## Local Chairs

| | |
|---|---|
| Gang Wu | Northeastern University, China |
| Junchang Xin | Northeastern University, China |
| Bin Wang | Northeastern University, China |
| Xiaochun Yang | Northeastern University, China |

## Publicity Chairs

| | |
|---|---|
| Xiaowang Zhang | Tianjin University, China |
| Xianling Mao | Beijing Institute of Technology, China |
| Yongpan Sheng | Southwest University, China |

## Publication Chairs

| | |
|---|---|
| Yongbin Liu | University of South China, China |
| Ningyu Zhang | Zhejiang University, China |

## Tutorial Chairs

| | |
|---|---|
| Xin Wang | Tianjin University, China |
| Tao Gui | Fudan University, China |

## Evaluation Chairs

| | |
|---|---|
| Lei Hou | Tsinghua University, China |
| Yuanzhe Zhang | Institute of Automation, Chinese Academy of Sciences, China |
| Tianxing Wu | Southeast University, China |

## Frontiers and Trends Forum Chair

| | |
|---|---|
| Zhixu Li | Fudan University, China |

## Young Scholar Forum Chair

| | |
|---|---|
| Wei Hu | Nanjing University, China |

## Poster/Demo Chairs

| | |
|---|---|
| Weiguo Zheng | Fudan University, China |
| Feiliang Ren | Northeastern University, China |

## Sponsorship Chairs

| | |
|---|---|
| Lei Zou | Peking University, China |
| Qiang Zhang | Zhejiang University, China |
| Tong Xiao | Northeastern University, China |

## Industry Track Chairs

| | |
|---|---|
| Jinlong Li | China Merchants Bank, China |
| Xiao Ding | Harbin Institute of Technology, China |

## Website Chairs



## Area Chairs

### Knowledge Representation and Knowledge Graph Reasoning



### Knowledge Acquisition and Knowledge Graph Construction



### Knowledge Integration and Knowledge Graph Management



### Natural Language Understanding and Semantic Computing



### Knowledge Graph Applications (Semantic Search, Question Answering, Dialogue, Decision Support, and Recommendation)

**Knowledge Graph Open Resources**

| | |
|---|---|
| Baotian Hu | Harbin Institute of Technology (Shenzhen), China |
| Gang Wu | KG Data, China |
| Fu Zhang | Northeastern University, China |

## Program Committee

| | |
|---|---|
| Ben Zheng | LinkedIn Corporation, USA |
| Bi Sheng | Southeast University, China |
| Bo Chen | Minzu University of China, China |
| Bo Xu | Donghua University, China |
| Chao Che | Dalian University, China |
| Chen Zhu | Baidu Talent Intelligence Center, China |
| Chengzhi Zhang | Nanjing University of Science and Technology, China |
| Chuanjun Zhao | Shanxi University of Finance and Economics, China |
| Cunli Mao | Kunming University of Science and Technology, China |
| Daifeng Li | Baidu, China |
| Fake Lin | University of Science and Technology of China, China |
| Fan Xu | Jiangxi Normal University, China |
| Fei Li | UMass, USA |
| Fu Zhang | Northeastern University, China |
| Gang Wu | Northeastern University, China |
| Hai Wan | Sun Yat-sen University, China |
| Haitao Cheng | Nanjing University of Posts and Telecommunications, China |
| Haiyun Jiang | Tencent, China |
| Haochun Wang | Harbin Institute of Technology, China |
| Hongye Tan | Shanxi University, China |
| Hongyu Wang | Wuhan University, China |
| Huaiyu Wan | Beijing Jiaotong University, China |
| Huiying Li | Southeast University, China |
| Huxiong Chen | Dalian University of Technology, China |
| Jia Li | Peking University, China |
| Jia-Chen Gu | University of Science and Technology of China, China |
| Jian Liu | Beijing Jiaotong University, China |

| | |
|---|---|
| Jiaoyan Chen | University of Oxford, UK |
| Jiaqi Li | Harbin Institute of Technology, China |
| Jie Hu | Southwest Jiaotong University, China |
| Jiezhong Qiu | Tsinghua University, China |
| Jifan Yu | Tsinghua University, China |
| Jihao Shi | HIT, China |
| Jing Zhang | Renmin University of China, China |
| Jingchang Chen | Harbin Institute of Technology, China |
| Jingping Liu | East China University of Science and Technology, China |
| Jingwei Cheng | Northeastern University, China |
| Jun Pang | Wuhan University of Science and Technology, China |
| Junshuang Wu | Beihang University, China |
| Junwen Duan | Central South University, China |
| Junyong Chen | Shanghai Jiao Tong University, China |
| Kai Zhang | Zhejiang University City College, China |
| Kang Xu | Nanjing University of Posts and Telecommunications, China |
| Li Du | Harbin Institute of Technology, China |
| Li Jin | Aerospace Information Institute, Chinese Academy of Sciences, China |
| Liang Hong | Wuhan University, China |
| Liang Pang | Institute of Computing Technology, Chinese Academy of Sciences, China |
| Linhao Luo | Monash University, Australia |
| Lishuang Li | Dalian University of Technology, China |
| Ma Chao | USTC, China |
| Maofu Liu | Wuhan University of Science and Technology, China |
| Meng Wang | Tongji University, China |
| Ming Jin | Monash University, Australia |
| Mingyang Chen | Zhejiang University, China |
| Muzhen Cai | Harbin Institute of Technology, China |
| Ningyu Zhang | Zhejiang University, China |
| Peng Bao | Beijing Jiaotong University, China |
| Peng Furong | Shanxi University, China |
| Peng Peng | Hunan University, China |
| Peng Zhang | Tianjin University, China |
| Qing Liu | CSIRO, Australia |
| Qu Shilin | Monash University, Australia, China |
| Ruifang He | Tianjin University, China |

| | |
|---|---|
| Ruijie Wang | University of Zurich, Switzerland |
| Shi Feng | Northeastern University, China |
| Sihang Jiang | Fudan University, China |
| Tianxing Wu | Southeast University, China |
| Tong Ruan | ECUST, China |
| Tong Xu | University of Science and Technology of China, China |
| Wang Zhiqiang | Shanxi University, China |
| Wei Du | Renmin University of China, China |
| Wei Shen | Nankai University, China |
| Weixin Zeng | National University of Defense Technology, China |
| Weizhuo Li | Nanjing University of Posts and Telecommunications, China |
| Wen Zhang | Zhejiang University, China |
| Wenliang Chen | Soochow University, China |
| Wenpeng Lu | Qilu University of Technology (Shandong Academy of Sciences), China |
| Wenqiang Liu | Tencent, China |
| Xiang Zhang | Southeast University, China |
| Xiang Zhao | National University of Defense Technology, China |
| Xiangyu Liu | Nanjing University, China |
| Xiaohui Han | Shandong Computer Science Center, China |
| Xiaowang Zhang | Tianjin University, China |
| Xiaoyu Guo | Beihang University, China |
| Xiaozhi Wang | Tsinghua University, China |
| Xin Chen | Taiyuan University of Science and Technology, China |
| Xin Xin | Beijing Institute of Technology, China |
| Xinhuan Chen | Tencent, China |
| Xinyi Wang | Nanjing University, China |
| Xiuying Chen | KAUST, China |
| Xu Lei | Wuhan University, China |
| Xue-Feng Xi | Suzhou University of Science and Technology, China |
| Yanan Cao | Institute of Information Engineering Chinese Academy of Science, China |
| Yang Liu | Nanjing University, China |
| Yanling Wang | Renmin University of China, China |
| Yao Li | Shanghai Jiao Tong University, China |
| Yaojia Lv | Harbin Institute of Technology, China |

| | |
|---|---|
| Yaojie Lu | Institute of Software, Chinese Academy of Sciences, China |
| Yi Zheng | Huawei Cloud, China |
| Yinglong Ma | NCEPU, China |
| Yixin Cao | National University of Singapore, Singapore |
| Yongpan Sheng | Southwest University, China |
| Youhuan Li | Hunan University, China |
| Yuan Yao | Tsinghua University, China |
| Yuanfang Li | Monash University, Australia |
| Yuanlong Wang | Shanxi University, China |
| Yuanning Cui | Nanjing University, China |
| Yuantong Li | UCLA, USA |
| Yufei Wang | Macquarie University, Australia |
| Yun Chen | Shanghai University of Finance and Economics, China |
| Yuncheng Hua | Southeast University, China |
| Yunxin Li | Harbin Institute of Technology, Shenzhen, China |
| Yuqiang Han | Zhejiang University, China |
| Yuxia Geng | Zhejiang University, China |
| Yuxiang Wang | Hangzhou Dianzi University, China |
| Yuxin Wang | Harbin Institute of Technology, China |
| Zequn Sun | Nanjing University, China |
| Zhenfang Zhu | Shandong Jiaotong University, China |
| Zhiliang Tian | National University of Defense Technology, China |
| Zhixu Li | Fudan University, China |
| Zhiyi Luo | Zhejiang Sci-Tech University, China |
| Zitao Wang | Nanjing University, China |
| Ziyang Chen | National University of Defense Technology, China |

## Sponsors

### Platinum Sponsor

## Gold Sponsors

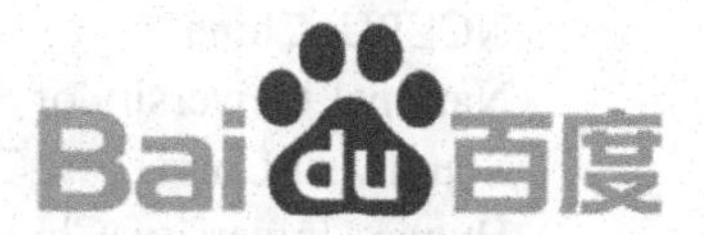

美团 Meituan

## Silver Sponsors

云孚科技

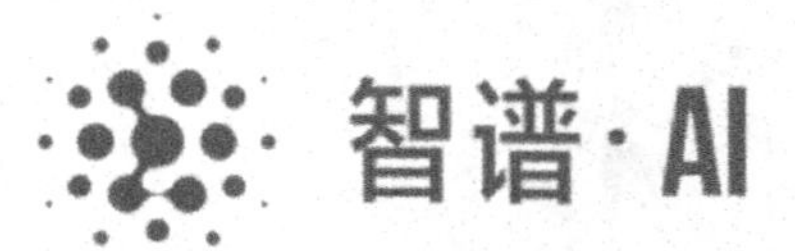
智谱·AI

小牛翻译
NiuTrans.com

QUTKE
况客科技

HashData

# Contents

**Knowledge Representation and Knowledge Graph Reasoning**



**Knowledge Acquisition and Knowledge Base Construction**



**Knowledge Integration and Knowledge Graph Management**

## Natural Language Understanding and Semantic Computing



## Knowledge Graph Applications



## Knowledge Graph Open Resources

# Knowledge Representation and Knowledge Graph Reasoning

# Dynamic Weighted Neural Bellman-Ford Network for Knowledge Graph Reasoning

Huanxuan Liao[1,2], Shizhu He[1,2(✉)], Yao Xu[1,2], Kang Liu[1,2], and Jun Zhao[1,2]

[1] The Laboratory of Cognition and Decision Intelligence for Complex Systems Institute of Automation, Chinese Academy of Sciences, Beijing, China
liaohuanxuan2023@ia.ac.cn, {shizhu.he,yao.xu,kliu,jzhao}@nlpr.ia.ac.cn
[2] School of Artificial Intelligence, University of Chinese Academy of Sciences, Beijing, China

**Abstract.** Recent studies have shown that subgraphs of the head entity, such as related relations and neighborhoods, are helpful for Knowledge Graph Reasoning (KGR). However, prior studies tend to focus solely on enhancing entity representations using related relations, with little attention paid to the impact of different relations on different entities and their importance in various reasoning paths. Meanwhile, conventional Graph Neural Networks (GNNs) utilized for KGR consider simultaneously neighboring nodes and connected relations of the head entity but typically use a standard message-passing paradigm over the entire Knowledge Graph (KG). This results in over-smoothed representations and limits efficiency. To address the above-mentioned limitations of existing methods, we propose a Dynamic Weighted Neural Bellman-Ford Network (DyNBF) for KGR, which utilizes relation weights generated from subgraphs to compute only the most relevant relations and entities. This way, we can integrate multiple reasoning paths more flexibly to achieve better interpretable reasoning, while scaling more easily to more complex and larger KGs. DyNBF consists of two key modules: 1) a transformer-based relation weights generator module, which computes the weights of different relations on the path with a sequence-to-sequence model, and 2) an NBFNet-based logic reasoner module, which obtains entity representations and conducts fact prediction with dynamic weights from the previous module. Empirical results on three standard KGR datasets demonstrate that the proposed approach can generate explainable reasoning paths and obtain competitive performance.

**Keywords:** Knowledge Graph Reasoning · NBFNet · Dynamic weights · Path interpretations

## 1 Introduction

Knowledge Graphs (KGs) store a large number of facts, containing entities and rich structural relations, which are usually in the form of ($h$, $r$, $t$), where $h$ is the head entity, $r$ is the relation, and $t$ is the tail entity. Popular public KGs


H. Wang et al. (Eds.): CCKS 2023, CCIS 1923, pp. 3–16, 2023.
https://doi.org/10.1007/978-981-99-7224-1_1

include Freebase [2], WordNet [12] etc. KGs are widely applied to a variety of applications such as question answering [4], recommender systems [11] etc.

However, due to the influence of artificial construction and information extraction technology, the currently constructed KGs suffer from incompleteness [21]. For example, 66% of people are lacking the relationship of birthplace in Freebase. To enhance the completeness of the KGs, researchers have proposed Knowledge Graph Reasoning (KGR) to discover missing facts. Recent advances in KGR primarily work on knowledge graph embedding (KGE) by mapping each entity and relation into a low-dimensional vector [3,18,23]. Despite steady progress in developing novel algorithms for KGE, researchers still face a common challenge: encoding all information (e.g. semantic and structural) about an entity into a single vector. To this end, several studies utilize Graph Neural Networks (GNNs) [15,26,27] or attention-based approaches [20,22] to learn representations for KGR based on both entities and their graph context. However, these methods have limited expressiveness due to their shallow network architectures and over-reliance on message passing throughout the entire KGs, which hampers its scalability for larger KGs. This also results in slower inference speed.

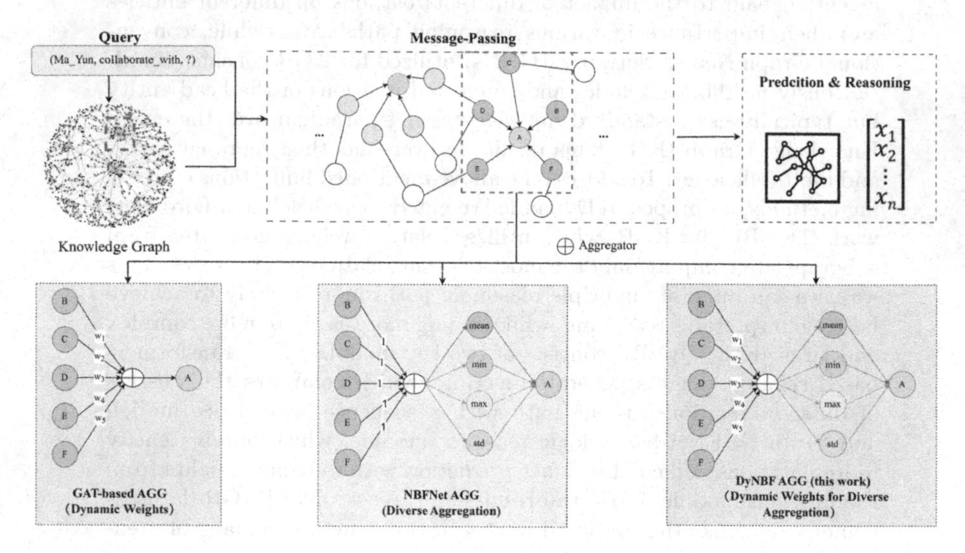


**Fig. 1.** Overview of the proposed DyNBF and comparison with GAT-Based and NBFNet in aggregation.

Intuitively, for each specific query, e.g., (*Ma_Yun*, *collaborate_with*, ?), only a small sub-graph [13] of the entire knowledge graph and specific relation types like *collaborate_with* may be relevant. Relations like *award* or *spouse* are not useful and may introduce bias. Using relevant subgraph information as context, more important relations for a given query can be identified and thus assigned higher weights, thereby facilitating accurate and efficient inference.

To make better use of relevant relations and subgraph information. GAT-based aggregation methods [25] use attention weights to learn node representations that are more informative than those generated by traditional GNNs without attention weights. However, these methods aggregate information in a single way, limiting their flexibility in understanding multiple properties of nodes. To solve the problem of a single aggregation method, the NBFNet [27] jointly learn the types and scales of the aggregation functions used to combine neighboring nodes' representations which are pair representations conditioned on the source node. This enables GNNs to capture multiple properties of nodes simultaneously and is critical for solving complex graph problems. But instead of selecting relevant relations for different queries for reasoning, it aggregates all the surrounding information. To address the above-mentioned issues, we explore the combination of Transformer architecture and GNNs for KGR with dynamic weights. The proposed **Dy**namic Weighted **N**eural **B**ellman-**F**ord Network (**DyNBF**) (see Fig. 1 for an overview and comparison with existing GNNs works above mentioned) consists of two main components: 1) a *weights generator* that identifies the relevant KG sub-graph as context and generate the weights dynamically, and 2) a *logic reasoner* that jointly considers the KGs and the relation weights for inferring answers. In such a way, we can simply leverage an optimization object for KGR with high efficiency, accuracy, and interpretability.

We evaluate DyNBF by conducting experiments on three popular benchmarks (WN18RR [6], UMLS [8], FB15k-237 [17]). The efficacy of DyNBF is demonstrated by the improvements observed in both performance and reduced time.

## 2 Related Work

Existing work on KGR can be generally classified into three main paradigms: path-based methods, embedding methods, and graph neural networks methods.

### 2.1 Path-Based Methods

Random walk inferences have been extensively researched in this type of model. For example, the PathRanking Algorithm (PRA) [9] derives the path-based logical rules under the constraints of the path. Rule mining methods such as AMIE [7] use predefined metrics such as confidence and support to prune incorrect rules. More recently, differentiable rule learning methods based on TensorLog [5] such as Neural-LP [24] and DRUM [14], learning probabilistic logical rules to weight different paths. Ruleformer [22] introduced a transformer-based rule mining approach to choose suitable rules for a query in a differentiable rule mining process. However, these approaches employed the Tensorlog framework resulting in inefficiency. Additionally, they require exploration of an exponentially large number of paths and are restricted to extremely brief paths, such as less than or equal to 3 edges.

### 2.2 Embedding Methods

Embedding methods learn a distributed representation for each node and edge by preserving the edge structure of the graph. Most of them like TransE [3],

ComplEx [18], DistMult [23] etc, design a scoring function to get a value for a triplet in the embedding space. Among them, TransE [3] is the most widely used KGE method, which views the relation as a translation from a head entity to a tail entity. DistMult [23] employs a rudimentary bilinear formulation and develops a novel method that leverages the acquired relation embeddings to extract logical rules. ComplEx [18] creates a composition of intricate embeddings that exhibit robustness towards a wide range of binary relations. Despite the simplicity of these works, they demonstrated impressive performance on reasoning tasks. However, these models often ignore the neighborhood sub-graphs of the entities and relations.

### 2.3 Graph Neural Networks Methods

GNNs encode topological structures of graphs and most of these frameworks encode node representations and decode edges as a function over node pairs. RGCN [15] deals with the highly multi-relational data characteristic of realistic KGs. The NBFnet [27] proposed a general and flexible representation learning framework for KGR based on the paths between two nodes to enhance generalization in the inductive setting, interpretability, high model capacity, and scalability. This framework's ability may be constrained by viewing every relation in the KGs equally for different queries. Such a constraint can impede capturing of rich information and be affected by biased training data, neglecting marginalized entities and leading to stereotyped predictions.

## 3 Methodology

**Knowledge Graph.** A knowledge graph $\mathcal{G}$ can be defined as a collection of facts $\mathcal{F}$, i.e., $\mathcal{G} = \{\mathcal{F}\}$. Each fact $f$ is a triplet like $(h, r, t)$, $h, t \in \mathcal{E}$, $r \in \mathcal{R}$, where $\mathcal{E}$ is a countable set of entities and $\mathcal{R}$ is a set of relations, respectively.

**Link Prediction.** The objective of link prediction is to deduce the absent relations amidst the entities in a given KG. To accomplish this, the model has to rank the target entity amid the group of candidate entities, in response to a link prediction question, either $(h, r, ?)$ or $(?, r, t)$.

We propose to reason answers by taking into account not only the relation weights but also the types and scales of the aggregation function used. To this end, we introduce our DyNBF to determine the weights of all relations at each step, which can be utilized as edge weights for aggregation, and Fig. 2 shows the details. Specifically, with regard to a query like $(h, r, t)$, the sub-graph related to the head entity $h$ is constructed from the KG. Then, the global relational weights for each aggregation are obtained using the Transformer weights generator, which incorporates a relational attention mechanism. Finally, these global relational weights are assigned to the edge weights in NBFNet to enable accurate logical inference and prediction. The details of the two components are respectively shown in Sect. 3.1 and Sect. 3.2.

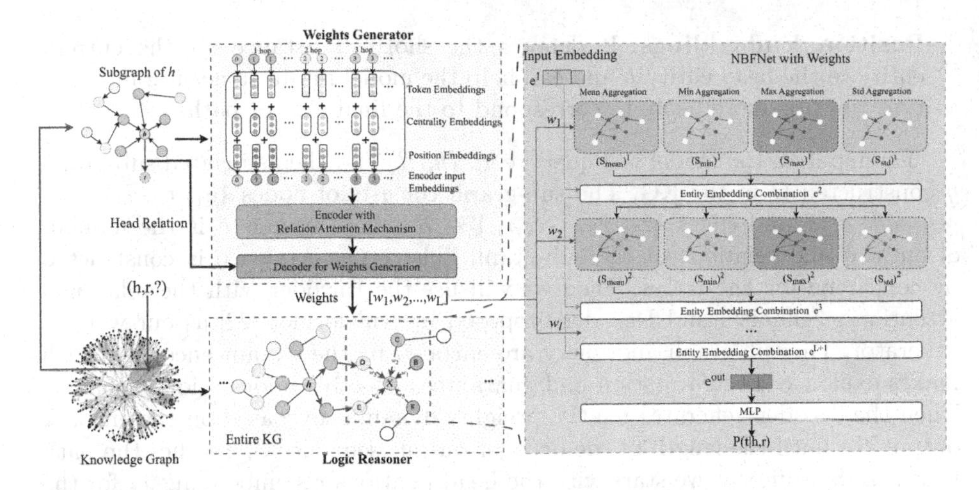


**Fig. 2.** The DyNBF framework consists of two primary components: a weights generator and a logic reasoner. The green node $X$ in the logic reasoner part of the figure represents our required representation, which aggregates information about the surrounding nodes, while the red arrows depict the different paths from the source node to $X$. Different colored nodes represent different distances from the source node, e.g., yellow nodes represent 2 hops away from source node $h$ and the purple node represents the distance from the $h$ greater than or equal to 3. (Color figure online)

### 3.1 Weights Generator

The role of the weights generator module is to generate the weights dynamically by leveraging the sub-graph of the head entity. Inspired by Ruleformer [22], we utilize contextualized sub-graph as input sequences to encode the local structural information.

We use a breadth-first approach to sample edges from the source entity until the specified context budget is reached, which is noteworthy for getting the sub-graph to enrich the query by considering relevant contextual information. To convert the graph into a sequence for Transformer input, we employ a summation method to combine the token embeddings, centrality embeddings, and position embeddings (see Fig. 2):

- **Token Embeddings**: We initialize randomly all entities and relations in the KG.
- **Centrality Embeddings**: The distribution of entity occurrences is characterized by a heavy-tailed pattern that poses a challenge to the development of suitable entity representations. In light of this, we use the degree of centrality as an additional signal to the neural network. To be specific, we develop a Centrality Encoding which assigns each node two real-valued embedding vectors according to its indegree and outdegree. As the centrality encoding is applied to each node, we simply add it to the node features as the input.

- **Position Embeddings**: It denotes the shortest distance for the current entity to the head entity $h$ and can help the model retain structural information which is more likely to correspond to the final answer paths.

The input to the model is a query, e.g., $(h,r,?)$ and the associated sub-graph $g$ constructed from the KG. The sub-graph consists of nodes $\{e_1, e_2, \cdots, e_m, \cdots pad\} \in \mathcal{E}$ and edges $\{r_1, r_2, \cdots, r_n\} \in \mathcal{R}$ where $m$ and $n$ is the number of entities and relations of the sub-graph. The input sequence is constructed by concatenating the nodes. Then we will use the Encoder with the relational attention mechanism and Decoder proposed by Ruleformer [22] as our weights generator. The sub-graph and query are encoded by the variant encoder, which makes exploit edge information and guarantees effective information exchange. Then the decoder generates relation weights dynamically based on the encoder output $S_e$, until the length of the decoder output sequence $S_r$ reaches the path length $L$. Specifically, we start with the head relations $r$'s embedding $x_r$ for the path sequence $S_r$ input to the decoder, which means $S_r^0 = x_r$. Consequently, a feed-forward neural network is employed to determine the weights $w_l^i$ of relation $r_i$ in step $l$. More formally, let $w_{l+1} \in \mathbb{R}^{|\mathcal{R}|+1}$ denote relation weights including the self-loop relation.

$$w_{l+1} = \sigma(f(RelationAttention(S_r^l, S_e))) \tag{1}$$

where $\sigma(\cdot)$ is the sigmoid function and $f(\cdot)$ is a feed-forward neural network. $S_r^l$ is the concatenation of the head relation and the relation with the greatest weight for each step before the l-step. Let the relation with maximum weight be expressed as $r_{l+1}$, then $S_r^{l+1} = [S_r^l, x_{r_{l+1}}]$.

### 3.2 Logic Reasoner

To effectively answer a link prediction query, we need to perform the prediction from $(h, r, ?)$ to $t$ or from $(?, r, t)$ to $h$. Drawing on previous work, we can think of this process as Knowledge Graph Completion (KGC) task and we can deploy the Tensorlog framework to get the prediction results like Ruleformeror or KGE methods to score triples $(h, r, t)$ like SE-GNN [10]. We choose the NBFNet [27] as our logic reasoner, which learns a pair representation $h_q(h,t)$ with a *generalized sum* of path representations $h_q(\mathcal{P})$ between $h$ and $t$ with a commutative *summation* operator $\oplus$. Additionally, each path representation $h_q(\mathcal{P})$ is defined as a *generalized product* of the edge representations in the path using the *multiplication* operator $\otimes$. After performing the above two operations, each node representation contains information about the cumulative sum of all paths from the source node to that node. Therefore, this information can be used to infer the probability of arrival. In other words, this representation captures the probability of arriving at each node, based on the paths traversed from the source node.

Following NBFNet, the relation embeddings are first randomly initialized, and then the source node $h$ is initialized to the sum of the embeddings of all relations connected to it, while the other nodes are initialized to 0. This initialization facilitates the transfer of information from the source node to each node on the path, which is consistent with our reasoning process. Then we perform message passing with entity embedding, relation embeddings, and relation weights using *Message* function and *Aggregate* function:

$$(S_{a_i})_t^{l+1} = Aggregate(\{Message((S_{a_i})_x^l, S_r, w^l) \mid (x,r,t) \in \mathcal{E}(t)\}) \quad (2)$$

where $a_i \in \{mean, min, max, std\}$ denotes the different aggregators; $(S_{a_i})_t^l$ is the representation of $t$ using $a_i$ aggregator in the $l$-th step; $w^l$ is a weight vector as the edge weights in the $l$-th step; $S_r$ is the representation for edge $e = (x, r, t)$ and $r$ is the relation of the edge, $\mathcal{E}(t)$ denotes the triplets with tail entity t in the KG.

After the first message passing, the representation of each node contains information about all paths of length 1 from the source node to that node. The process is repeated twice to retrieve information up to two hops (the path length from the source node is 2) and so on. Then the combination function combines the node representations of different operators to obtain sufficient information from nodes' neighborhoods, thus enhancing expressive power learning abilities:

$$e^{l+1} = Combine(\{(S_{a_i})_t^l \mid a_i \in \{min, max, mean, std\}\}) \quad (3)$$

where $e^l$ is the node representation after aggregating $l$ hop information. *Combine* is the combination strategy where we concatenate messages from high to low dimensions.

Finally, we can use the final representation obtained after a previously predetermined number of $L$ message passing, and then reason for the query.

### 3.3 Optimizing

We now demonstrate the process of utilizing the learned pair representations $e^{out}$ to solve the link prediction task. Specifically, we aim to estimate the conditional likelihood of the tail entity $t$ given the head entity $h$ and relation $r$ as $p(t|h,r) = \sigma(f(e^{out}))$. Here, $\sigma(\cdot)$ denotes the sigmoid function, while $f(\cdot)$ is a feed-forward neural network. Similarly, the conditional likelihood of the head entity $h$ can be predicted by $p(h|r^{-1}, t) = \sigma(f(e^{out'}))$ with the same model. As prior research [3,16], we minimize the negative log-likelihood of positive and negative triplets (Eq. 4). The negative samples are generated through the corruption of one entity in a positive triplet to produce a negative example.

$$\mathcal{L} = -log\ p(h,r,t) - \sum_{i=1}^{n} \frac{1}{n} log(1 - p(h_i', r, t_i')) \quad (4)$$

where $n$ is the number of negative samples per positive sample and $(h_i', r, t_i')$ is the $i$-th negative samples.

# 4 Experiments

## 4.1 Datasets

We utilize well-established link prediction benchmarks: UMLS [8], FB15K-237 [17], and WN18RR [6] to assess the performance of our model. The summary statistics of the datasets are shown in Table 1.

**Table 1.** This summary of statistics for the three datasets used in this study: WN18RR, FB15K-237, and UMLS.

| Dataset | UMLS | FB15K-237 | WN18RR |
|---|---|---|---|
| Entities | 135 | 14541 | 40943 |
| Relations | 46 | 237 | 11 |
| Train Triplets | 5216 | 272115 | 86835 |
| Valid Triplets | 652 | 17535 | 3034 |
| Test Triplets | 661 | 20466 | 3034 |

## 4.2 Experiment Setup

The experiments are implemented with the PyTorch framework and are trained on RTX3090 GPU for 30 epochs. The Adam optimizer is used with 5 steps of warmups for parameter tuning and the learning rate is set to 0.0001/0.0003/0.0015 respectively for the FB15K-237, UMLS, and WN18RR.

**Implementation Details.** The weights generator model comprises an encoder and a decoder block, both of which are implemented as two-layer Transformers with default settings of six heads per layer. The entity dimension, as well as the position encoding dimension, is set to 200. Dropout is applied with a possibility $p = 0.1$. And the reasoner is the NBFNet with 6 layers, each with 32 hidden units. The feed-forward network $f(\cdot)$ is set to a 2-layer MLP with 64 hidden units. The ReLU activation function was implemented for all hidden layers in order to optimize performance. We select models based on their performance on the validation set.

**Evaluation.** We adopt the filtered ranking protocol proposed in [3] for the Knowledge Graph Completion task. We evaluate the performance using three widely-used metrics: mean rank (MR), mean reciprocal rank (MRR), and Hits at N (Hits@N). Hits@N measures the proportion of correct entities ranked in the top N.

**Baselines.** We compare our model DyNBF against path-based methods including Path Ranking [9], Neural-LP [24], DRUM [14], Ruleformer [22]; Embedding methods like TransE [3], Distmult [23], ComplEx [18] etc.; and GNNs methods such as RGCN [15], CompGCN [19] and NBFNet [27]. There are a total of 13 baselines above for KGR.

### 4.3 Main Results

Table 2 summarizes the results of the KGR. As the table shows, we observe that our proposed approach can achieve competitive performance on all metrics and both datasets compared with baselines. DyNBF has obvious improvement compared to Ruleformer, which is a typical recent path-based method model. This suggests that NBFNet can combine a greater amount of path information in order to perform inference. The results demonstrate a significant improvement over NBFNet, particularly on the FB15-237 dataset, when attention weights are incorporated. This emphasizes that incorporating attention weights can assist NBFNet in focusing on relevant information and improving inferences, rather than simply gathering all information without discrimination.

**Table 2.** Knowledge graph reasoning results. The best results are in bold. † denotes that we run these methods with the same evaluation process in our way. ‡ means that the results are from [22]. Other results are from the published paper.

| Class | Method | UMLS | | | | | FB15K-237 | | | | | WN18RR | | | | |
|---|---|---|---|---|---|---|---|---|---|---|---|---|---|---|---|---|
| | | MR | MRR | Hits | | | MR | MRR | Hits | | | MR | MRR | Hits | | |
| | | | | @1 | @3 | @10 | | | @1 | @3 | @10 | | | @1 | @3 | @10 |
| Path-based | Path Ranking [9] ‡ | - | 0.197 | 0.147 | 0.256 | 0.376 | 3521 | 0.174 | 0.119 | 0.186 | 0.285 | 22438 | 0.324 | 0.276 | 0.360 | 0.406 |
| | NeuralLP(l = 3) [24] ‡ | - | 0.735 | 0.627 | 0.820 | 0.923 | - | 0.239 | 0.160 | 0.261 | 0.399 | - | 0.425 | 0.394 | 0.432 | 0.492 |
| | DRUM(l = 3) [24] ‡ | - | 0.784 | 0.643 | 0.912 | 0.972 | - | 0.328 | 0.247 | 0.362 | 0.499 | - | 0.441 | 0.412 | 0.456 | 0.516 |
| | Ruleformer(l = 3) [22] | - | 0.857 | 0.752 | 0.958 | 0.984 | - | 0.342 | 0.255 | 0.374 | 0.513 | - | 0.452 | 0.417 | 0.465 | 0.530 |
| Embeddings | TransE [3] | 1.84 | 0.668 | 0.468 | 0.845 | 0.930 | 357 | 0.294 | - | - | 0.465 | 3384 | 0.226 | - | - | 0.501 |
| | DistMult [23] | 5.52 | 0.753 | 0.651 | 0.821 | 0.930 | 254 | 0.241 | 0.155 | 0.263 | 0.419 | 5110 | 0.430 | 0.390 | 0.440 | 0.490 |
| | ComplEx [18] | 2.59 | 0.829 | 0.748 | 0.897 | 0.961 | 339 | 0.247 | 0.158 | 0.275 | 0.428 | 5261 | 0.440 | 0.410 | 0.460 | 0.510 |
| | ConvE [6] | 1.51 | 0.908 | 0.862 | 0.944 | 0.981 | - | 0.325 | 0.237 | 0.356 | 0.501 | - | 0.430 | 0.400 | 0.440 | 0.520 |
| | TuckER [1] | - | - | - | - | - | - | 0.358 | 0.266 | 0.394 | 0.544 | - | 0.470 | 0.443 | 0.482 | 0.526 |
| | RotatE [16] | - | 0.948 | 0.914 | **0.960** | **0.994** | 177 | 0.338 | 0.241 | 0.375 | 0.553 | 3340 | 0.476 | 0.428 | 0.492 | 0.571 |
| GNNs | RGCN [15] | - | - | - | - | - | 221 | 0.273 | 0.182 | 0.303 | 0.456 | 2719 | 0.402 | 0.345 | 0.437 | 0.494 |
| | CompGCN [24] | - | 0.735 | 0.627 | 0.820 | 0.923 | 197 | 0.355 | 0.204 | 0.390 | 0.535 | 3533 | 0.479 | 0.443 | 0.494 | 0.546 |
| | NBFNet(l = 3) [27] † | 2.21 | 0.902 | 0.857 | 0.937 | 0.976 | 123 | 0.407 | 0.314 | 0.446 | 0.592 | 1195 | 0.519 | 0.473 | 0.540 | 0.609 |
| | NBFNet(l = 6) [27] ‡ | 1.60 | 0.933 | 0.900 | 0.956 | 0.987 | 114 | 0.415 | 0.321 | 0.454 | 0.599 | **636** | **0.551** | **0.497** | **0.573** | **0.666** |
| | **DyNBF(l = 3)** | 1.87 | 0.916 | 0.875 | 0.952 | 0.979 | 84 | 0.458 | 0.349 | 0.508 | 0.677 | 1552 | 0.499 | 0.445 | 0.529 | 0.599 |
| | **DyNBF(l = 6)** | **1.50** | **0.956** | **0.933** | 0.973 | 0.992 | **76** | **0.557** | **0.455** | **0.617** | **0.752** | 691 | 0.543 | 0.481 | 0.568 | 0.660 |

On UMLS, the existing state-of-the-art (SOTA) performance sets a high standard. However, we still outperform RotatE by at least 0.02 units in Hits@1 and 0.008 units in MRR. This is mainly because UMLS is a relatively small dataset, thus the embedding methods' simple modeling approach achieves very high performance, and large models can suffer from overfitting. DyNBF performs the best on FB15K-237, which outperforms the best-compared method NBFNet(l=6) by 14% in MRR, 13% in Hits@1 and 15% in Hits@10. The primary reason for such significant improvements is that the weights generated from the sub-graph of the query are capable of instructing the model, providing the appropriate path and relation to obtain the answer. Compared to the improvements made on FB15k-237, it only achieves a performance similar to SOTA NBFNet on WN18RR. One potential explanation is that despite the inclusion of word relationships in WordNet, the number of relationships provided is relatively low, with more emphasis on abstract semantics, such as hypernyms and hyponyms. And the FB15k-237 dataset's graph is relatively denser with more than 200 types of relations and each relation contains ample information (see Table 1). In this case, the relation weights will play a more important role for larger and more complex KG to help KGR.

### 4.4 Fine-Grained Analysis

**Impact of Weights.** We explore the influence of weights on the performance of our model. Specifically, by adding weights to the initial k layers, we simulate the initial k hops, thereby applying an attention weight restriction to identify and select pertinent relational paths for effective information aggregation. Firstly, Fig. 3 highlights a noticeable decrease in performance when random weights are assigned to each relation in contrast to the improvement when attention weights generated from the sub-graph are applied as demonstrated by our adopted model. This observation indicates the effectiveness of the attention weights approach. Additionally, Fig. 4 exhibits a linear increase in attention weights assigned to relational edges at varying distances. This progressive increase promotes the selection of more relevant and favorable node information during the aggregation process in each hop, resulting in enhanced inference accuracy.

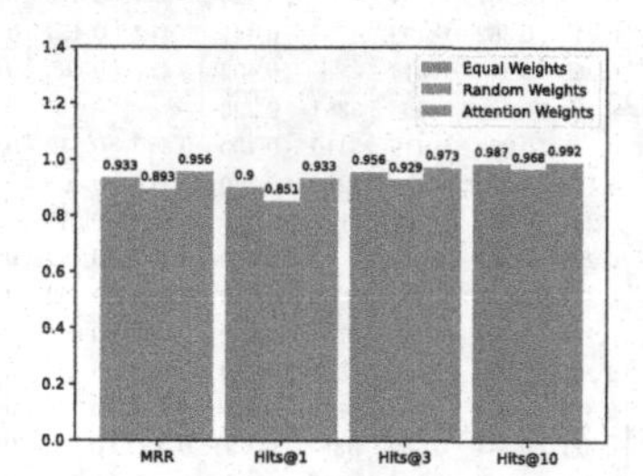


**Fig. 3.** Performance impact of different weighting approaches on KGR.

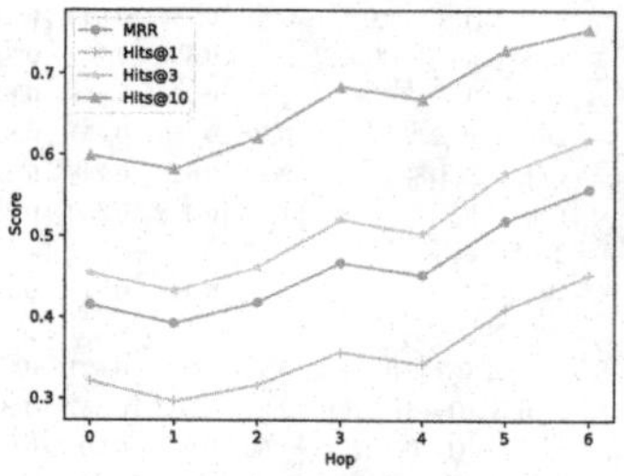

**Fig. 4.** Performance impact of adding attention weights to the first K hops on KGR.

**Performance by Relation Category.** We classify all relations into four categories based on the cardinality of head and tail arguments following the rules by [3]: one-to-one (1-1), one-to-many (1-N), many-to-one (N-1), and many-to-many (N-N). We compute the average number of tails per head and the average number of heads per tail. The category is *one* if the average number is smaller than 1.5 and *many* otherwise. As shown in Table 3, our DyNBF not only enhances simple one-to-one cases but also improves performance in hard cases where there are several correct solutions for a given query. Predicting the "N" side is generally more difficult, since numerous plausible options may lead to model confusion. In addition, the incompleteness of the knowledge graph is a challenge. Some predicted triples could be accurate following human evaluation, especially for *instance of* and *place of birth* 1-N relations in head entity prediction, and others.

**Table 3.** Performance w.r.t. relation category. The two scores are the rankings over-heads and tails respectively.

| Method | Relation Category | | | |
|---|---|---|---|---|
| | 1-1 | 1-N | N-1 | N-N |
| TransE [3] | 0.498/0.488 | 0.455/0.071 | 0.079/0.744 | 0.224/0.330 |
| RotatE [16] | 0.487/0.484 | 0.467/0.070 | 0.081/0.747 | 0.234/0.338 |
| NBFNet [27] | 0.578/0.600 | 0.499/0.122 | 0.165/0.790 | 0.348/0.456 |
| DyNBF | **0.588/0.656** | **0.634/0.293** | **0.261/0.813** | **0.376/0.719** |

### 4.5 Efficiency Analysis

Table 4 shows the comparison of training time and space occupation of our method to two prominent baselines Ruleformer and NBFNet. All experiments were performed on a single RTX3090. From the table, we can find that our approach is clearly more efficient than Ruleformer. While our training space occupation has been reduced, the training time has been greatly reduced, from 15 h per epoch to 2 h per epoch. For comparison with NBFNet, because our handling of subgraphs and the generation of weights increases the space occupation and training time, it is a measure of utilizing resources in exchange for better performance. However, when used for larger KGs, the overhead of the subgraph processing part is negligible, and the training time and space occupation gap between the two will gradually decrease, so our method is potentially more efficient than NBFNet for large-scale KGs.

**Table 4.** Comparison of training time and space occupation of our approach to baseline methods on FB15K-237. The training time indicates the time required for each epoch of training and the training batch size indicates the training space occupation.

| Method | Training Time | Training batchsize |
|---|---|---|
| Ruleformer (l=3) [22] | 15 h | 32 |
| NBFNet (l=3) [27] | 0.7 h | 128 |
| NBFNet (l=6) [27] | 1.5 h | 64 |
| DyNBF (l=3) (Ours) | 2 h | 42 |
| DyNBF (l=6) (Ours) | 3 h | 28 |

### 4.6 Path Interpretations of Prediction

Followed by [27], the importance of each path is estimated by summing up the importance of its edges, where we obtain the edge importance through auto-differentiation. However, we calculate the differential using weights generated by the weight generator, in contrast to the edge weights which are all equal to 1. We obtain the top-K path interpretations by identifying the top-k paths on the edge importance graph, which is solvable by using Bellman-Ford-style beam search.

Table 5 visualizes path interpretations from the FB15k-237 test set. Examining the table reveals that DyNBF's discovered rule paths are not only shorter but also have higher path weights. These results suggest that adding relational weights to specific queries during the inference process helps the model focus on more pertinent information and make more precise predictions. For example, when presented with the first query, "What is the language of Pearl Harbor (film)", DyNBF quickly identifies the country and region to which it belongs and infers that the language is Japanese, rather than venturing further like NBFNet to obtain less relevant information. Additionally, this approach yields more credible mining rules, as evidenced by the improved final path weights.

**Table 5.** Path interpretations and comparison of predictions on FB15k-237 test set. For each query triplet, we visualize the Top-1 path interpretations and their weights. Inverse relations are denoted with a superscript $^{-1}$.

| Query $(h, r, t)$ | NBFNet [Weight] | DyNBF [Weight] |
|---|---|---|
| (*Pearl Harbor (film)*, *language*, *Japanese*) | (*Pearl Harbor(film)*, *film actor*, *C.-H. Tagawa*) ∧ (*C.-H. Tagawa*, *nationality*, *Japan*) ∧ (*Japan*, *country of origin*, *Yu-Gi-Oh!*) ∧ (*Yu-Gi-Oh!*, *language*, *Japanese*) **0.211** | (*pearl harbor (film)*, *film release region*, *usofa*) ∧ (*usofa*, *country of origin*$^{-1}$, *transformers the rebirth*) ∧ (*transformers the rebirth*, *languages*, *Japanese*) **0.467** |
| (*disney cartoon studios*, *production companies*, *hopper (a bug's life)*) | (*hopper (a bug's life)*, *film*$^{-1}$, *andrew stanton*) ∧ (*andrew stanton*, *film*, *nemo (finding nemo)*) ∧ (*nemo (finding nemo)*, *production companies*, *disney cartoon studios*)**0.189** | (*hopper (a bug's life)*, *titles*$^{-1}$, *the walt disney studio*) ∧ (*the walt disney studio*, *child*, *disney cartoon studios*) **0.319** |

## 5 Conclusion

In this paper, we draw attention to using context-aware weights as the edge significance to instruct knowledge graph reasoning. We leverage the Transformer with a relational attention mechanism, enabling the dynamic generation of attention weights from subgraphs and serving as a filter for relational importance during query reasoning, which helps improve the accuracy and expediency of the inference process. The conducted experiments show that our proposed approach exceeds the performance of the majority of existing methods. This outcome provides empirical evidence of the effectiveness and significance of removing irrelevant relations during the inference phase. Despite this approach showing promise, utilizing subgraphs would require significant memory resources and training time. Thus, future research will be devoted to developing an efficient method for generating weights from subgraphs while exclusively utilizing them for information aggregation and inference.

**Acknowledgement.** This work was supported by the Strategic Priority Research Program of Chinese Academy of Sciences (No. XDA27020100) and the National Natural Science Foundation of China (No. U1936207, No. 61976211). This work was supported by the Youth Innovation Promotion Association CAS and Yunnan Provincial Major Science and Technology Special Plan Projects (No. 202202AD080004).

## References

1. Balažević, I., Allen, C., Hospedales, T.M.: Tucker: tensor factorization for knowledge graph completion. In: Empirical Methods in Natural Language Processing (2019)
2. Bollacker, K.D., Evans, C., Paritosh, P.K., Sturge, T., Taylor, J.: Freebase: a collaboratively created graph database for structuring human knowledge. In: SIGMOD Conference (2008)
3. Bordes, A., Usunier, N., Garcia-Durán, A., Weston, J., Yakhnenko, O.: Translating embeddings for modeling multi-relational data. In: Proceedings of the 26th International Conference on Neural Information Processing Systems, NIPS 2013, vol. 2, pp. 2787–2795. Curran Associates Inc., Red Hook (2013)
4. Chen, M., et al.: Meta-knowledge transfer for inductive knowledge graph embedding. In: Proceedings of the 45th International ACM SIGIR Conference on Research and Development in Information Retrieval, SIGIR 2022, pp. 927–937. Association for Computing Machinery, New York (2022). https://doi.org/10.1145/3477495.3531757
5. Cohen, W.W.: Tensorlog: a differentiable deductive database. CoRR abs/1605.06523 (2016). https://arxiv.org/abs/1605.06523
6. Dettmers, T., Pasquale, M., Pontus, S., Riedel, S.: Convolutional 2D knowledge graph embeddings. In: Proceedings of the 32th AAAI Conference on Artificial Intelligence, pp. 1811–1818 (2018). https://arxiv.org/abs/1707.01476
7. Galárraga, L.A., Teflioudi, C., Hose, K., Suchanek, F.: Amie: association rule mining under incomplete evidence in ontological knowledge bases. In: Proceedings of the 22nd International Conference on World Wide Web, WWW 2013, pp. 413–422. Association for Computing Machinery, New York (2013). https://doi.org/10.1145/2488388.2488425
8. Kok, S., Domingos, P.: Statistical predicate invention. In: Proceedings of the 24th International Conference on Machine Learning, ICML 2007, pp. 433–440. Association for Computing Machinery, New York (2007). https://doi.org/10.1145/1273496.1273551
9. Lao, N., Cohen, W.W.: Relational retrieval using a combination of path-constrained random walks. Mach. Learn. **81**, 53–67 (2010)
10. Li, R., et al.: How does knowledge graph embedding extrapolate to unseen data: a semantic evidence view (2022)
11. Liu, D., Lian, J., Liu, Z., Wang, X., Sun, G., Xie, X.: Reinforced anchor knowledge graph generation for news recommendation reasoning. In: Proceedings of the 27th ACM SIGKDD Conference on Knowledge Discovery; Data Mining, KDD 2021, pp. 1055–1065. Association for Computing Machinery, New York (2021). https://doi.org/10.1145/3447548.3467315
12. Miller, G.A.: WordNet: a lexical database for English. In: Speech and Natural Language: Proceedings of a Workshop Held at Harriman, New York, 23–26 February 1992 (1992). https://aclanthology.org/H92-1116
13. Rui, Y.: Knowledge mining: a cross-disciplinary survey. Mach. Intell. Res. **19**, 89–114 (2022). https://doi.org/10.1007/s11633-022-1323-6. www.mi-research.net/en/article/doi/10.1007/s11633-022-1323-6

14. Sadeghian, A.R., Armandpour, M., Ding, P., Wang, D.Z.: Drum: end-to-end differentiable rule mining on knowledge graphs. arXiv abs/1911.00055 (2019)
15. Schlichtkrull, M., Kipf, T., Bloem, P., van den Berg, R., Titov, I., Welling, M.: Modeling relational data with graph convolutional networks. In: Extended Semantic Web Conference (2017)
16. Sun, Z., Deng, Z.H., Nie, J.Y., Tang, J.: Rotate: knowledge graph embedding by relational rotation in complex space. In: International Conference on Learning Representations (2019). https://openreview.net/forum?id=HkgEQnRqYQ
17. Toutanova, K., Chen, D., Pantel, P., Poon, H., Choudhury, P., Gamon, M.: Representing text for joint embedding of text and knowledge bases. In: Proceedings of the 2015 Conference on Empirical Methods in Natural Language Processing, Lisbon, Portugal, pp. 1499–1509. Association for Computational Linguistics (2015). https://doi.org/10.18653/v1/D15-1174. https://aclanthology.org/D15-1174
18. Trouillon, T., Welbl, J., Riedel, S., Gaussier, E., Bouchard, G.: Complex embeddings for simple link prediction. In: Balcan, M.F., Weinberger, K.Q. (eds.) Proceedings of the 33rd International Conference on Machine Learning. Proceedings of Machine Learning Research, New York, USA, vol. 48, pp. 2071–2080. PMLR (2016). https://proceedings.mlr.press/v48/trouillon16.html
19. Vashishth, S., Sanyal, S., Nitin, V., Talukdar, P.: Composition-based multi-relational graph convolutional networks (2020)
20. Wang, X., He, X., Cao, Y., Liu, M., Chua, T.S.: KGAT: knowledge graph attention network for recommendation. In: Proceedings of the 25th ACM SIGKDD International Conference on Knowledge Discovery; Data Mining, KDD 2019, pp. 950–958. Association for Computing Machinery, New York (2019). https://doi.org/10.1145/3292500.3330989
21. Wang, X.: Large-scale multi-modal pre-trained models: a comprehensive survey. Mach. Intell. Res. **20**, 1 (2023). https://doi.org/10.1007/s11633-022-1410-8. https://www.mi-research.net/en/article/doi/10.1007/s11633-022-1410-8
22. Xu, Z., Ye, P., Chen, H., Zhao, M., Chen, H., Zhang, W.: Ruleformer: context-aware rule mining over knowledge graph. In: Proceedings of the 29th International Conference on Computational Linguistics, Gyeongju, Republic of Korea, pp. 2551–2560. International Committee on Computational Linguistics (2022). https://aclanthology.org/2022.coling-1.225
23. Yang, B., tau Yih, W., He, X., Gao, J., Deng, L.: Embedding entities and relations for learning and inference in knowledge bases. CoRR abs/1412.6575 (2014)
24. Yang, F., Yang, Z., Cohen, W.W.: Differentiable learning of logical rules for knowledge base reasoning. In: Proceedings of the 31st International Conference on Neural Information Processing Systems, NIPS 2017, pp. 2316–2325. Curran Associates Inc., Red Hook (2017)
25. Zhang, X.: Transfer hierarchical attention network for generative dialog system. Int. J. Autom. Comput. **16**, 720 (2019). https://doi.org/10.1007/s11633-019-1200-0. https://www.mi-research.net/en/article/doi/10.1007/s11633-019-1200-0
26. Zhang, Z., Wang, J., Ye, J., Wu, F.: Rethinking graph convolutional networks in knowledge graph completion. In: Proceedings of the ACM Web Conference 2022, WWW 2022, pp. 798–807. Association for Computing Machinery, New York (2022). https://doi.org/10.1145/3485447.3511923
27. Zhu, Z., Zhang, Z., Xhonneux, L.P., Tang, J.: Neural bellman-ford networks: a general graph neural network framework for link prediction. In: Advances in Neural Information Processing Systems, vol. 34 (2021)

# CausE: Towards Causal Knowledge Graph Embedding

Yichi Zhang and Wen Zhang(✉)

Zhejiang University, Hangzhou, China
{zhangyichi2022,zhang.wen}@zju.edu.cn

**Abstract.** Knowledge graph embedding (KGE) focuses on representing the entities and relations of a knowledge graph (KG) into the continuous vector spaces, which can be employed to predict the missing triples to achieve knowledge graph completion (KGC). However, KGE models often only briefly learn structural correlations of triple data and embeddings would be misled by the trivial patterns and noisy links in real-world KGs. To address this issue, we build the new paradigm of KGE in the context of causality and embedding disentanglement. We further propose a **Caus**ality-enhanced knowledge graph **E**mbedding (**CausE**) framework. CausE employs causal intervention to estimate the causal effect of the confounder embeddings and design new training objectives to make stable predictions. Experimental results demonstrate that CausE could outperform the baseline models and achieve state-of-the-art KGC performance. We release our code in https://github.com/zjukg/CausE.

**Keywords:** Knowledge Graph Embedding · Knowledge Graph Completion · Causal Inference

## 1 Introduction

Knowledge graphs (KGs) [2] modeling the world knowledge with structural triples in the form of *(head entity, relation, tail entity)*, which portrays the relation between the head and tail entity. Expressive KGs have become the new infrastructure of artificial intelligence (AI), which have been widely used in question answering [18], recommender systems [20], and fault analysis [6].

KGs are usually inherently incomplete due to their vast diversity and complexity. To address this issue, knowledge graph completion (KGC) has become a popular research topic, aimed at identifying undiscovered triples in KGs. A mainstream solution to KGC is knowledge graph embedding (KGE), which utilizes low-dimensional continuous space to embed entities and relations from the KG. The triple structure is modeled through a score function [3,12,17] that measures the plausibility of each triple, forming the basis for predictions in KGC tasks.

However, in KGs, various confounding factors (such as trivial structural patterns, noisy links, etc.) may mislead KGE models, resulting in spurious correlations [11] being learned and non-causal predictions being made. Figure 1 provides an intuitive view of such a situation. While many existing methods propose


H. Wang et al. (Eds.): CCKS 2023, CCIS 1923, pp. 17–28, 2023.
https://doi.org/10.1007/978-981-99-7224-1_2

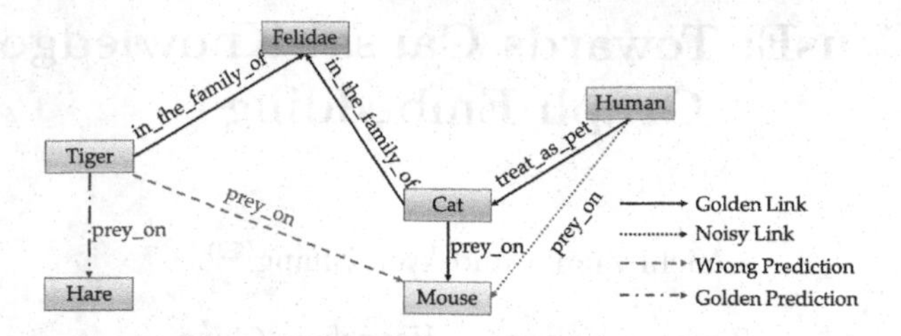


**Fig. 1.** A simple example to explain that the confounding factors like noisy links e.g. (Human, prey_on, Mouse) and trivial patterns (Both Tiger and Cat are in the family of Felidae) might mislead the link prediction. In this case, the prediction result of (Tiger, prey_on, ?) would be misled to Mouse.

scoring functions to model different relationship patterns, they overlook the possibility that the knowledge graph data itself may contain information that could mislead the model.

To address the mentioned problem, We decouple the embeddings of entities and relations into causal and confounder embeddings. Then we introduce the theory of causal inference [9] to model and analyze this problem. We construct the structural causal model (SCM) [10] to analyze the KGE task in the context of causality. Meanwhile, we propose a **Caus**ality-enhanced knowledge graph **E**mbedding (CausE) framework to guide the KGE models to learn causal features in the KG. In CausE, we design the intervention operator to implement the backdoor adjustment [10], which would combine the two kinds of embeddings to estimate the effect of the causal and confounder embeddings. Besides, we design two auxiliary training objectives to enhance the model. We conduct comprehensive experiments on two public benchmarks with the link prediction task to demonstrate the effectiveness of CausE on KGC and make further explorations. The main contribution of this paper can be summarized as follows:

- We are the first work to introduce causality theory into the field of KGE.
- We propose a new learning paradigm for KGE in the context of causality and design a **Caus**ality-enhanced knowledge graph **E**mbedding (**CausE** for short) framework to learn causal embeddings for KGE models.
- We conduct comprehensive experiments on public benchmarks to demonstrate the effectiveness of CausE. We also make further exploration to understand it deeply.

## 2 Related Works

### 2.1 Knowledge Graph Embedding

Knowledge graph embedding [15] usually represent the entities and relations of a KG into the low dimensional continuous space to learn the structural features in the KG. A score function is defined in the KGE model to model the triple structure and discriminate the plausibility of triples.

Existing KGE methods [3–5,12,14,17] focus on design elegant and expressive score functions to modeling the triples. Translation-based methods [3,5,12] modeling the relation as a translation from head to tail in the representation space. TransE [3] treats the translation as a vector addition. RotatE [12] represents the relation as a rotation in the complex space. PairRE [5] employs two vectors for relation representation and designs a more complicated score function in the Euclidean space. Besides, semantic matching [14,17] models employ latent semantic matching to score the triples, which could be regarded as implicit tensor factorization. DistMult [17] treats the process as 3D tensor factorization and ComplEx [14] further extends it to the complex space. Although various KGE methods are proposed and achieve state-of-the-art knowledge graph completion results, no existing methods are concerned with learning the causality of triple structure and making knowledge graph completion better.

### 2.2 Causal Inference-Enhanced Graph Learning

Causal inference [9,10] is a popular statistical research topic which aims to discovering causality between data. In recent years, it is becoming increasingly visible to combine causal inference and machine learning to learn the causality from data rather then the correlation for stable and robust prediction. As for graph learning (GL), causal inference also brings a different perspective to the learning paradigm of graphs. CGI [8] employs causal theory to select trustworthy neighbors for graph convolution networks. CAL [11] proposes a causal attention learning framework to learn the causal feature of graphs to enhance the graph classification task. However, there is no existing work to introduce causal theory into the knowledge graph community.

## 3 Preliminary

A knowledge graph can be denoted as $\mathcal{G} = (\mathcal{E}, \mathcal{R}, \mathcal{T})$, where $\mathcal{E}$ is the entitiy set, $\mathcal{R}$ is the relation set, and $\mathcal{T} = \{(h, r, t)|h, t \in \mathcal{E}, r \in \mathcal{R}\}$ is the triple set.

A KGE model would embed each entity $e \in \mathcal{E}$ and each relation $r \in \mathcal{R}$ into the continuous vector space and represent each of them with an embedding. We denote $\mathbf{E}^{|\mathcal{E}|\times d_e}$ and $\mathbf{R}^{|\mathcal{R}|\times d_r}$ as the embedding matrix of entity and relation respectively, where $d_e, d_r$ are the dimensions of the entity embeddings and the relation embeddings. Besides, a score function $\mathcal{F}(h, r, t)$ is defined to measure the triple plausibility. The overall target of the KGE model is to give positive triples higher scores and give negative triples lower scores. During training, negative triples are generated by randomly replacing the head or tail entity for positive-negative contrast. We denote the negative triple set as $\mathcal{T}' = \{(h', r, t)|(h, r, t) \in \mathcal{T}, h' \in \mathcal{E}, h' \neq h\} \cup \{(h, r, t')|(h, r, t) \in \mathcal{T}, t' \in \mathcal{E}, t' \neq t\}$. Sigmoid loss proposed by [12] is widely used by recent state-of-the-art KGE methods, which could be denoted as:

$$\mathcal{L} = \frac{1}{|\mathcal{T}|} \sum_{(h,r,t)\in\mathcal{T}} \Big( -\log\sigma(\gamma - \mathcal{F}(h, r, t)) - \sum_{i=1}^{K} p_i \log\sigma(\mathcal{F}(h'_i, r'_i, t'_i) - \gamma) \Big) \quad (1)$$

where $\sigma$ is the sigmoid function, $\gamma$ is the margin, and $K$ is the number of negative triples generated for each positive triple. The negative triples for $(h, r, t)$ is denoted as $(h'_i, r'_i, t'_i), i = 1, 2, \ldots, K$. Besides, $p_i$ is the self-adversarial weight [12] for each negative triple $(h'_i, r'_i, t'_i)$. It could be denoted as $p_i = \frac{\exp(\alpha \mathcal{F}(h'_i, r'_i, t'_i))}{\sum_{j=1}^{K} \exp(\alpha \mathcal{F}(h'_j, r'_j, t'_j))}$, where $\alpha$ is the temperature of self-adversarial weight.

# 4 Methodology

In this section, we first present the structural causal model (SCM) for the KGE task. Then we further propose our causality-enhanced KGE framework CausE to learn causal and confounder embeddings with carefully designed objectives.

## 4.1 SCM for KGE Task

In KGE models described in Sect. 3, each entity and relation has a single embedding that encodes both the useful (causal) and harmful (confounder) features. However, as discussed in Sect. 1, this approach is not robust enough since some local structural information in the KG (e.g. trivial patterns, noisy links) can mislead embedding learning. To develop better embeddings that account for structural causality and make accurate predictions, we introduce the structural causal model (SCM) [10] for KGE, as shown in Fig. 2.

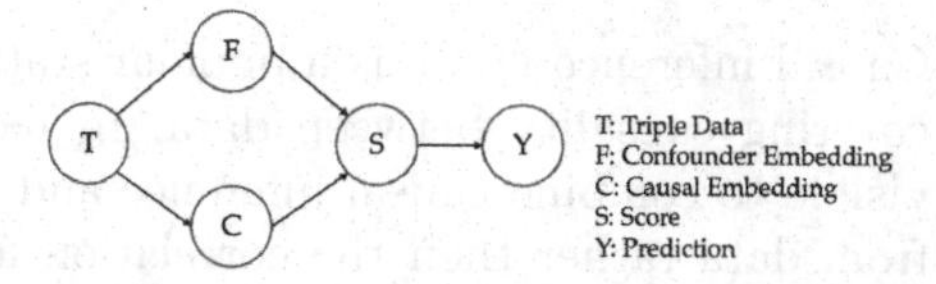


**Fig. 2.** Our SCM for KGE models.

The SCM defines variables: the triple data $T$, the confounder embeddings $F$, the causal embeddings $C$, the triple score $S$, and the prediction result $Y$. Besides, the SCM demonstrates several causal relations among those variables:

- $F \leftarrow T \rightarrow C$. The causal embeddings $C$ encode the implicit knowledge about the triple structure. The confounder embeddings $F$, however, have no contribution to the prediction. As both of them could be learned from the KG data $T$, such causal relations exist in the SCM.
- $F \rightarrow S \leftarrow C$. $S$ represents the score of a triple, which is based on both the causal embeddings and confounder embeddings.
- $S \rightarrow Y$. We denote $Y$ as the prediction results. The overall target of a KGE model is to predict the proper results $Y$ based on the triple scores $S$ in the inference stage.

In the original KGE paradigm, the causal and confounder embedding of each entity or relation co-exist in one embedding. With SCM, we explicitly disentangle the structural embeddings from the causal and confounder embeddings and analysis their effects on the prediction results in $Y$. The next question is how to mitigate the impact of $F$ on the final prediction $Y$ to make causal predictions.

### 4.2 Causal Intervention

According to the SCM, both the confounder embeddings $C$ and causal embeddings $F$ could be learned from the triple data, which would be all considered in the triple score $S$. Thus, $F \leftarrow T \rightarrow C \rightarrow S \rightarrow Y$ is a backdoor path [9] and $F$ is the confounder between $C$ and $Y$.

To make causal predictions based on causal embeddings $C$, we need to model $P(Y|C)$. However, the backdoor path creates a confounding effect of $F$ on the probability distribution $P(Y|C)$, opening a backdoor from $F$ to $Y$. Therefore, it is crucial to block the backdoor path and reduce the impact of confounder embeddings. This will allow KGE models to make predictions by utilizing the causal embeddings fully. Causality theory [9,10] provides powerful tools to solve the backdoor path problem.

We employ do-calculus [9,10] to make the causal intervention on the variable $C$, which could **cut off the backdoor path** $F \leftarrow T \rightarrow C \rightarrow S \rightarrow Y$. With the help of do-calculus, the influence from the confounder $F$ to $C$ is manually cut off, which means $C, F$ are independent. Our target turns to estimate $P(Y|do(C))$ instead of the confounded $P(Y|C)$. Combined with Bayes Rule and the causal assumptions [9,10], we could deduce as follows:

$$P(Y|do(C)) = P(Y|S) \sum_{d \in \mathcal{D}} P(S|C,d)P(d) \tag{2}$$

The above derivation shows that to estimate the causal effect of $C$ on $Y$, it is necessary to consider the scores with both causal and counfounder embeddings. This can be understood as re-coupling the decoupled embeddings and using them to calculate the score of the triple. In the next section, we would propose our **Caus**ality-enhanced knowledge graph **E**mbedding (CausE) framework and implement the backdoor adjustments mentioned above.

### 4.3 CausE Framework

In this section, we would demonstrate our **Caus**ality-enhanced knowledge graph **E**mbedding (CausE) framework. We would first describe the basic settings of CausE and emphasize how we implement the backdoor adjustment in the CausE.

**Basic Definition.** The overall framework of CausE is shown in Fig. 3. In the embedding layer, we define two embeddings called causal embedding and confounder embedding for each entity and relation in the KG, aiming to achieve the disentanglement of causal and confounder features. Specifically, for each entity $e \in \mathcal{E}$, we define a causal embedding $\boldsymbol{e}_{caus}$ and a confounder embedding $\boldsymbol{e}_{conf}$ for it. Similarly, for each relation $r \in \mathcal{R}$, the two embeddings are $\boldsymbol{r}_{caus}$ and $\boldsymbol{r}_{conf}$. Such design is consistent with the SCM in Fig. 2.

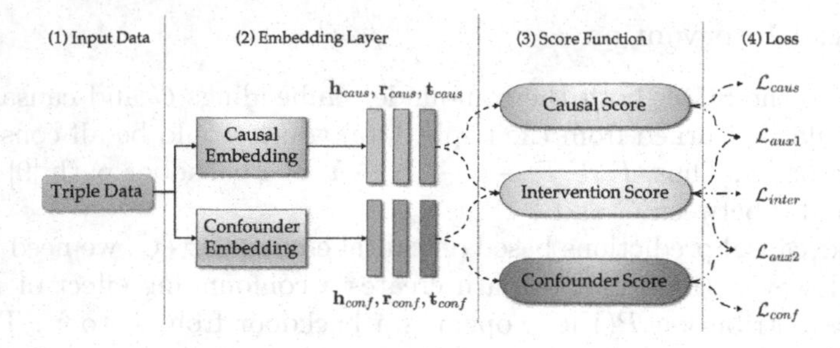


**Fig. 3.** The overall architecture of CausE. We disentangle the embeddings into two parts called causal and confounder embeddings respectively while applying three score functions. We also design five loss functions to train these embeddings while the causal intervention is integrated into them.

As for the score function, we employ three score functions $\mathcal{F}_{caus}, \mathcal{F}_{conf}, \mathcal{F}_{inter}$, which are called causal score, confounder score, and intervention score respectively. The three score functions are in the same form but can be any general score functions proposed by the existing KGE models. Besides, we design several loss functions to guide the training process of CausE. We would describe the details of the score functions and their corresponding loss functions.

**Causal and Confounder Scores.** The causal score function $\mathcal{F}_{caus}(h, r, t)$ takes the causal embeddings $\boldsymbol{h}_{caus}, \boldsymbol{r}_{caus}, \boldsymbol{t}_{caus}$ of $h, r, t$ as input and calculate the causal score the triple. According to our assumption, the causal embeddings are expected to make reasonable and causal predictions. Thus, the causal score $\mathcal{F}_{caus}(h, r, t)$ should still follow the general rule of KGE models: positive triple should have higher scores. We apply sigmoid loss function with self-adversarial negative sampling as the loss function to train the causal embeddings. The causal loss $\mathcal{L}_{caus}$ has the same form as Eq. 1, which is based on $\mathcal{F}_{caus}$.

Meanwhile, the confounder score function $\mathcal{F}_{conf}(h, r, t)$ would calculate the confounder score of the confounder embeddings $\boldsymbol{h}_{conf}$, $\boldsymbol{r}_{conf}$, $\boldsymbol{t}_{conf}$. Different from the causal embeddings, we assume that confounder embeddings learn the harmful features from the KGs and they make no positive contribution to the reasonable prediction. Hence, the confounder score $\mathcal{F}_{caus}(h, r, t)$ should be close to the confounder score of negative triples, which means the KGE model is misled by the harmful features and could not distinguish the positive triple from high plausibility from the negative triples. Therefore, we apply the mean squared error (MSE) loss to train the confounder embeddings. The training objective can be denoted as:

$$\mathcal{L}_{conf} = \frac{1}{|\mathcal{T}|} \sum_{(h,r,t)\in\mathcal{T}} \left(\mathcal{F}_{caus}(h, r, t) - \sum_{i=1}^{K} p_i \mathcal{F}_{caus}(h'_i, r'_i, t'_i)\right)^2 \tag{3}$$

By the two loss functions proposed above, we could achieve the disentanglement of the causal and confounder embeddings.

**Intervention Scores.** As shown in Eq. 2, we need to implement the backdoor adjustment. As we mentioned above, the formula for backdoor adjustment can be understood as jointly measuring the triple plausibility with both causal and confounder embeddings, while considering all possible confounder embedding. This is equivalent to recombining the two decoupled embeddings into the original embeddings and computing the score.

We call this score an intervention score $\mathcal{F}_{inter}(h,r,t)$. Besides, we propose a **intervetion operator** $\Phi$ to recombine the two embeddings and get the intervention embeddings as the output. This process can be denoted as:

$$\boldsymbol{e}_{inter} = \Phi(\boldsymbol{e}_{caus}, \boldsymbol{e}_{conf}), \boldsymbol{e} \in \{\boldsymbol{h}, \boldsymbol{t}\} \quad \boldsymbol{r}_{inter} = \Phi(\boldsymbol{r}_{caus}, \boldsymbol{r}_{conf}) \tag{4}$$

We employ the addition operation as the intervention operation. Hence, we could calculate the intervetion score $\mathcal{F}_{inter}(h,r,t)$ with the intervention embedding $\boldsymbol{h}_{inter}, \boldsymbol{r}_{inter}, \boldsymbol{t}_{inter}$. From another perspective, causal intervention is such a process that employs the confounder embeddings to disrupt the prediction of the causal embeddings to estimate the causal effect of the confounder embeddings. We expect the intervention scores could still lead to reasonable predictions. Thus, the training objective $\mathcal{L}_{inter}$ is also a sigmoid loss like 1 based on $\mathcal{F}_{inter}$.

**Auxiliary Objectives.** To further improve the performance of CausE, we utilize the intervention score and propose two auxiliary training objectives.

As we mentioned above, the intervention embeddings can be regarded as the causal embeddings perturbed by the confounder embeddings. Therefore, the effectiveness of the causal scores should be worse than the causal scores but better than the confounder scores. Based on such an assumption, we design two auxiliary training objectives. The first auxiliary objective is between the causal and intervention scores. We apply the sigmoid loss function to make the contrast between them and push the causal scores higher than the intervention scores:

$$\mathcal{L}_{aux1} = \frac{1}{|\mathcal{T}|} \sum_{(h,r,t)\in\mathcal{T}} \Big( -\log\sigma(\gamma - \mathcal{F}_{caus}(h,r,t)) - \log\sigma(\mathcal{F}_{inter}(h,r,t) - \gamma) \Big) \tag{5}$$

The second auxiliary objective $\mathcal{L}_{aux2}$ is similarly designed as $\mathcal{L}_{aux1}$ to push the intervention scores higher than the confounder scores. In summary, the overall training objective of CausE is:

$$\mathcal{L} = \mathcal{L}_{caus} + \mathcal{L}_{conf} + \mathcal{L}_{inter} + \mathcal{L}_{aux1} + \mathcal{L}_{aux2} \tag{6}$$

## 5 Experiments

In this section, we will demonstrate the effectiveness of our methods with comprehensive experiments. We first detailedly introduce our experimental settings in Sect. 5.1. Then we would demonstrate our results to answer the following questions:

- **RQ1**: Could CausE outperform the existing baseline methods in the knowledge graph completion task?
- **RQ2**: How does CausE perform in the noisy KGs?
- **RQ3**: How much does each module of CausE contribute to the performance?
- **RQ4**: Do the learned embeddings achieve our intended goal?

### 5.1 Experiment Settings

**Datasets/Tasks/Evaluation Protocols.** In the experiments, we use two benchmark datasets FB15K-237 [13] and WN18RR [7].

We evaluate our method with link prediction task, which is the main task of KGC. Link prediction task aims to predict the missing entities for the given query $(h, r, ?)$ or $(?, r, t)$. We evaluate our method with mean reciprocal rank (MRR), and Hit@K (K = 1,3,10) following [12]. Besides, we follow the filter setting [3] which would remove the candidate triples that have already appeared in the training data to avoid their interference.

**Baselines.** As for the link prediction task, we select several state-of-the-art KGE methods, including translation-based methods (TransE [3], RotatE [12], PairRE [5]), semantic matching methods (DistMult [17], ComplEx [14]), quaternion-based methods (QuatE [19], DualE [4]), and neural network based methods (ConvE [7], MurP [1]). We report the baseline results from the original paper.

**Parameter Settings.** We implement CausE framework to five representative score functions: TransE [3], DistMult [16], ComplEx [14], PairRE [5], and DualE [4]. We apply grid search to tune the hyper-parameters to find the best results of CausE. We search the embedding dimension of the KGE model $d_e, d_r \in \{256, 512, 1024\}$, the margin $\gamma \in \{0, 4, 6, 8\}$, the training batch size $\in \{512, 1024\}$, the temperature $\alpha \in \{1.0, 2.0\}$, the negative sample number $N_k \in \{64, 128, 256\}$, and the learning rate $\eta \in \{1e^{-3}, 1e^{-4}, 2e^{-5}\}$. We conduct all the experiments on Nvidia GeForce 3090 GPUs with 24 GB RAM.

### 5.2 Main Results (RQ1)

Our main experiment results are in Table 1. From the results, we could find that The CausE could outperform the baseline methods on the two benchmarks. For example, CausE can achieve a relatively 1.4% Hit@1 improvement on the

**Table 1.** Link prediction results on FB15K-237 and WN18RR. The best results are **bold** and the second best results are <u>underlined</u> for each metrics.

| Model | FB15K-237 | | | | WN18RR | | | |
|---|---|---|---|---|---|---|---|---|
| | MRR | Hit@10 | Hit@3 | Hit@1 | MRR | Hit@10 | Hit@3 | Hit@1 |
| TransE [3] | 0.279 | 0.441 | 0.376 | 0.198 | 0.224 | 0.520 | 0.390 | 0.022 |
| DistMult [17] | 0.281 | 0.446 | 0.301 | 0.199 | 0.444 | 0.504 | 0.470 | 0.412 |
| ComplEx [14] | 0.278 | 0.450 | 0.297 | 0.194 | 0.449 | 0.530 | 0.469 | 0.409 |
| ConvE [7] | 0.312 | 0.497 | 0.341 | 0.225 | 0.456 | 0.531 | 0.470 | 0.419 |
| RotatE [12] | 0.338 | 0.533 | 0.375 | 0.241 | 0.476 | 0.571 | 0.492 | 0.428 |
| MurP [1] | 0.336 | 0.521 | 0.370 | 0.245 | 0.475 | 0.554 | 0.487 | 0.436 |
| QuatE [19] | 0.311 | 0.495 | 0.342 | 0.221 | 0.481 | **0.564** | 0.500 | 0.436 |
| DualE [4] | 0.330 | 0.518 | 0.363 | 0.237 | <u>0.482</u> | 0.561 | <u>0.500</u> | <u>0.440</u> |
| PairRE [5] | <u>0.351</u> | <u>0.544</u> | <u>0.387</u> | <u>0.256</u> | - | - | - | - |
| CausE (TransE) | 0.332 | 0.517 | 0.368 | 0.234 | 0.227 | 0.536 | 0.391 | 0.023 |
| CausE (DistMult) | 0.298 | 0.473 | 0.327 | 0.212 | 0.447 | 0.517 | 0.452 | 0.415 |
| CausE (ComplEx) | 0.324 | 0.504 | 0.357 | 0.234 | 0.467 | 0.527 | 0.482 | 0.436 |
| CausE (SOTA) | **0.355** | **0.547** | **0.392** | **0.259** | **0.486** | <u>0.562</u> | **0.502** | **0.446** |

WN18RR dataset. Such results demonstrate that CausE becomes a new state-of-the-art KGE method.

Meanwhile, CausE is a universal framework and can be applied in various KGE models. The results in Table 1 also demonstrate that CausE could enhance the performance of various KGE models, compared with the corresponding baselines trained w/o CausE. For example, the MRR results on the FB15K-237 dataset of the TransE/DistMult/ComplEx models get relative improvement by 18.9%, 13.5%, and 16.5% respectively. We speculate that this is due to the design defects in the early KGE models, which would mislead the model to learn the confounder features in the KG and make non-causal predictions in the inference stage. Overall, we show that CausE can outperform the baseline methods in various score functions. Thus, the **RQ1** is solved.

### 5.3 Link Prediction on Noisy KG (RQ2)

To answer the **RQ2**, we make further exploration on the noisy link prediction task, aiming to validate the robustness of CausE on noisy KGs. We set a parameter called noisy rate $\lambda$, it is defined as $\lambda = \frac{|\mathcal{T}_{noisy}|}{|\mathcal{T}_{train}|}$, where $\mathcal{T}_{noisy} \subset \mathcal{T}_{train}$ is the noisy link set of the training set. We generate noisy KGs by randomly replacing the positive triples and setting the noisy rate $\lambda$ from 1% to 10%. We conduct experiments on these noisy datasets with DistMult [17] and ComplEx [14]. The results are shown in Fig. 4.

According to the noisy link prediction results, we could first observe that the performance of KGE models is gradually declining as the noisy links in the training data increase. Further, the models enhanced with CausE outperform the baseline models on different benchmarks and score functions. Such experimental results show that our design is effective to counter the noise in the data set and achieve better link prediction performance.

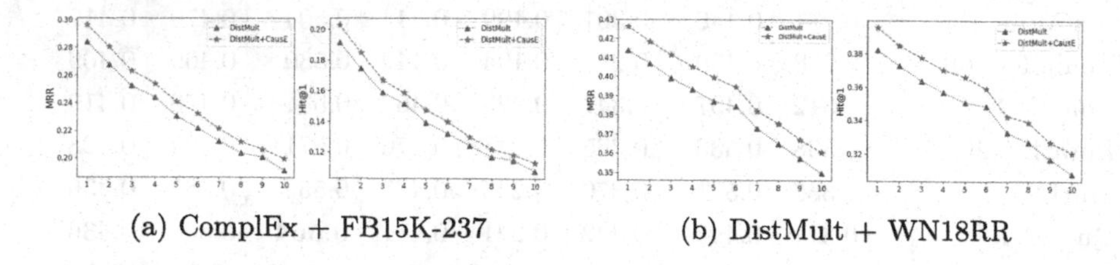


(a) ComplEx + FB15K-237 (b) DistMult + WN18RR

**Fig. 4.** The noisy link prediction results. We report the Hit@1 and MRR results for different experiment settings. The x-axis represents the noisy rate (%) of the training dataset.

**Table 2.** Ablation Study result on WN18RR dataset with ComplEx score.

| | Model | MRR | Hit@10 | Hit@3 | Hit@1 |
|---|---|---|---|---|---|
| | CausE-ComplEx | 0.467 | 0.527 | 0.482 | 0.436 |
| $\mathcal{L}$ | w/o $\mathcal{L}_{caus}$ | 0.458 | 0.525 | 0.479 | 0.421 |
| | w/o $\mathcal{L}_{conf}$ | 0.453 | 0.509 | 0.467 | 0.424 |
| | w/o $\mathcal{L}_{inter}$ | 0.427 | 0.494 | 0.452 | 0.407 |
| | w/o $\mathcal{L}_{aux1}$ | 0.454 | 0.508 | 0.466 | 0.426 |
| | w/o $\mathcal{L}_{aux2}$ | 0.446 | 0.497 | 0.460 | 0.419 |
| $\Phi$ | subtraction | 0.454 | 0.507 | 0.464 | 0.426 |
| | multiple | 0.439 | 0.494 | 0.454 | 0.409 |
| | concatenation | 0.433 | 0.482 | 0.442 | 0.409 |

### 5.4 Ablation Study (RQ3)

To explore the **RQ3**, we conduct ablation studies on different components of CausE in this section. We mainly verify the effectiveness and necessity of module design from two aspects.

First, we remove each of the five training objectives and conduct link prediction experiments. Secondly, we validate the effectiveness of the intervention operator by replacing the addition operation $\Phi$ with other common operators.

Our ablation studies are conducted in the mentioned settings with ComplEx score and WN18RR dataset, while keeping other hyper-parameters same. The results are shown in Table 2. The experiment results show that all five parts of the training objective are of great significance, as the model performs worse when any of them is removed. The performance of the model degrades most when $\mathcal{L}_{inter}$ is removed. Hence, the results emphasize that causal intervention plays a very important role in CausE. Meanwhile, when the intervention operator is changed to other settings, the performance of the model has also decreased. Thus, we could conclude that the addition operation is a pretty good choice, as it is simple but effective enough.

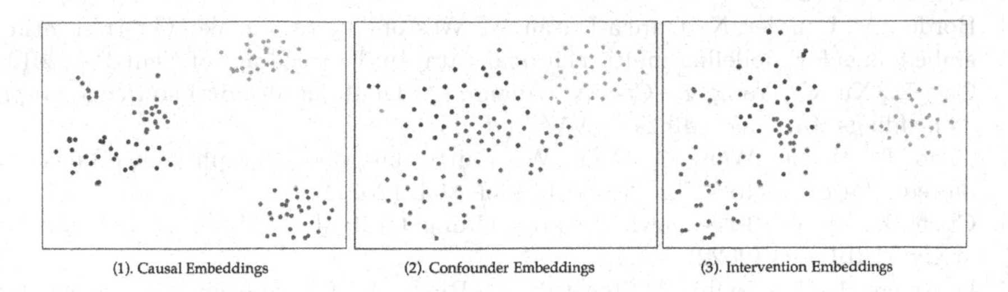

(1). Causal Embeddings (2). Confounder Embeddings (3). Intervention Embeddings

**Fig. 5.** Embedding visualization results with t-SNE, we assign different colors for the entities with different types.

### 5.5 Visualization

To answer **RQ4** and to illustrate the effectiveness of CausE intuitively, we selected entities with several different types and visualize their embeddings with t-SNE, which is shown in Fig. 5. We can find that the causal embedding distribution of different types can be clearly distinguished, while the confounder embedding are relatively mixed and closer together. The distribution of the intervention embeddings which could represent the original embeddings without disentanglement lies between the two. This shows that our approach make causal embeddings learn more distinguishable and achieve the designed goal.

## 6 Conclusion

In this paper, we emphasis that learning correlation in knowledge graph embedding models might mislead the models to make wrong predictions. We resort to causal inference and propose the new paradigm of knowledge graph embedding. Further, we propose a novel framework called CausE to enhance the knowledge graph embedding models. CausE would disentangle the causal and confounder features to different embeddings and train those embeddings guided by the causal intervention. Comprehensive experiments demonstrate that CausE could outperform the baseline methods achieve new state-of-the-art results. In the future, we plan to introduce more causality theory into knowledge graph embeddings and we attempt to apply the causal theory in more complex scenarios such as multi-modal knowledge graphs, and temperal knowledge graphs.

**Acknowledgement.** This work is funded by Zhejiang Provincial Natural Science Foundation of China (No. LQ23F020017), Yongjiang Talent Introduction Programme (2022A-238-G), and NSFC91846204/U19B2027.

## References

1. Balazevic, I., Allen, C., Hospedales, T.M.: Multi-relational poincaré graph embeddings. In: Proceedings of NeurIPS (2019)
2. Bollacker, K.D., Evans, C., Paritosh, P.K., Sturge, T., Taylor, J.: Freebase: a collaboratively created graph database for structuring human knowledge. In: Proceedings of SIGMOD (2008)
3. Bordes, A., Usunier, N., García-Durán, A., Weston, J., Yakhnenko, O.: Translating embeddings for modeling multi-relational data. In: Proceedings of NeurIPS (2013)
4. Cao, Z., Xu, Q., Yang, Z., Cao, X., Huang, Q.: Dual quaternion knowledge graph embeddings. In: Proceedings of AAAI (2021)
5. Chao, L., He, J., Wang, T., Chu, W.: Pairre: knowledge graph embeddings via paired relation vectors. In: Proceedings of ACL (2021)
6. Chen, Z., et al.: Tele-knowledge pre-training for fault analysis. arXiv preprint arXiv:2210.11298 (2022)
7. Dettmers, T., Minervini, P., Stenetorp, P., Riedel, S.: Convolutional 2D knowledge graph embeddings. In: Proceedings of AAAI (2018)
8. Feng, F., Huang, W., He, X., Xin, X., Wang, Q., Chua, T.: Should graph convolution trust neighbors? A simple causal inference method. In: Proceedings of SIGIR (2021)
9. Pearl, J.: Interpretation and identification of causal mediation. Psychol. Methods **19**(4), 459 (2014)
10. Pearl, J., et al.: Models, Reasoning and Inference. Cambridge University Press, Cambridge (2000)
11. Sui, Y., Wang, X., Wu, J., Lin, M., He, X., Chua, T.: Causal attention for interpretable and generalizable graph classification. In: Proceedings of KDD (2022)
12. Sun, Z., Deng, Z., Nie, J., Tang, J.: Rotate: knowledge graph embedding by relational rotation in complex space. In: Proceedings of ICLR (2019)
13. Toutanova, K., Chen, D., Pantel, P., Poon, H., Choudhury, P., Gamon, M.: Representing text for joint embedding of text and knowledge bases. In: Proceedings of EMNLP (2015)
14. Trouillon, T., Dance, C.R., Gaussier, É., Welbl, J., Riedel, S., Bouchard, G.: Knowledge graph completion via complex tensor factorization. J. Mach. Learn. Res. (2017)
15. Wang, Q., Mao, Z., Wang, B., Guo, L.: Knowledge graph embedding: a survey of approaches and applications. IEEE Trans. Knowl. Data Eng. **29**(12), 2724–2743 (2017)
16. Wang, Z., Zhang, J., Feng, J., Chen, Z.: Knowledge graph embedding by translating on hyperplanes. In: Proceedings of AAAI (2014)
17. Yang, B., Yih, W., He, X., Gao, J., Deng, L.: Embedding entities and relations for learning and inference in knowledge bases. In: Proceedings of ICLR (2015)
18. Yasunaga, M., Ren, H., Bosselut, A., Liang, P., Leskovec, J.: QA-GNN: reasoning with language models and knowledge graphs for question answering. In: Proceedings of NAACL (2021)
19. Zhang, S., Tay, Y., Yao, L., Liu, Q.: Quaternion knowledge graph embeddings. In: Proceedings of NeurIPS (2019)
20. Zhu, Y., et al.: Knowledge perceived multi-modal pretraining in e-commerce. In: MM 2021: ACM Multimedia Conference, Virtual Event, China, 20–24 October 2021 (2021)

# Exploring the Logical Expressiveness of Graph Neural Networks by Establishing a Connection with $\mathcal{C}_2$

Zhangquan Zhou[1(✉)], Chengbiao Yang[2], Qianqian Zhang[1], and Shijiao Tang[1]

[1] School of Information Science, Nanjing Audit University Jinshen College, Nanjing, China
quanzz1129@163.com

[2] School of Computer Science and Engineering, Southeast University, Nanjing, China

**Abstract.** Graph neural networks (GNNs) have gained widespread application in various real-world scenarios due to their powerful ability to handle graph-structured data. However, the computational power and logical expressiveness of GNNs are still not fully understood. This work explores the logical expressiveness of GNNs from a theoretical view and establishes a connection between them and the fragment of first-order logic, known as $\mathcal{C}_2$, which servers as a logical language for graph data modeling. A recent study proposes a type of GNNs called ACR-GNN, demonstrating that GNNs can mimic the evaluation of unary $\mathcal{C}_2$ formulas. Working upon this, we give a variant of GNN architectures capable of handling general $\mathcal{C}_2$ formulas. To achieve this, we leverage a mechanism known as message passing to reconstruct GNNs. The proposed GNN variants enable the simultaneous updating of node and node pair features, allowing for the handling of both unary and binary predicates in $\mathcal{C}_2$ formulas. We prove that the proposed models possess the same expressiveness as $\mathcal{C}_2$. Through experiments conducted on synthetic and real datasets, we validate that our proposed models outperform both ACR-GNN and a widely-used model, GIN, in the tasks of evaluating $\mathcal{C}_2$ formulas.

**Keywords:** Graph neural networks · logical expressiveness · $\mathcal{C}_2$

## 1 Introduction

Graph neural networks (GNNs), a class of neural network architectures designed to process graph-structured data, have been introduced to enhance a wide range of real-world applications in fields such as social science [1], drug discovery [2] and knowledge graphs [3,4]. While these applications benefit greatly from the fault-tolerance and high-performance of GNN models, their inner computational properties and logical expressive power are not yet fully understood [5,6]. Even the currently popular large language models (LLMs) also exhibit suboptimal performance in logical reasoning [7]. This motivates researchers to explore in depth *the expressive power* of GNNs by establishing connections between GNNs and *deterministic models* [8], such as classical graph algorithms [5,9] and first-order logic languages [10]. From these connections, researchers can gain insights


H. Wang et al. (Eds.): CCKS 2023, CCIS 1923, pp. 29–40, 2023.
https://doi.org/10.1007/978-981-99-7224-1_3

into the inherent behavior of specific GNNs, which further leads to potential improvements in the performance and interpretability of GNNs.

A recent advancement in exploring the expressive power of GNNs has established a close connection between GNNs and a fragment of first-order logic, known as $\mathcal{C}_2$ [10,11]. The logic language $\mathcal{C}_2$ allows a maximum of two *free variables* and *counting qualifiers* within logical formulas [12]. The work of [10] proposes a class of GNNs, referred to as ACR-GNNs, which can evaluate formulas of a limited version of $\mathcal{C}_2$. We refer to this limited version as *unary-*$\mathcal{C}_2$, where only unary predicates (e.g., $\texttt{Red}(x)$) and statements of node connectivity are permitted. The proposed approach is based on the finding that unary predicates can be considered as features of graph nodes, and evaluating formulas can thus be transferred to node classification. The authors of [10] prove that ACR-GNN is equivalent to *unary-*$\mathcal{C}_2$ in terms of logical expressiveness. However, ACR-GNNs have not been shown to be capable of handling user-defined binary predicates (e.g., **hasFather(x,y)**). It remains unknown whether a specific type of GNNs or its variants exist for mimicking the evaluation of general $\mathcal{C}_2$ formulas, taking into account various binary predicates as well.

Our work takes a step forward in addressing the question of whether a class of GNNs exists for evaluating general $\mathcal{C}_2$ formulas. Our focus on $\mathcal{C}_2$ stems from its remarkable expressive power in modeling a wide range of graph theory problems [12,13]. Hence investigating the connection between GNNs and well-formed deterministic models from the perspective of $\mathcal{C}_2$ offers a valuable perspective.

It is challenging to evaluate general $\mathcal{C}_2$ formulas directly using widely-used GNN models, such as GIN [5], GCN [14], and ACR-GNN. The difficulty arises from the fact that these GNN models are primarily designed to classify local nodes or global structures of graphs. In contrast, evaluating binary predicates within $\mathcal{C}_2$ formulas requires handling node pairs, which are actually *sub-graph structures*. This issue prompts us to consider two questions: Can we reconstruct GNNs to deal with node pairs, and are the reconstructed GNNs capable of evaluating general $\mathcal{C}_2$ formulas? In this paper, we aim to answer these two questions. Our contributions are as follows:

- We propose a novel GNN architecture, named NP-GNN, to handle node pairs in graph-structured data. NP-GNN is built upon the framework of ACR-GNN model and incorporates a pair-wise *message passing scheme* to enable effective information exchange among both nodes and nodes pairs. We show that NP-GNN is capable of evaluating binary predicates within $\mathcal{C}_2$ formulas whose variables are *guarded*.
- While NP-GNNs lack readout functions for global checking, we establish their logical expressiveness by showing their equivalence to ACR-GNNs. We further introduce NPR-GNNs, a variant of NP-GNN that incorporates readout functions. By integrating readout functions into the architecture, NPR-GNNs can evaluate general $\mathcal{C}_2$ formulas.
- We conduct experiments on synthetic and real datasets to validate the performance of NP-GNNs and NPR-GNNs in evaluating $\mathcal{C}_2$ formulas. The exper-

imental results show that our proposed models can evaluate $\mathcal{C}_2$ formulas and outperform both ACR-GNN and a widely-used GNN model, GIN.

The rest of the paper is organized as follows. In Sect. 2, we introduce basic notions about $\mathcal{C}_2$ and GNNs. We then discuss the approach of evaluating $\mathcal{C}_2$ formulas Sect. 3. We give NP-GNN model and NPR-GNN model in Sects. 4 and 5 respectively, and discuss their logical expressiveness in these two sections. We report the experimental results in Sect. 6 and conclude the work in Sect. 7. The technical report and the source code of our implementations can be found at this address https://gitlab.com/p9324/ccks2023.

## 2 Background Knowledge

### 2.1 The Logic Basics

We study graph-structured datasets in this paper, defined as $G = \langle V, \mathbf{C}, \mathbf{R}\rangle$, where $V$ is a set of graph nodes, $\mathbf{C}$ is a set of *unary predicates* on $V$, and $\mathbf{R}$ is a set of *binary predicates* on $V^2$. We also use $v \in G$ to represent that $v \in V$. We build first-order logic for modeling graph datasets by using $V$ as the *universe set*, with the predicate symbols in $\mathbf{C}$ and $\mathbf{R}$, the variables $x_1, x_2, ..., x_n$, logical operators ($\wedge$, $\vee$, $\neg$, and $\rightarrow$), and quantifiers ($\forall$ and $\exists$). A fragment of first-order logic, $\mathcal{C}_2$, has been studied and used for tasks built on graph-structured datasets [12]. The formulas in $\mathcal{C}_2$ satisfy that (i) *counting qualifiers* ($\exists^{\geq k}$) are allowed, (ii) the appearing predicates are from $\mathbf{C}$ and $\mathbf{R}$, and (iii) one or two *free variables*[1] occur in formulas. A formula $\varphi$ is called a *unary* (resp. *binary*) *formula* if $\varphi$ has one (resp. two) free variables.

We apply model theory for logical semantics by mapping the node set $V$ to the universe set in a model $\mathcal{M}$. We say that a graph $G$ satisfies a unary formula $\varphi$ if $\langle \mathcal{M}, v\rangle$ exists such that $\varphi(v)$ holds, denoted by $(G, v) \models \varphi$. Similarly, a graph $G$ satisfies a binary formula $\varphi$ if $\langle \mathcal{M}, (v, w)\rangle$ exists such that $\varphi(v, w)$ holds, denoted by $(G, (v, w)) \models \varphi$. Evaluation of $\mathcal{C}_2$ formulas on graphs can be seen as *logical classification* [10], where a logical classifier $\mathcal{L}$ is set to decide whether an assignment $\omega$ exists for $\varphi$ such that $\mathcal{L}(G, \omega) = 1$ if $(G, \omega) \models \varphi$ and $\mathcal{L}(G, \omega) = 0$ otherwise, where $\omega$ corresponds to a node $v$ (resp. a node pair $(v, w)$) if $\varphi$ is a unary (resp. binary) formula.

### 2.2 Graph Neural Network

The classical architecture of GNNs [15] involves multiple iterations that perform *aggregation* and *combination* functions. Formally, a GNN classifier $\mathcal{A}$ has $T$ iterations, denoted by $t \in \{1, 2, ..., T\}$, with the input dimension $d_t$. Given a graph $G = \langle V, \mathbf{C}, \mathbf{R}\rangle$ as the input, the computation in each iteration can be described by the following formula:

$$\boldsymbol{x}_v^{t+1} = \textsc{Comb}^{t+1}(\boldsymbol{x}_v^t, \textsc{Agg}^{t+1}(\{\boldsymbol{x}_u^t | u \in \mathcal{N}_G(v)\})) \quad (1)$$

[1] A variable $x$ is free if it is not qualified by $\exists$ and $\forall$. See the formula $\exists y.\varphi(x, y)$, where variable $x$ is free and variable $y$ is not.

where $\boldsymbol{x}_v^t$ is the feature vector of node $v \in V$ in iteration $t$, and $\mathcal{N}_G(v)(:= \{u | \forall u\ (v, u) \in \mathbf{R}\ or\ (u, v) \in \mathbf{R}\})$ is a set of the neighborhoods of $v$. The aggregation function $\text{AGG}^t$ maps a multiset of input vectors in $\mathbb{R}^{d_t}$ to one vector, while the combination function $\text{COMB}^{t+1}$ takes input vectors in $\mathbb{R}^{d_t}$ and outputs one vector in $\mathbb{R}^{d_{t+1}}$. Note that the output $\boldsymbol{x}_v^T$ of the GNN $\mathcal{A}$ is actually the feature vector of the node $v$ after all the $T$ iterations finish computation.

Suppose we have constructed a GNN model $\mathcal{A}_\varphi$ for evaluating the target formula $\varphi$ given a graph $G$. We can then use the following formal definition to describe the equivalent logical expressiveness between $\mathcal{C}_2$ and GNNs.

**Definition 1.** ***(Formula Capturing)*** *Given a graph $G$, we say that a GNN model $\mathcal{A}_\varphi$* captures *the logical classifier $\mathcal{L}_\varphi$ with respect to the $\mathcal{C}_2$ formula $\varphi$ if it holds that, for any assignment $\omega$, $\mathcal{A}_\varphi(G, \omega) = 1$ if and only if $\mathcal{L}_\varphi(G, \omega) = 1$.*

## 3 The Basic Idea of Evaluating Formulas via GNNs

From Definition 1, one can use GNNs to evaluate $\mathcal{C}_2$ formulas through building a connection between $\mathcal{C}_2$ logical classifiers and GNNs. This connection is based on the following observation: evaluating a formula from its innermost subformula to the outmost one can be viewed as a process of attaching node labels. To illustrate it, consider the following unary formula (see (2)) as an example, which describes all boys who have a friend.

$$\texttt{boyWithFreind}(x) \equiv \texttt{child}(x) \wedge \underbrace{\overbrace{\texttt{male}(x) \wedge \underbrace{\exists y.\texttt{friend}(y, x)}_{\phi_1}}^{\phi_2}}_{\phi_3} \tag{2}$$

A logical classifier of handling this formula evaluates the innermost subformula $\phi_1$, then the secondary subfomula $\phi_2$, and finally the outmost one $\phi_3$. It can also be seen as a repeating process of *attaching formulas* ($\phi_1, \phi_2, \phi_3$ in the example), as new labels, to graph nodes and node pairs. The following example shows a case of how this process works for Formula 2.

*Example 1.* Suppose a GNN model $\mathcal{A}$ handles a graph dataset containing two nodes $v_1$ and $v_2$, and a node pair $(v_2, v_1)$. Initially, node $v_1$ holds the two labels '`child`' and '`male`', while the node pair $(v_2, v_1)$ is labeled '`friend`'. The attached labels correspond to the predicates in Formula 2. After model $\mathcal{A}$ verifies that node $v_1$ holds the labels '`child`' and '`male`' and that a '`friend`'-labeled node pair $(v_2, v_1)$ exists, it then attaches the formula notation '`boyWithFreind`' as a label to node $v_1$.

From the perspective of GNNs, the process of attaching formulas can be implemented by mapping node labels to feature vectors and locally visiting nodes to aggregation and combination functions. However, directly using GNNs presents an issue. Unary formulas allow for the statement of isolated variables that are not *guarded* by any free variable. Consider the formula $\varphi(x) \wedge \exists y.\phi(y)$, for example. The qualified variable $y$ is not guarded by the free variable $x$. Consequently, when evaluating this formula, variable $y$ could be assigned to any node in the given graph. This necessitates global checking in addition to the local aggregation and combination functions. In their work, the authors of [10] address this issue by using a global function called readout (denoted by READ), which is placed between the aggregation and combination functions to collect global information after each iteration of GNNs (see (3)).

$$\boldsymbol{x}_v^{t+1} = \text{COMB}^{t+1}(\boldsymbol{x}_v^t, \text{AGG}^{t+1}(\{\boldsymbol{x}_u^t | u \in \mathcal{N}_G(v)\}), \text{READ}^{t+1}(\{\boldsymbol{x}_u^t | u \in G\})) \quad (3)$$

The class of GNNs consisting of *aggregation*, *combination* and *readout* functions are referred to as ACR-GNNs. It has been shown by the following theorem that, ACR-GNNs evaluate formulas of a limited version of $\mathcal{C}_2$. We refer to this limited version as *unary-*$\mathcal{C}_2$, where only unary predicates and statements of node connectivity are permitted, and various user-defined binary predicates are not allowed.

**Theorem 1** (*[10]*)**.** *For any unary-*$\mathcal{C}_2$ *formula* $\varphi$ *whose logical classifier is* $\mathcal{L}_\varphi$*, there exists an ACR-GNN classifier* $\mathcal{A}$ *that captures the logical classifier* $\mathcal{L}_\varphi$*.*

Our objective is to evaluate general $\mathcal{C}_2$ formulas that involve both unary and binary predicates. However, it is apparent that ACR-GNNs cannot achieve this goal because they solely deal with graph nodes, while evaluating binary predicates necessitates GNNs to classify node pairs.

$$\texttt{bloodBrother}(x, y) \equiv \texttt{boy}(x) \wedge \texttt{boy}(y) \wedge \exists z.(\texttt{father}(z, x) \wedge \texttt{father}(z, y)) \quad (4)$$

Consider Formula (4), which represents all blood brothers. Suppose a logical classifier $\mathcal{L}$ evaluates this formula given a graph $G$. This classifier deals with each node pair $(v, u)$ in $G$ and satisfies $\mathcal{L}(G, (v, u)) = 1$ if and only if $(G, (v, u)) \models$ `bloodBrother`. Similarly, A GNN model evaluating Formula (4) should attach the corresponding label `bloodBrother` to node pairs. This can hardly be implemented using aforementioned aggregation, combination and readout functions, as they are restricted to dealing with nodes. This raises two questions:

(1) Can we reconstruct GNN models to classify node pairs and further evaluate general $\mathcal{C}_2$ formulas?
(2) If such GNNs exist, do they offer greater logical expressiveness than ACR-GNNs?

In the following sections, we provide our answers to these questions.

## 4 Evaluating General $\mathcal{C}_2$ Formulas via GNNs

We will now address the first question: how to reconstruct GNNs to enable them to handle node pairs. This necessitates the development of GNNs that are capable of processing features associated with node pairs. One possible approach is to treat node pairs as graph edges and consider the utilization of *edge features*. A notable variant of GNNs, referred to as *message-passing* GNNs (MP-GNNs), has been proposed to facilitate the update of edge features [16]. MP-GNNs have also been employed in the early stages of research in the field of quantum chemistry [17]. In this work, we adopt the methodology given in [16] to construct our own GNNs, leveraging the existing advancements made in ACR-GNNs as well.

One issue arises when applying MP-GNNs directly to evaluate general $\mathcal{C}_2$ formulas. MP-GNNs are designed to update features of existing graph edges, but in a $\mathcal{C}_2$ graph dataset, it is possible for two nodes to not be connected by any edge. For instance, consider the scenario where two boys, denoted as $b_1$ and $b_2$, are discovered to be blood brothers after applying Formula (4). In the original graph, their corresponding nodes are not connected by any edge. In the following work, we address this issue by converting the original input graphs to a *completely connected graph* during the initial stage. This ensures that each node pair has a corresponding edge. Although this conversion results in a theoretical space complexity of $O(n^2)$, the practical computing efficiency can be guaranteed by employing *sparse matrix scheme* [18].

We call our GNN models capable of handling node pairs, NP-GNNs, which are fed with sparsely stored completely connected graphs. Specifically, given a graph $G = \langle V, \mathbf{C}, \mathbf{R}\rangle$, for each node $v \in V$ and node pair $(v, w) \in V^2$, let $\boldsymbol{x}_v$ and $\boldsymbol{e}_{vw}$ be the feature vectors of $v$ and $(v, w)$ respectively. Similar to classical GNNs, NP-GNNs operate through several iterations. The message-passing scheme for updating $\boldsymbol{x}_v^{t+1}$ and $\boldsymbol{e}_{vw}^{t+1}$ in iteration $t+1$ $(t \geq 0)$ can be formalized as follows:

$$\boldsymbol{x}_v^{t+1} = \textsc{Comb}_1^{t+1}(\boldsymbol{x}_v^t, \textsc{Agg}_1^{t+1}(\{\boldsymbol{x}_w^t, \boldsymbol{e}_{vw}^t, \boldsymbol{e}_{wv}^t | w \in \mathcal{N}(v)\})) \quad (5)$$

$$\boldsymbol{e}_{vw}^{t+1} = \textsc{Comb}_2^{t+1}(\boldsymbol{e}_{vw}^t, \textsc{Agg}_2^{t+1}(\{\boldsymbol{e}_{wv}^t, \boldsymbol{e}_{vu}^t, \boldsymbol{e}_{uv}^t, \boldsymbol{e}_{uw}^t, \boldsymbol{e}_{wu}^t, \boldsymbol{x}_v^t, \boldsymbol{x}_w^t | u \in V\})) \quad (6)$$

where the functions $\textsc{Agg}_i^{t+1}$ and $\textsc{Comb}_i^{t+1}$ $(i = 1, 2)$ represent the corresponding aggregation and combination functions. Comparatively, NP-GNNs differ from ACR-GNNs (3) in that both node and node pair feature vectors are simultaneously updated within each iteration. The functions $\textsc{Agg}_i$ $(i = 1, 2)$ are specially designed for this purpose. To be specific, for a given node $v$, the function $\textsc{Agg}_1$ collects information not only from the node's neighbor (i.e., $\boldsymbol{x}_w$) but also from its related node pairs (i.e., $\boldsymbol{e}_{vw}^t$ and $\boldsymbol{e}_{wv}^t$). Similarly, for a given node pair $(v, w)$, $\textsc{Agg}_2$ collects information from all node pairs involving $v$ or $w$ (i.e., $\boldsymbol{e}_{wv}^t, \boldsymbol{e}_{vu}^t, \boldsymbol{e}_{uv}^t, \boldsymbol{e}_{uw}^t$ and $\boldsymbol{e}_{wu}^t$), as well as its related nodes (i.e., $\boldsymbol{x}_v^t$ and $\boldsymbol{x}_w^t$).

Returning to our question, we address it by constructing a specific class of NP-GNNs. We can achieve this by specifying the parameters of these models to ensure their ability to capture logical classifiers that evaluate $\mathcal{C}_2$ formulas in which *all variables are guarded.*

**Theorem 2.** *For any $\mathcal{C}_2$ formula $\varphi$ in which variables are guarded, with its corresponding logical classifier denoted as $\mathcal{L}_\varphi$, there exists an NP-GNN model $\mathcal{A}$ that captures $\mathcal{L}_\varphi$.*

The proof can be found in the technique report. The basic idea of the construction of the specific NP-GNNs is that the components in the feature vectors used for labelling graph nodes (resp. node pairs) represent sub-formulas of the given unary formula (resp. binary formula). Further, if a feature component is set 1 for a node (resp. a node pair) then the corresponding sub-formula is satisfied under the node (resp. the node pair).

## 5 The Logical Expressiveness of NP-GNN

While NP-GNN does not incorporate readout functions, it exhibits competitive logical expressiveness compared to ACR-GNN as demonstrated in Proposition 1. Consider the formula $\varphi(x) \wedge \exists y.\phi(y)$ again, where variable $y$ is unguarded. ACR-GNN is capable of capturing this formula by globally collecting information on the predicate $\phi$ through its readout functions. However, we can rewrite the formula by introducing a *watching variable* $z$ as follows, and utilize NP-GNN to handle the revised formula without relying on readout functions:

$$\psi(x,z) \equiv \varphi(x) \wedge \exists y.\phi'(y,z) \tag{7}$$

Here, $z$ is the newly introduced watching variable. The predicate $\phi'$ is defined based on the original predicate $\phi$. Specifically, for any node pair $(v,w)$, $\phi'(v,w)$ holds if $\phi(v)$ holds. NP-GNNs operate by labeling node pairs with the predicate $\phi'$. The presence of such labels implies the satisfiability of the original sub-formula $\exists y.\phi(y)$. In a more general sense, any unary predicate with unguarded variables can be transformed into an equivalent binary predicate by utilizing watching variables, making it amenable to be handled by NP-GNNs. The feasibility of watching variables lies in the fact that fully connected graphs, serving as input to NP-GNNs, naturally distribute the global information among all nodes through node pairs.

**Proposition 1.** *The NP-GNNs are logically expressive as the ACR-GNNs.*

We will now discuss formulas with unguarded variables in binary predicates. These types of formulas are commonly used in multi-agent systems to determine whether a local event leads to a special situation, contingent upon the satisfaction of a global condition. As an example, we consider the multi-agent system called *Capture The Flag* [19], in which each robot endeavors to capture flags and deliver them to designated coordinates. The termination condition of this game can be expressed by the following formula (see 8): If any robot $x$ delivers a flag $y$ to the destination, and no robot $z$ (including $x$ itself) can move any flag $z'$, then the game round concludes with $x$ gaining a point with flag $y$.

$$\texttt{gainPoint}(x,y) \equiv \texttt{capture}(x,y) \wedge \forall z,z'.\neg\texttt{move}(z,z') \tag{8}$$

As we can see, the variables in the sub-formula $\neg\texttt{move}(z,z')$ are both unguarded. To evaluate such a formula, the corresponding logical classifier must perform a global check to verify whether, for each node pair $(w,v)$, $\neg\texttt{move}(w,v)$ holds. In order to facilitate this global checking by GNNs, we utilize the ACR-GNN method and incorporate two readout functions to update the features of nodes and node pairs, respectively, during each iteration of NP-GNNs.

$$\begin{aligned}
\boldsymbol{x}_v^{t+1} &= \textsc{Comb}_1^{t+1}(\boldsymbol{x}_v^t, \\
&\qquad \textsc{Agg}_1^{t+1}(\{\boldsymbol{x}_w^t, \boldsymbol{e}_{vw}^t, \boldsymbol{e}_{wv}^t | w \in \mathcal{N}(v)\}), \\
&\qquad \textsc{Read}^{t+1}(\{\boldsymbol{x}_v^t | v \in V\})) \\
\boldsymbol{e}_{vw}^{t+1} &= \textsc{Comb}_2^{t+1}(\boldsymbol{e}_{vw}^t, \\
&\qquad \textsc{Agg}_2^{t+1}(\{\boldsymbol{e}_{wv}^t, \boldsymbol{e}_{vu}^t, \boldsymbol{e}_{uv}^t, \boldsymbol{e}_{uw}^t, \boldsymbol{e}_{wu}^t, \boldsymbol{x}_v^t, \boldsymbol{x}_w^t | u \in V\}), \\
&\qquad \textsc{Read}^{t+1}(\{\boldsymbol{e}_{vw}^t | (v, w) \in V^2\}))
\end{aligned} \tag{9}$$

We refer to the modified models as NPR-GNNs, where each layer integrates *aggregation*, *combination* and *readout* functions. It can be checked that NPR-GNNs can capture general $\mathcal{C}_2$ formula classifiers, as shown by the following result.

**Theorem 3.** *For any $\mathcal{C}_2$ formula $\varphi$, with its logical classifier as $\mathcal{L}_\varphi$, there exists an NPR-GNN classifier $\mathcal{A}$ that captures $\mathcal{L}_\varphi$.*

# 6 Experiments

We conducted experiments on both synthetic and real datasets to validate the performance of the proposed models, namely NP-GNN and NPR-GNN, in evaluating $\mathcal{C}_2$ formulas. In order to compare the expressiveness of our models with others, we also tested GIN [5] and ACR-GNN [10] on the same datasets. All of the tested models were implemented using PyTorch Geometric Library (PyG). The synthetic dataset used in the first set of experiments is called *Family Tree*, which serves as a benchmark for inductive logic programming [20]. In the second set of experiments, we used *Planetoid*, a widely-used benchmark that integrates real-world datasets of paper citations [21]. We further evaluated the performance of our models in node classification on the *PPI* dataset [22]. The accuracy results were reported by counting the number of correctly classified nodes and node pairs in all graphs within a dataset. In every experiment, we set up 200 epochs with the Adam optimizer.

**Family Tree.** A manually generated graph of Family Tree consists of $n$ nodes representing different family members, and it includes six types of binary predicates: `hasHusband`, `hasWife`, `hasFather`, `hasMother`, `hasSon`, and `hasDaughter`. The objective is to deduce additional relationships among family members. One of the reasons why we selected this dataset is that it solely contains binary predicates, enabling us to transform node pairs to *high-order graph nodes* and binary predicates to labels for high-order nodes according to the method given in [9]. This transformation allows us to compare our models with GIN and ACR-GNN since they can only handle graph nodes.

We investigated four implicit relationships, which can be described using the original binary predicates as follows:

$$\texttt{hasParent}(x, y) \equiv \texttt{hasFather}(x, y) \wedge \texttt{hasMother}(x, y)$$
$$\texttt{hasGrandParent}(x, y) \equiv \exists z(\texttt{hasParent}(x, z) \wedge \texttt{hasParent}(z, y))$$
$$\texttt{hasSister}(x, y) \equiv \exists z(\texttt{hasParent}(x, z) \wedge \texttt{hasDaughter}(z, y) \wedge x \neq y)$$
$$\texttt{hasUncle}(x, y) \equiv \exists z(\texttt{hasGrandParent}(x, z) \wedge \texttt{hasSon}(z, y) \wedge \neg\texttt{hasFather}(x, y))$$

In each experiment of evaluating the above four relationships, we generated 500 graphs with the numbers of family members randomly set between 10 and 20. We divided the graphs into training and testing sets with an 80%/20% split. The accuracy results are given in Table 1. All models exhibited high performance for the `hasParent` and `hasGrandParent` relationships. However, for `hasSister` and `hasUncle`, the accuracies significantly decreased for GIN and ACR-GNN, whereas NP-GNN and NPR-GNN maintained high performance levels. This indicates that it is more challenging for GIN and ACR-GNN to generalize negative expressions (i.e., $x \neq y$ and $\neg\texttt{hasFather}(x, y)$) compared to NP-GNN and NPR-GNN. The primary reason is that GIN and ACR-GNN treat high-order nodes transformed from node pairs in isolation, whereas NPR-GNN and NP-GNN can perceive the complete connections among nodes. Furthermore, as observed from the `hasSister` and `hasUncle` results, NPR-GNN outperforms NP-GNN. Although NP-GNN can handle the guarded variable $z$ in the corresponding expressions, the readout functions of NPR-GNN improve the learning effectiveness to some extent.

**Table 1.** The Experimental Results on Family Tree Datasets.

| | GIN | | ACR-GNN | | NP-GNN | | NPR-GNN | |
|---|---|---|---|---|---|---|---|---|
| | train | test | train | test | train | test | train | test |
| `hasParent` | 0.93 | 0.92 | 1.00 | 1.00 | 1.00 | 1.00 | 1.00 | 1.00 |
| `hasGrandParent` | 0.97 | 0.96 | 0.98 | 0.94 | 1.00 | 0.97 | 1.00 | 0.98 |
| `hasSister` | 0.87 | 0.84 | 0.91 | 0.89 | 0.97 | 0.94 | 1.00 | 1.00 |
| `hasUncle` | 0.86 | 0.85 | 0.94 | 0.90 | 0.96 | 0.93 | 1.00 | 0.97 |

**Planetiod.** In the second series of experiments, we evaluated the performance of NP-GNN and NPR-GNN on the Planetoid benchmark, which comprises three graph datasets: *Cora*, *Citeseer*, and *Pub*. In these datasets, each node represents an academic paper and has a category label indicating its research direction, while each edge represents a citation link between two papers.

To evaluate the models' performance on tasks that incorporate both unary and binary predicates, we extracted two types of sub-graphs from each dataset using the following formulas, denoted as ($\alpha$) and ($\beta$), respectively:

$$\texttt{cat}_i(y) \equiv \texttt{cites}(x, y) \wedge \texttt{cat}_i(x) \tag{$\alpha$}$$
$$\texttt{cites}(x, z) \equiv \texttt{cites}(x, y) \wedge \texttt{cites}(y, z) \wedge \texttt{cat}_i(x) \wedge \texttt{cat}_i(z) \tag{$\beta$}$$

In the sub-graphs extracted based on formula $\alpha$ (denoted by sub-graphs ($\alpha$)), for each paper $x$ that cites some paper $y$ and $x$ is labeled by the category $\mathtt{cat}_i$, paper $y$ has to be labeled by $\mathtt{cat}_i$ as well. On the other hand, the sub-graphs extracted based on formula $\beta$ (denoted by sub-graphs ($\beta$)) describe a transitive relation among citation links. Specifically, if paper $x$ cites paper $y$ and $y$ cites another paper $z$, and if paper $x$ and paper $z$ are labeled by the same category, then the citation relation between $x$ and $z$ should exist. The statistics of all the sub-graphs for Cora, Citeseer, and Pub are presented in the following table, including the number of nodes, edges and paper categories.

**Table 2.** The Statistics of Extracted Sub-graphs.

| | sub-graph($\alpha$) | | | sub-graph($\beta$) | | |
|---|---|---|---|---|---|---|
| | ♯node | ♯edge | ♯cat | ♯node | ♯edge | ♯cat |
| Cora | 2,551 | 4,418 | 7 | 1,470 | 2,964 | 7 |
| Citeseer | 2,733 | 3,518 | 6 | 1,182 | 2,126 | 6 |
| Pub | 17,196 | 35,578 | 3 | 4,835 | 15,219 | 3 |

**Table 3.** The Experimental results of the Extracted Sub-graphs.

| | | sub-graph($\alpha$) | | sub-graph($\beta$) | |
|---|---|---|---|---|---|
| | | train | test | train | test |
| NP-GNN | Cora | 0.97 | 0.94 | 0.87 | 0.86 |
| | Citeseer | 0.95 | 0.92 | 0.85 | 0.83 |
| | Pub | 0.96 | 0.93 | 0.89 | 0.87 |
| NPR-GNN | Cora | 0.97 | 0.97 | 0.91 | 0.90 |
| | Citeseer | 0.95 | 0.91 | 0.89 | 0.86 |
| | Pub | 0.98 | 0.95 | 0.92 | 0.89 |

We partitioned all sub-graphs into training and testing sets, with an 80%/20% split, respectively. In the testing data, we excluded all instances of $\mathtt{cat}_i(y)$ (respectively, $\mathtt{cites}(x, z)$) that appear on the left-hand side of formula $\alpha$ (respectively, $\beta$). This was done to validate whether the models are capable of deducing the correct facts. It is worth noting that GIN and ACR-GNN are unable to simultaneously handle unary and binary predicates. Consequently, we only present the performance results of NP-GNN and NPR-GNN here.

As shown in Table 3, both models have similar performance for sub-graphs ($\alpha$). NPR-GNN, however, demonstrates a noticeable increase of 3 to 4% points compared to NP-GNN for sub-graphs ($\beta$), which can be attributed to the readout functions. It should be noted that neither model achieves a test accuracy exceeding 0.90. Both NP-GNN and NPR-GNN models tend to quickly learn transitivity expressions. However, in the case of formula $\beta$, it represents a conditional expression rather than a strictly defined transitivity expression. Specifically, paper $x$ cites paper $z$ only when they belong to the same category, as indicated by the presence of the sub-formula $(\mathtt{cat}_i(x) \wedge \mathtt{cat}_i(z))$, which impedes the learning process of NP-GNN and NPR-GNN. Additional experiments were conducted by modifying parameters such as increasing the number of epochs and changing optimizers, but these changes did not yield significant improvements.

**PPI.** In our final set of experiments, we assessed the node classification capabilities of NP-GNN and NPR-GNN using the well-known benchmark dataset, *Protein-Protein Interaction* (PPI) [22], where 4,997 nodes and 129,764 edges are included. Each node in this dataset is represented by a 50-dimensional feature vector and assigned one of 210 classes.

Although the results obtained by NP-GNN and NPR-GNN were competitive with those of GIN and ACR-GNN (NP-GNN: 0.76, NPR-GNN: 0.74, GIN: 0.72, ACR-GNN: 0.76), NP-GNN and NPR-GNN did not demonstrate a significant improvement. This can be attributed to the nature of the PPI dataset, which primarily involves node connectivities rather than diverse binary predicates among nodes. As a result, NP-GNN and NPR-GNN were unable to utilize additional information from node pairs to enhance prediction accuracy.

## 7 Conclusion

We aimed at exploring a variant of GNN architectures capable of capturing general $\mathcal{C}_2$ formulas. By introducing the message passing mechanism, we discovered a type of GNNs called NP-GNNs that can handle both unary and binary predicates. NP-GNNs were found to be comparable to ACR-GNNs in terms of logical expressiveness. To handle formulas with unguarded variables, we proposed NPR-GNNs by extending NP-GNNs with readout functions. Through various experiments, we validated the logical expressiveness of NP-GNNs and NPR-GNNs. The results showed that our models outperformed GIN and ACR-GNN on Family Tree Datasets but did not exhibit significant improvement in PPI classification tasks. Additionally, comparing NP-GNNs and NPR-GNNs revealed that readout functions indeed enhance global checking capabilities.

**Acknowledgement.** Zhangquan Zhou and Shijiao Tang were supported by The Natural Science Foundation of the Jiangsu Higher Education Institutions of China under grant 22KJB520003. Qianqian Zhang was supported by the same foundation under grant 22KJD510009.

## References

1. Yongji, W., Lian, D., Yiheng, X., Le, W., Chen, E.: Graph convolutional networks with Markov random field reasoning for social spammer detection. Proc. AAAI Conf. Artif. Intell. **34**, 1054–1061 (2020)

2. Sun, M., Zhao, S., Gilvary, C., Elemento, O., Zhou, J., Wang, F.: Graph convolutional networks for computational drug development and discovery. Brief. Bioinform. **21**(3), 919–935 (2020)
3. Hamaguchi, T., Oiwa, H., Shimbo, M., Matsumoto, Y.: Knowledge transfer for out-of-knowledge-base entities: a graph neural network approach. arXiv preprint arXiv:1706.05674 (2017)
4. Ji, S., Pan, S., Cambria, E., Marttinen, P., Philip, S.Y.: A survey on knowledge graphs: representation, acquisition, and applications. IEEE Trans. Neural Networks Learn. Syst. **33**(2), 494–514 (2021)
5. Xu, K., Hu, W., Leskovec, J., Jegelka, S.: How powerful are graph neural networks? arXiv preprint arXiv:1810.00826 (2018)
6. Chen, T., Bian, S., Sun, Y.: Are powerful graph neural nets necessary? a dissection on graph classification. arXiv preprint arXiv:1905.04579 (2019)
7. Liu, H., Ning, R., Teng, Z., Liu, J., Zhou, Q., Zhang, Y.: Evaluating the logical reasoning ability of ChatGPT and GPT-4. arXiv preprint arXiv:2304.03439 (2023)
8. Sato, R.: A survey on the expressive power of graph neural networks. arXiv preprint arXiv:2003.04078 (2020)
9. Morris, C., et al.: Weisfeiler and leman go neural: higher-order graph neural networks. Proc. AAAI Conf. Artif. Intell. **33**, 4602–4609 (2019)
10. Barceló, P., Kostylev, E.V., Monet, M., Pérez, J., Reutter, J., Silva, J.-P.: The logical expressiveness of graph neural networks. In: 8th International Conference on Learning Representations (ICLR 2020) (2020)
11. Grohe, M.: The logic of graph neural networks. In: 2021 36th Annual ACM/IEEE Symposium on Logic in Computer Science (LICS), pp. 1–17. IEEE (2021)
12. Immerman, N., Lander, E.: Describing graphs: a first-order approach to graph canonization. In: Complexity theory retrospective, pp. 59–81. Springer (1990). https://doi.org/10.1007/978-1-4612-4478-3_5
13. Cai, J.-Y., Fürer, M., Immerman, N.: An optimal lower bound on the number of variables for graph identification. Combinatorica **12**(4), 389–410 (1992)
14. Kipf, T.N., Welling, M.: Semi-supervised classification with graph convolutional networks. CoRR, abs/1609.02907 (2016)
15. Scarselli, F., Gori, M., Tsoi, A.C., Hagenbuchner, M., Monfardini, G.: The graph neural network model. IEEE Trans. Neural Networks **20**(1), 61–80 (2008)
16. Kearnes, S., McCloskey, K., Berndl, M., Pande, V., Riley, P.: Molecular graph convolutions: moving beyond fingerprints. J. Comput. Aided Mol. Des. **30**(8), 595–608 (2016). https://doi.org/10.1007/s10822-016-9938-8
17. Gilmer, J., Schoenholz, S.S., Riley, P.F., Vinyals, O., Dahl, G.E.: Neural message passing for quantum chemistry. In: International Conference on Machine Learning, pp. 1263–1272. PMLR (2017)
18. Langr, D., Tvrdik, P.: Evaluation criteria for sparse matrix storage formats. IEEE Trans. Parallel Distrib. Syst. **27**(2), 428–440 (2015)
19. Zhou, Z., Huang, J., Xu, J., Tang, Y.: Two-phase jointly optimal strategies and winning regions of the capture-the-flag game. In: IECON 2021–47th Annual Conference of the IEEE Industrial Electronics Society, pp. 1–6. IEEE (2021)
20. Dong, H., Mao, J., Lin, T., Wang, C., Li, L., Zhou, D.: Neural logic machines. arXiv preprint arXiv:1904.11694 (2019)
21. Yang, Z., Cohen, W., Salakhudinov, R.: Revisiting semi-supervised learning with graph embeddings. In: International Conference on Machine Learning, pp. 40–48. PMLR (2016)
22. Zitnik, M., Leskovec, J.: Predicting multicellular function through multi-layer tissue networks. Bioinformatics **33**(14), i190–i198 (2017)

# Research on Joint Representation Learning Methods for Entity Neighborhood Information and Description Information

Le Xiao[1], Xin Shan[1], Yuhua Wang[1], and Miaolei Deng[2(✉)]

[1] School of Information Science and Engineering, Henan University of Technology, Zhengzhou 450001, China

[2] Graduate School, Henan University of Technology, Zhengzhou 450001, China

dml_1978@foxmail.com

**Abstract.** To address the issue of poor embedding performance in the knowledge graph of a programming design course, a joint representation learning model that combines entity neighborhood information and description information is proposed. Firstly, a graph attention network is employed to obtain the features of entity neighboring nodes, incorporating relationship features to enrich the structural information. Next, the BERT-WWM model is utilized in conjunction with attention mechanisms to obtain the representation of entity description information. Finally, the final entity vector representation is obtained by combining the vector representations of entity neighborhood information and description information. Experimental results demonstrate that the proposed model achieves favorable performance on the knowledge graph dataset of the programming design course, outperforming other baseline models.

**Keywords:** Knowledge Graph · Representation Learning · Entity Neighborhood · Entity Description

## 1 Introduction

Knowledge graph representation learning is a method that transforms entities and relationships in a knowledge graph into low-dimensional vectors, enabling efficient computation of complex semantic associations [1]. It serves as the foundation for downstream tasks such as knowledge reasoning and knowledge base construction. In the knowledge graph of a programming design course, a target entity is connected to other relevant entities through relationships, and the relationships and connected entities starting from the target entity are referred to as its structural neighborhood. Different entities within the neighborhood have varying degrees of importance to the tar-get entity, and entities typically possess rich description information.

As illustrated in Fig. 1, the target entity “Array” is connected to neighboring entities such as “Memory” and “One-dimensional Array” through relationships like “require” and “include”. Additionally, it is accompanied by the description information “Definition


H. Wang et al. (Eds.): CCKS 2023, CCIS 1923, pp. 41–53, 2023.
https://doi.org/10.1007/978-981-99-7224-1_4

of an array". Traditional representation learning methods for course knowledge graphs typically focus on embedding individual entities without considering their neighborhood information and description information, leading to suboptimal embedding performance.

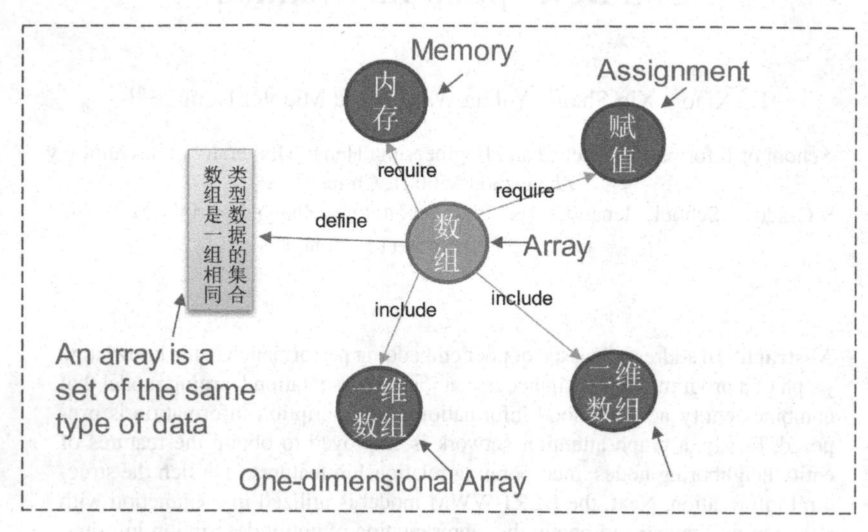


**Fig. 1.** Sample Knowledge Graph of Programming Course

This paper proposes a representation learning model, named NDRL (A Representation Learning Model for Joint Entity Neighborhood Information and Description Information), based on the characteristics of the programming design course knowledge graph. The model aims to effectively integrate entity neighborhood information and description information, utilizing the information within the knowledge graph to obtain high-quality embedding representations. This model plays a significant role in subsequent tasks such as knowledge reasoning [2, 3], completion [4, 5], and applications [6, 7] based on the course knowledge graph.

## 2 Related Work

Knowledge graphs are often represented symbolically, which can lead to issues such as low computational efficiency and data sparsity [8]. With the development and application of deep learning, there is a growing desire for more simple and efficient representations of knowledge graphs. This has given rise to knowledge graph representation learning methods, which aim to map the elements of knowledge graphs, including entities and relationships, into a continuous low-dimensional vector space. The goal is to learn vector representations for each element in the vector space, thereby mapping the triplets from a high-dimensional one-hot vector space to a continuous low-dimensional dense real-valued vector space [9]. This approach addresses the problem of data sparsity in knowledge bases and enables efficient computation.

Among the existing research, translation models are the most representative and classic methods [10]. The fundamental idea behind translation models is to map entities and relationships into a shared vector space, where the semantic information of entities and relationships can be represented by the similarity between vectors. However, these methods typically independently learn the structural features of each triplet, without incorporating the semantic information present in the knowledge graph.

To address this issue, several knowledge graph representation learning methods based on semantic information have been proposed in recent years. KG-BERT [11] represent entities and relations as their name/description textual sequences and turn the knowledge graph completion problem into a sequence classification problem. The RotatE [12] model defines each relation as a rotation from the source entity to the target entity in the complex vector space, which is able to model and infer various relation patterns. The methods based on Graph Convolutional Neural Networks (GCN) [13] are one of the most commonly used approaches. GCN learns the semantic relationships between entities and relationships by performing convolutions on the graph, thereby integrating this information into vector representations.

In recent years, researchers have proposed improved graph neural network models, such as Graph Attention Network (GAT) [14]. GAT utilizes attention mechanisms to learn the interaction between neighbor nodes. Compared to GCN, GAT not only captures the complex relationships between nodes more comprehensively but also exhibits better interpretability and generalizability.

In addition to the structural information of the triplets themselves, knowledge graphs often contain rich additional information such as entity descriptions and attribute information. Xie et al. [15] incorporated entity description information from the knowledge graph into knowledge graph representation learning and proposed the DKRL model. The model used both convolutional neural networks and continuous bag-of-words models to encode entity description information. It leveraged both factual triplets and entity description information for learning and achieved good inference performance. However, since these representations did not include the entire semantic information of entity descriptions, there might be some loss of semantic information. Additionally, in many large-scale knowledge graphs, there is a lack of entity descriptions for many entities. To address this, Wang et al. [16] introduced an external text corpus and used the semantic structures of entities in the text corpus as part of the entity representation, further improving the accuracy of knowledge inference in cases of missing entity descriptions. Reference [17] proposed a rule-guided joint embedding learning model for knowledge graphs. It utilized graph convolutional networks to fuse context information and textual information into the embedding representation of entities and relationships, further enhancing the representation capability of entities and relationships. Inspired by translation-based graph embeddings designed for structural learning, Wang et al.[18] apply a concatenated text encoder. Then, a scoring module is proposed based on these two representations, in which two parallel scoring strategies are used to learn contextual and structural knowledge.

In this work, we address the following issues based on existing work [15, 19]. The following research and improvements have been conducted:

1. A representation learning model is proposed that jointly considers entity neighborhood information and description information. Improved GAT and BERT models are

used to obtain vector representations of entity neighborhood information and description information, respectively, to fully utilize the hidden complex entity relationship feature vectors in knowledge graph triplets and represent the target entity.
2. Joint learning is performed on the vector representations of entity neighborhood information and description information, training both types of representations in the same continuous low-dimensional vector space to better utilize the two different types of information.
3. For entities that already have rich neighborhood information, to avoid noise interference caused by the addition of description information, the concept of "entity structure richness" is defined. Based on the magnitude of entity structure richness, different representation learning methods are selected to obtain the optimal vector representation. Experiments on a dataset of programming course knowledge graph demonstrate that the proposed model outperforms other baseline models.

## 3 Our Model

In the entity neighborhood information representation module, we employ an improved Graph Attention Network (GAT) model to obtain the embedding representation. In the entity description information module, we use the BERT-WWM model and attention mechanism to obtain the corresponding embedding representation. Different approaches are selected to obtain the final entity vector representation based on the entity structure richness. The model framework is illustrated in Fig. 2.

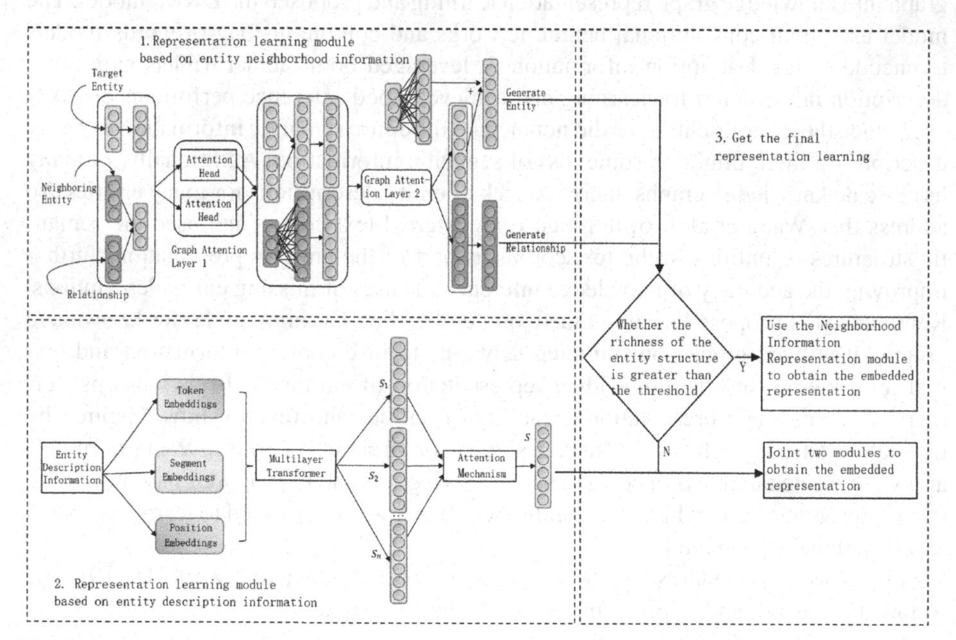


**Fig. 2.** Representation learning model with joint entity neighborhood information and description information

### 3.1 Representation Learning Based on Entity Neighborhood Information

Traditional Graph Attention Network (GAT) models learn the weights of neighboring nodes to perform weighted summation of their features, but they do not consider the importance of relationships for entity representation. To incorporate relationships as important information during training and combine them with the structural features of triplets in the knowledge graph, we enhance the GAT model by adding relationships as significant information to the graph attention mechanism. Specifically, to apply the attention mechanism to the target node and its neighboring nodes, we first compute weighted sums of entities and relationships in the neighborhood. The construction methods for the target node and its neighboring nodes are defined as Eqs. (1) and (2):

$$h_i = t_s \tag{1}$$

$$h_j = \rho h_s + (1 - \rho) r_s \tag{2}$$

Among them, $h_s$, $t_s$ and $r_s$ represent the initial vector representations of the head entity, tail entity, and relationship, respectively. The weight parameter $\rho \in (0,1)$ is used to adjust the proportion of the relationship vector compared to the entity vector when constructing neighboring nodes. This allows both the entity and the relationship of each triplet to participate in the computation of the graph attention model. To calculate the influence weight of $h_j$ on the target node $h_i$ we define their attention value $v_{ij}$ as shown in Eq. (3):

$$v_{ij} = a\left(Wh_i, Wh_j\right) \tag{3}$$

whereas: parameter $W$ represents the projection matrix, and the attention mechanism a is a single-layer feed-forward neural network. Expanding Eq. (3) yields the specific calculation formula:

$$v_{ij} = LeakyRelu\left(z^T\left[Wh_i \| Wh_j\right]\right) \tag{4}$$

After multiplying the projection matrix with the feature vectors and concatenating them together, a linear transformation is applied using the weight vector $z$. Then, a nonlinear activation is performed using the LeakyReLU function. Finally, the Softmax function is applied to normalize the attention values between each node and all its neighboring nodes. The normalized attention weights serve as the final attention coefficients, as shown in Eq. (5):

$$\alpha_{ij} = Softmax_j\left(v_{ij}\right) = \frac{\exp(v_{ij})}{\sum_{k \in N_i} \exp(v_{ik})} \tag{5}$$

where $N_i$ represents the neighboring nodes of the target node $h_i$, which consists of the entities $h_s$ adjacent to the target node $t_s$ and the relations $r_s$ between them as defined in Eq. (2). The attention coefficients calculated are then weighted and summed up as shown in Eq. (6):

$$h_i^{'} = \sigma\left(\sum_{j \in N_i} \alpha_{ij} Wh_j\right) \tag{6}$$

where $h_i^{'}$ represents the new feature vector for each node $i$ based on the output of GAT, which integrates the neighborhood information of entities in the knowledge graph. The function $\sigma$ is the activation function, and the output of the target node is related to the feature vectors of all neighboring nodes. To enable the model to learn the features of neighboring nodes more stably, a multi-head attention mechanism is used to obtain different features for integration. To prevent overfitting, the vectors obtained from K-independent attention mechanisms are concatenated. The specific representation is given by Eq. (7):

$$h_i^{'} = \overset{K}{\underset{k=1}{\|}} \sigma\left(\sum_{j \in N_i} \alpha_{ij}^k W^k h_j\right) \tag{7}$$

In the last layer of the graph attention model, the obtained vector representations are averaged instead of concatenated. This can be expressed by Eq. (8):

$$h_i^{'} = \sigma\left(\frac{1}{K}\sum_{j \in N_i} \alpha_{ij}^k W^k h_j\right) \tag{8}$$

To ensure that relation vector representations have the same output dimension as entity vector transformations, they will share the output dimension. Following a graph attention calculation, the relation vectors undergo a linear transformation, as depicted in Eq. (9):

$$R^{'} = RW^R \tag{9}$$

The input set of relation vectors is denoted as R, and $W^R \in R^{T \times T'}$ represents the linear transformation matrix. Here, T corresponds to the dimension of the original vectors, and T^' represents the dimension after transformation. However, in the process of obtaining new entity vector representations, there is a potential loss of the original structural features. To tackle this issue, the initial entity vectors undergo a linear transformation and are then added to the final entity representations in the following manner:

$$E^{'} = EW^E + E^f \tag{10}$$

$W^E \in R^{T^i \times T^f}$, The parameter $T^i$ represents the initial dimension of entity vectors, while $T^f$ represents the final dimension. $E$ represents the set of initial input entity vectors, and $E^f$ denotes the set of entity vector representations learned through GAT.

### 3.2 Representation Learning Based on Entity Description Information

In this paper, the BERT model is introduced to represent the complete entity description information. The entity description information serves as the direct input to the BERT model, minimizing information loss and capturing the full semantic representation of the entity description. The model employs a multi-layer Transformer structure, which captures bidirectional relationships within sentences. However, since the BERT model masks individual characters during training and does not consider Chinese word segmentation conventions, this paper utilizes the BERT-WWM model, an upgraded version of BERT specifically optimized for Chinese tasks. It incorporates the whole-word masking technique, allowing for better handling of the complex language structure in Chinese. As shown in Table 1, through tokenization, the input text "数组经常被用作实际参数" (Arrays are often used as actual parameters) is segmented into several words, such as " 数组" (arrays), "经常" (often), "被" (are), "用作" (used as), "实际" (actual), and "参数" (parameters). Traditional BERT masking randomly selects words for masking, for example, replacing "组" (group) and "参" (participation) with the [MASK] token. In contrast, according to the BERT-WWM model, the "数" (number) in "数组" (arrays) and the "数" in "参数" (parameters) would also be replaced by the [MASK] token. This enhancement aims to improve the model's performance.

**Table 1.** Example table of whole word MASK

| Input text | 数组经常被用作实际参数 |
|---|---|
| word segentation | 数组 经常 被 用作 实际 参数 |
| BERT masking mechanism | 数[MASK] 经常 被 用作 实际 [MASK]数 |
| BERT-WWM masking mechanism | [MASK][MASK] 经常 被 用作 实际 [MASK][MASK] |

Firstly, the entity description information is transformed into word embedding, segmentation embedding, and location embedding; then it is vector stitched as the input of the BERT-WWM model, and the sentence vector $S_i$ $(i = 1,2,\ldots,n)$ of this entity description information is obtained by multi-layer Transformer structure, which is represented as the sentence vector of the $i$-th sentence.

After that, using the vector representation $h_i^{'}$ of the target entity obtained in the entity neighborhood representation module, the influence weight of each sentence vector $S_i$ of the description information of the target entity is calculated in the same way as Eqs. (3)–(8), and the attention weight of $S_i$ depends on the correlation between $S_i$ and $h_i^{'}$, from which the attention weight distribution of each sentence vector is calculated, and the weighted aggregation of each sentence vector representation is obtained to obtain the entity vector representation $S$ of descriptive information.

### 3.3 Obtaining the Final Embedding Representation

After obtaining the vector representations of the above two modules separately, this paper takes two approaches to the two vector representations, one is to perform a joint representation of the two, by combining the triadic structural information of the knowledge graph and entity descriptions for the training of the model in an integrated manner. The vector representations of entities and relations are learned in the same continuous low-dimensional vector space. The energy function of the synthesis is defined as shown in Eq. (11)

$$d = d_g + d_w \tag{11}$$

where $d_g = \|h_g + r - t_g\|_2$ is the GAT-based energy function and the $h_g$ and $t_g$ sub-tables are the GAT-based representations of the head entity and tail entity; $d_w$ is the energy function based on the description information. To achieve unity between the two in the same vector space, the relational representation r in GAT is used in training, and $d_w$ is defined as in Eq. (12)

$$d_w = d_{ww} + d_{wg} + d_{gw} \tag{12}$$

where $d_{ww} = \|h_w + r - t_w\|_2$, $h_w$ and $t_w$ are the description information-based representations of the head entities and tail entities, respectively. In $d_{wg}$ and $d_{gw}$, one of the head and tail entities is represented using a vector based on entity description; the other is represented using a vector obtained from GAT training as $d_{wg} = \|h_w + r - t_g\|_2$ and $d_{gw} = \|h_g + r - t_w\|_2$, respectively. The two types of representations are jointly trained in the above way to obtain the final vector representation.

However, after our experiments, we found that when an entity has more neighboring entities, joint representation learning is not yet as effective as using only the neighborhood information module. This is because when the entity has richer neighbors, these neighbors are already able to provide enough information to the entity, and then joined with the description information at this time will bring some noise instead, which affects the embedding effect. Therefore, this leads to the second treatment, which is to use only the vector representation obtained in the neighborhood information representation module as the final vector representation, and therefore, this involves the question of a threshold, i.e., which entity is represented by the joint representation and which entity is represented by the neighborhood information representation module. Therefore, we define a concept called "entity structure richness" to measure the size of entity neighborhood information, and select different representation learning methods according to the size of entity structure richness, which is defined as shown in Eq. (13).

$$N(e) = n_e + kn_{N_e} \tag{13}$$

where $n_e$ is the entity degree, $n_{N_e}$ is the degree of entity neighbor nodes, and $k$ is the hyperparameter in the range of 0–1.

### 3.4 Loss Function

Based on the above computational analysis, the loss function is further constructed. A boundary-based optimization method is defined and used as the training objective to optimize this model by minimizing the loss function L. Both vector representations in the model use this loss function.

$$L = \sum_{(h,r,t)\in T} \sum_{(h',r',t')\in T'} max\left(\gamma + d(h,r,t) - d\left(h',r,t'\right), 0\right) \tag{14}$$

where $\gamma$ is the boundary parameter measuring the correct and incorrect triples. $T$ is the set of positive examples consisting of the correct triples $(h, r, t)$ and $T'$ is the set of negative examples consisting of the incorrect triples $\left(h', r, t'\right)$, and $T'$ is defined as shown in Eq. (15):

$$T' = \left\{\left(h', r, t\right) | h' \in \varepsilon\right\} \cup \left\{\left(h, r, t' | t' \in \varepsilon\right)\right\} \tag{15}$$

$T'$ in Eq. (15) is obtained by randomly replacing the head entity, tail entity, or relationship in the set of positive examples to obtain the corresponding set of negative examples. During the training of the model, the optimization operation is performed using stochastic gradient descent to minimize the value of its loss function.

# 4 Experiment

## 4.1 Dataset

At present, there is no public authoritative knowledge graph of programming classes in the field of knowledge mapping representation learning. In this paper, the course knowledge obtained from the "C Programming" textbook, teacher's courseware, Baidu encyclopedia, etc. are used as data sources, and the course knowledge points are used as entities, the description information of knowledge points as attributes and three entity relationships between knowledge points and knowledge points are set, namely, " include": indicates the inclusion relationship between entities of similar knowledge points, "require": indicates the logical dependency relationship between knowledge points, "relate": indicates the relates": denotes the related relationship between knowledge points. Thus, the programming course knowledge graph dataset (PDCKG) was constructed. A total of 2685 entities and 9869 triples were obtained, and the training set, validation set, and test set were divided according to the ratio of 7:1.5:1.5.

## 4.2 Parameter Settings

In obtaining the GAT-based representation, the TransE training vector based on Xavier initialization is used as the initialized structural feature vector representation of entities and relationships. The alpha parameter of LeakyReLU is set to 0.2. In order to prevent overfitting of the model, L2 regularization is used; in order to obtain the BERT-WWM-based representation of entity description information, the alpha parameter of LeakyReLU is set to 0.2. The hyperparameter k in the entity richness expression is set to 0.5, and the entity structure richness threshold is set to 12, i.e., when the entity structure richness is not less than 12, the vector representation obtained based on the neighborhood information module is used as the final representation, and vice versa, the vector representation obtained from the joint representation is used as the final representation. Let the learning rate $\lambda$ be 0.004, the boundary value $\gamma$ be 1.0, and the size of the batch be 512 during the model training.

## 4.3 Link Prediction Task Experiment

Link prediction aims to test the inferential prediction ability of the model. For a given correct triple $(h, r, t)$, after the missing head entity h or tail entity $t$, the head and tail entities are randomly selected in the original entity set to complete the set, and for the missing positions, the scores of the reconstituted triples are calculated by the model, and then sorted in ascending order, and the ranking of the final correct triple is recorded. That is, the tail entity t in the missing triple $(h, r, )$ or the head entity $h$ in the missing triple (, $r$, $t$) is predicted. Mean Rank (MR), Mean Reciprocal Rank (MRR), Hits@1, and Hits@10 are standard evaluation measures for the dataset and are evaluated in our experiments.

In addition, there is a problem that when constructing negative samples, the new triple formed after replacing the head entity or tail entity may already exist in the knowledge graph, which may interfere with the actual ranking of the correct triple and have some influence on the evaluation results. Therefore, in this paper, the link prediction experiments are divided into "raw" and "filter" according to whether to filter the existing triples or not. The experimental results of each model on PDCKG are shown in Table 2.

**Table 2.** Effectiveness of link prediction for each model

| Model | hits@1/% | | hits@10/% | | MR | | MRR | |
|---|---|---|---|---|---|---|---|---|
| | Filter | Raw | Filter | Raw | Filter | Raw | Filter | Raw |
| TransE | 8.39 | 5.85 | 32.34 | 20.54 | 556.98 | 804.51 | 0.126 | 0.091 |
| DKRL | 10.11 | 7.64 | 35.79 | 22.66 | 387.25 | 519.55 | 0.185 | 0.156 |
| R-GCN | 10.41 | 7.28 | 28.09 | 17.85 | 679.99 | 906.79 | 0.173 | 0.148 |
| RotatE | 25.68 | 20.35 | 51.84 | 40.12 | 201.36 | 341.66 | 0.346 | 0.296 |
| KG-BERT | 22.07 | 17.26 | 48.25 | 39.16 | 190.58 | 301.52 | 0.365 | 0.287 |
| KBGAT | 26.87 | 20.96 | 53.25 | 44.15 | 185.62 | 310.19 | 0.304 | 0.245 |
| StAR | 21.06 | 17.34 | 44.38 | 37.91 | 180.55 | 292.36 | 0.283 | 0.251 |
| **NDRL(Ours)** | **28.64** | **21.58** | **64.13** | **50.39** | **105.63** | **198.68** | **0.387** | **0.332** |

It can be seen that on the filtered dataset, each model performs better than the original dataset, which illustrates the necessity of performing filtering operations, and NDRL outperforms other comparable models in all indexes. Compared with the DKRL model, our model uses BERT to represent the entity description information, which can minimize the loss of information and obtain the entity description as much as possible. Compared with the R-GCN model, our model uses the method of constructing neighbor nodes by combining entities and relations so that the target node combines more information about entities and relations in the neighborhood, which improves the inference ability of the model. Compared with KBGAT, because the structure-based model only considers the structured information of the triad, when the corresponding information is missing, it will not be able to make predictions, the entity description information can be used as a favorable supplement to the structure-based model, thus improving the ability of knowledge graph representation learning and the performance of prediction, and therefore, the model has a richer representation capability.

### 4.4 Ablation Experiment

To further verify the effectiveness of the model in this paper, some modules in the model in this paper are removed and compared with the model in this paper, Hits@1, Hits@10, and MR are used as evaluation indexes, and the comparison models are as follows:

1. (NDRL-r): In the entity neighborhood information representation module, the traditional GAT is used to obtain the entity vector representation without adding the relationship into the model.
2. (NDRL-a): In the entity description information representation module, after obtaining the sentence vectors, the attention mechanism is not used to obtain the final vector representation, and the traditional averaging summation method is used.
3. (NDRL-s): In the module of obtaining the final vector representation, the final result is obtained directly by using the joint representation learning method without doing the discriminant of entity structural richness.

The experimental results are shown in Table 3.

**Table 3.** Ablation experiment

| Model | hits@1/% | | hits@10/% | | MR | | MRR | |
|---|---|---|---|---|---|---|---|---|
| | Filter | Raw | Filter | Raw | Filter | Raw | Filter | Raw |
| NDRL-r | 25.58 | 18.69 | 58.65 | 44.98 | 150.64 | 272.36 | 0.364 | 0.305 |
| NDRL-a | 26.21 | 19.36 | 60.36 | 48.68 | 125.65 | 223.85 | 0.371 | 0.311 |
| NDRL-s | 25.28 | 19.10 | 57.01 | 45.54 | 161.27 | 279.35 | 0.349 | 0.296 |
| NDRL | **28.64** | **21.58** | **64.13** | **50.39** | **105.63** | **198.68** | **0.387** | **0.332** |

As can be seen from the table, for each metric, removing any of the modules from the model makes the performance of the model decrease compared to the NDRL model, so the effect of all modules is positive.

### 4.5 Error Analysis

During the experiment, we found that there are many ORC (one-relation-circle) structures on the PDCKG dataset, i.e., structures composed of some special relations such as symmetry and propagation relations, etc. After statistics, the ORC structure data in PDCKG accounted for 15.38% of the total data. And the translation model based on TransE cannot handle such ORC structures. For example, for the symmetric relation $r$, there exist $(h, r, t) \in G$ and $(t, r, h) \in G$. In TransE, the corresponding vectors then satisfy $h + r \approx t$ and $t + r \approx h$, which can be mathematically introduced: $h \approx t$ and $r \approx 0$. This leads to the inability to model, which has a certain impact on the experimental effect. In the subsequent work, other more effective modeling approaches will be explored to deal with the ORC structure present in the data and make the model obtain better performance.

# 5 Summary

In this paper, we proposed a representation learning model, NDRL, for the union of entity domain information and description information, which integrates the entity neighborhood information and entity description information in the knowledge graph of programming courses. First, a combined representation of the relationships of entities and neighboring entities using the GAT-based representation learning model is used to obtain the corresponding neighborhood information representation; then, the entity description information is encoded and represented by the BERT-WRM model to obtain the entity description information representation corresponding to the entities; finally, it is integrated into a joint model for joint training and learning, and the experimental on the PDCKG dataset The results show that the NDRL model proposed in this paper can improve the performance of link prediction and triad classification tasks well compared with other benchmark models.

In our future work, we will further investigate the representation learning method for the knowledge graph of programming courses and hope to improve it in the following 2 aspects: 1) The model in this paper focuses on using the neighborhood information and entity description information of entities, but it has not been utilized for the knowledge graph such as category information and other knowledge base information, so the joint knowledge representation learning method of multiple sources is still future research and improvement. 2) For the ORC structures existing in the knowledge graph of programming courses, we will further explore other more effective modeling approaches to deal with such structures so that the model can obtain better performance.

# References


1. Zhang, T.C., Tian, X., Sun, X.H., et al.: Overview on knowledge graph embedding technology research. Ruan Jian Xue Bao/Journal of Software **34**(1), 277–311 (2023). (in Chinese)
2. Tian, L., Zhou, X., Wu, Y.P., et al.: Knowledge graph and knowledge reasoning: a systematic review. J. Electronic Science and Technol. **20**(2), 100159 (2022)
3. Li, P., Wang, X., Liang, H., et al.: A fuzzy semantic representation and reasoning model for multiple associative predicates in knowledge graph. Inf. Sci. **599**, 208–230 (2022)
4. Chou, Y.Y., Wu, P.F., Huang, C.Y., et al.: Effect of digital learning using augmented reality with multidimensional concept map in elementary science course. Asia Pac. Educ. Res. **31**(4), 383–393 (2022)
5. Kejriwal, M.: Knowledge graphs: constructing, completing, and effectively applying knowledge graphs in tourism. Applied Data Science in Tourism: Interdisciplinary Approaches, Methodologies, and Applications. Springer International Publishing, Cham, pp. 423-449 (2022). https://doi.org/10.1007/978-3-030-88389-8_20
6. Su, X., Hu, L., You, Z., et al.: Attention-based knowledge graph representation learning for predicting drug-drug interactions. Briefings in Bioinformatics, **23**(3), bbac140 (2022)
7. Zhao, B.W., Hu, L., You, Z.H., et al.: Hingrl: predicting drug–disease associations with graph representation learning on heterogeneous information networks. Briefings in Bioinformatics, **23**(1), bbab515 (2022)
8. Feng, H., Ma, J.L., Xu, S.J., et al.: Multi-label text classification method combining label embedding and knowledge-aware. J. Nanjing University (Natural Science), **59**(02), 273–281 (2023) (in Chinese)

9. Pinhanez, C., Cavalin, P.: Exploring the Advantages of Dense-Vector to One-Hot Encoding of Intent Classes in Out-of-Scope Detection Tasks. arXiv preprint arXiv:2205.09021 (2022)
10. Karetnikov, A., Ehrlinger, L., Geist, V.: Enhancing transe to predict process behavior in temporal knowledge graphs. Database and Expert Systems Applications-DEXA 2022 Workshops: 33rd International Conference, DEXA 2022, Vienna, Austria, August 22–24, 2022, Proceedings. Springer International Publishing, Cham, pp. 369-374 (2022). https://doi.org/10.1007/978-3-031-14343-4_34
11. Yao, L., Mao, C., Luo, Y.: Kg-Bert: BERT for Knowledge Graph Completion. arXiv preprint arXiv:1909.03193 (2019)
12. Sun, Z., Deng, Z.H., Nie, J.Y., et al.: Rotate: Knowledge Graph Embedding by Relational Rotation in Complex Space. arXiv preprint arXiv:1902.10197 (2019)
13. Ren, H., Lu, W., Xiao, Y., et al.: Graph convolutional networks in language and vision: a survey. Knowledge-Based Syst. 109250 (2022)
14. Veličković, P., Cucurull, G., Casanova, A., et al.: Graph Attention Networks. arXiv preprint arXiv:1710.10903 (2017)
15. Xie, R., Liu, Z., Jia, J., et al.: Representation learning of knowledge graphs with entity descriptions. Proceedings of the AAAI Conference on Artificial Intelligence **30**(1) (2016)
16. Wang, Z., Li, J., Liu, Z., et al.: Text-enhanced representation learning for knowledge graph. Proceedings of International Joint Conference on Artificial Intelligent (IJCAI), pp. 4–17 (2016)
17. Yao, S.Y., Zhao, T.Z., Wang, R.J., et al.: Rule-guided joint embedding learning of knowledge graphs. J. Computer Research and Dev. **57**(12), 2514–2522 (2020) (in Chinese)
18. Wang, B., Shen, T., Long, G., et al.: Structure-augmented text representation learning for efficient knowledge graph completion. Proceedings of the Web Conference 2021, pp. 1737–1748 (2021)
19. Lin, Y., Liu, Z., Luan, H., et al.: Modeling Relation Paths for Representation Learning of Knowledge Bases. arXiv preprint arXiv:1506.00379 (2015)

# Knowledge Acquisition and Knowledge Base Construction

# Harvesting Event Schemas from Large Language Models

Jialong Tang[1,3], Hongyu Lin[1], Zhuoqun Li[1,3], Yaojie Lu[1], Xianpei Han[1,2], and Le Sun[1,2](✉)

[1] Chinese Information Processing Laboratory, Beijing, China
[2] State Key Laboratory of Computer Science, Institute of Software, Chinese Academy of Sciences, Beijing, China
[3] University of Chinese Academy of Sciences, Beijing, China
{jialong2019,hongyu,xianpei,sunle}@iscas.ac.cn

**Abstract.** Event schema provides a conceptual, structural and formal language to represent events and model the world event knowledge. Unfortunately, it is challenging to automatically induce high-quality and high-coverage event schemas due to the open nature of real-world events, the diversity of event expressions, and the sparsity of event knowledge. In this paper, we propose a new paradigm for event schema induction – knowledge harvesting from large-scale pre-trained language models, which can effectively resolve the above challenges by discovering, conceptualizing and structuralizing event schemas from PLMs. And an **E**vent **S**chema **H**arvest**er** (**ESHer**) is designed to automatically induce high-quality event schemas via in-context generation-based conceptualization, confidence-aware schema structuralization and graph-based schema aggregation. Empirical results show that ESHer can induce high-quality and high-coverage event schemas on varying domains.

**Keywords:** Event schema induction · Pre-trained language models

## 1 Introduction

Event is one of the basic units for human beings to understand and experience the world. An event is a specific occurrence involving multiple participants, such as *bombing*, *election*, and *marriage*. To represent events and model the world event knowledge, event schema provides a conceptual, structural and formal language which can describe the types of events and the semantic roles (slots) of specific events. Specifically, an event schema is a frame such as "***Type***: *bombing*, ***Slots***: *perpetrator, victm, target, instrument*", which is central in event extraction, event relationship understanding, and event knowledge base construction. Due to its importance, it is critical to automatically discover and construct large-scale, high-quality, and high-coverage event schemas.

Event schema induction, unfortunately, is a non-trivial task due to the open nature of real-world events, the diversity of event expressions, and the sparsity of event knowledge. Firstly, in real-world applications, the size of event types


H. Wang et al. (Eds.): CCKS 2023, CCIS 1923, pp. 57–69, 2023.
https://doi.org/10.1007/978-981-99-7224-1_5

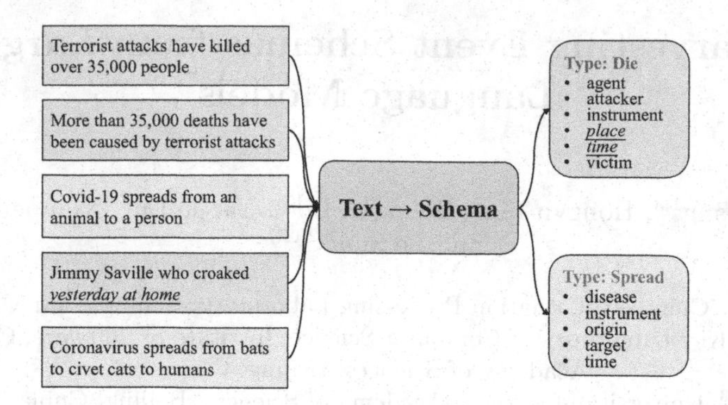


**Fig. 1.** The event harvest paradigm for event schema induction, which induces high-quality event schemas from diverse event expressions and dispersed event knowledge on open domains, e.g., the top two texts on the left use different utterances to describe the same event while the fourth text complements the slots "*place*" and "*time*".

is very large and new types of events are constantly emerging. To address this open problem, event schemas must be induced automatically and with a high coverage on varying domains. Secondly, as shown in Fig. 1, events are usually expressed using very different natural language utterances, therefore it is critical to normalize diverse event expressions by conceptualizing and structuralizing them into formal event schemas. Finally, due to the economic principle of language [3], event expressions are mostly incomplete and many event arguments are missing. To resolve this sparsity problem, an event schema induction method must aggregate dispersed event knowledge across different expressions.

Up to recently, all most event schemas are still hand-engineered by human experts, which are expensive and labour-intensive (e.g., schemas in ACE [4]). On the other hand, traditional automatic event schema induction methods still cannot overcome the open, diversity, and sparsity challenges. For instance, bottom-up concept linking methods [5] discover event types/slots by parsing and linking event expressions to external schema resources such as FrameNet, which are limited by the quality and the coverage of external schema resources. Top-down clustering methods [11] cluster event expressions according to pre-defined schema templates (e.g., the 5W1H template, or the predefined number of event types/slots), which are highly constrained by the pre-defined templates. To sum up, it remains a critical challenge to automatically discover schemas on open domains, normalise event schemas from diverse expressions, and aggregate dispersed knowledge from sparse descriptions.

In this paper, we propose a new paradigm for event schema induction – harvesting knowledge from large pre-trained language models (PLMs), which can effectively address the open, diversity, and sparsity challenges. The main idea is to leverage the strong text generation and in-context learning abilities of PLMs for discovering, conceptualizing, and structuralizing event schemas.

Specifically, we design an **E**vent **S**chema **H**arvester (**ESHer**) to automatically discover and normalize event types and their semantic roles via the fol-

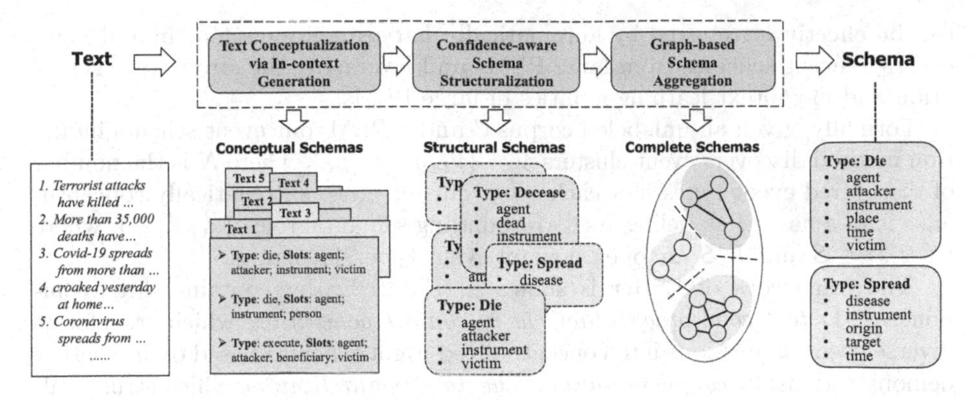


**Fig. 2.** An overview of ESHer, which automatically discovers and normalizes event types and their semantic roles via (1) *text conceptualization via in-context generation*, (2) *confidence-aware schema structuralization* and (3) *graph-based schema aggregation*.

lowing components: 1) *text conceptualization via in-context generation*, which can unsupervised-ly transform diverse event expressions into conceptualized event schema candidates based on in-context demonstrations; 2) *confidence-aware schema structuralization*, which structuralizes event schemas by selecting and associating event types with their salient, reliable and consistent slots; 3) *graph-based schema aggregation*, which aggregates dispersed event schemas via graph-based clustering. In this way, the open, diversity, and sparsity challenges can be effectively resolved via schema conceptualization, structuralization and aggregation.

We conducted experiments on ERE-EN [9] and additional datasets in multiple domains including finance (ChFinAnn [14]), pandemic (Cov-19 [11]), and daily news (New York Time and People's Daily). Empirical results show that ESHer surpasses the traditional methods in discovering high-quality and high-coverage event schemas and can be quickly extended to varying domains and emerging event types.

In general, this paper's main contributions are:

- We propose a new event schema induction paradigm, knowledge harvesting from large-scale PLMs, which can effectively resolve the open, diversity, and sparsity challenges.
- We design ESHer, which can automatically induce event schemas via in-context generation-based text conceptualization, confidence-aware schema structuralization, and graph-based schema aggregation.
- Experiments show ESHer can induce high-quality and high-coverage event schemas on varying domains. And we believe the induced event schemas are valuable resources which can benefit many downstream NLP tasks.

## 2 Event Schema Harvester

This section describes how to discover, conceptualize, and structuralize event schemas from large PLMs so that the open, diversity and sparsity challenges can be effectively resolved by automatically harvesting open-domain and high-coverage event schemas from large PLMs and leveraging the strong text generation and in-context learning abilities of large PLMs.

Formally, given an unlabeled corpus $\mathcal{C}$ and a PLM, our event schema induction method discovers event clusters $\mathcal{Y} = \{y_1, y_2, ..., y_N\}$, where $N$ is the number of discovered event types. For each event cluster $y$, we automatically conceptualize it to a name $t$ as well as its corresponding semantic roles $\{s_1^t, s_2^t, ...\}$, where $t \in \mathcal{T}$, $s \in \mathcal{S}$ and $\mathcal{T}/\mathcal{S}$ are open domain event type/slot names.

The framework of ESHer is shown in Fig. 2. ESHer contains three components: 1) *text conceptualization via in-context generation*, which transforms diverse event expressions into conceptualized event schemas based on in-context demonstrations; 2) *confidence-aware schema structuralization*, which structuralizes event schemas by selecting and associating event types with their salient, reliable and consistent slots; 3) *graph-based schema aggregation*, which aggregates dispersed sparse event knowledge across individual event schemas via graph-based clustering. In follows we describe these components in detail.

## 2.1 Text Conceptualization via In-Context Generation

Events are usually expressed in diverse natural language utterances, which poses a critical challenge for schema induction. For example, "*Terrorist attacks have killed over 35,000 people*" and "*More than 35,000 deaths have been caused by terrorist attacks*" convey the same event, but with quite different words and syntactic structures. To address this challenge, we conceptualize diverse utterances into schema candidates, which can distil event schema knowledge and uniformly represent them. For example, our method will distil and represent the event types and the semantic roles in the above two examples as the same schema "***Type**: die, **Slots**: agent; attacker; instrument; victim*" (as shown in Fig. 2).

To this end, this section proposes an unsupervised text-to-schema framework by leveraging the strong in-context generation ability of PLMs. Specifically, we model text conceptualization as an in-context generation process:

$$[Demonstrations; Text] \rightarrow Schema$$

where "*Demonstrations*" is a list of examples used to instruct PLMs how to conceptualize text to the schema, and each demonstration is a <text, schema> pair', "*Text*" is the event utterance we want to conceptualize, "*Schema*" is the conceptualized schema represented as "***Type***: $t$, ***Slots***: $s_1^t$; $s_2^t$ ...", and "$\rightarrow$" is a special token that separates the text and the event schema.

We can see that, our method is unsupervised so it can effectively resolve open and emerging events in real-world applications, and it is in-context instructed so it can be generalized to different domains/languages by instructing PLMs with appropriate in-context demonstrations. There are many ways to select appropriate in-context demonstrations. This paper directly samples them from existing human-annotated event datasets (e.g., ACE [4], DuEE [8], etc.) and we found text conceptualization benefits from high-quality and diverse demonstrations.

We believe this is because diverse demonstrations can help PLMs to better generalize to different domains, event types and event semantic roles. Furthermore, to recall more event knowledge from an instance, we generate $n$ schema candidates $c_1, c_2, ...c_n$ for each text, where $n$ is a hyperparameter.

### 2.2 Confidence-Aware Schema Structuralization

The text-to-schema component distils and conceptualizes event knowledge in diverse expressions. This section describes how to structuralize these conceptualized event schemas by selecting and associating the salient, reliable, and consistent slots for each event type. For instance, we can structuralize a "*die*" event frame by evaluating the association between event type "*die*" and slots "*agent; attacker; instrument; victim*" (as shown in Fig. 2).

Formally, we use $\mathcal{O}$ to denote the results of text conceptualization, in which $j$-th instance is $(text^j, \{c_1^j, c_2^j, ...c_n^j\})$ and $\{c_1^j, c_2^j, ...c_n^j\}$ are $n$ generated schema candidates, and we use $SlotSet^j$ to denote the union of all generated slots of instance j by summarizing slots in $\{c_1^j, c_2^j, ...c_n^j\}$. To select high-quality slots for event types, we design a set of metrics to estimate the quality of slots and type-slot associations, including ***Salience***, ***Reliability*** and ***Consistency***:
***Salience*** - a salient slot of an event type $t$ should appear frequently in the generated schemas of $t$, but less frequent in other events. For example, in Fig. 2, the slots "*attacker*" is more salient than "*person*" for the "*die*" event. Following the TF-IDF idea, the salience of a slot $s$ in $j$-th instance is computed as:

$$\begin{aligned} Salience(s)^j &= (1 + log(freq(s)^j)^2) \\ &* log(\frac{|\mathcal{O}|}{\sum_k^{|\mathcal{O}|} freq(s)^k}) \end{aligned} \tag{1}$$

where $freq(s)^j$ is the frequency of the slot $s$ in $SlotSet^j$, $|\mathcal{O}|$ is the total number of instances in the outputs $\mathcal{O}$.

***Reliability*** - a slot is reliable if it co-occurs frequently with other slots in multiple candidates of one instance. For example, in Fig. 2, the slot "*agent*" is considered reliable to "*die*" event because it co-occurs with all other slots. We use PageRank algorithm to compute the slot reliability as follows:

$$\begin{aligned} Reliability(s)^j = \beta \sum_k^{|SlotSet^j|} \frac{Reliability(s^k)}{d(s^k)} \\ + (1-\beta)\frac{1}{|SlotSet^j|} \end{aligned} \tag{2}$$

where $\beta$ is a hyper-parameter, $|SlotSet^j|$ is the number of slots in $SlotSet^j$ and $d(s^k) = \sum_{k(s \leftrightarrow s^k)}^{|SlotSet^j|} Reliability(s^k)$, $s \leftrightarrow s^k$ means that slots $s$ and $s^k$ co-occur in the same candidate. We initialize the reliability score for all slots as $\frac{1}{|SlotSet^j|}$ and run PageRank $T$ iterations or the change is less than $\epsilon$.

***Consistency*** – because PLMs may generate unfaithful schemas which are unfaithful to input event expressions, we evaluate the consistency between the generated event schemas and event expressions using semantic similarities based on WordNet, HowNet and BERT. And the consistency score of a slot in $j$-th instance is:

$$Consistency(s)^j = Sim(\hat{t^j}, text^j | s, \hat{t^j} \in c) \tag{3}$$

where $Sim(\cdot)$ is a semantic similarity function, $s, \hat{t^j} \in o$ denotes that slot $s$ and event type $\hat{t^j}$ are in the same schema candidate $c$.

The final confidence of a slot is computed by combining the salience, reliability, and consistency scores:

$$\begin{aligned} Score(s)^j = &(\lambda_1 * Salience(s)^j \\ &+ \lambda_2 * Reliability(s)^j) \\ &* Consistency(s)^j \end{aligned} \tag{4}$$

where $\lambda_1$ and $\lambda_2$ are two hyperparameters.

Finally, we only retain the top-1 consistent event type for each instance and filter all slots in that instance if their confidence scores are below a certain threshold. In this way, we obtained structuralized event schemas such as "***Type:*** *die,* ***Slots:*** *agent; attacker; instrument; victim*" (as shown in Fig. 2).

### 2.3 Graph-Based Schema Aggregation

As described above, event knowledge is sparse in event expressions due to the economical principle of language [3]. This section describes how to address the sparsity issue by aggregating dispersed semantic roles across different schemas. For example, we can obtain a more complete schema for the "*die*" event by combining "***Type:*** *die,* ***Slots:*** *agent; attacker; instrument; victim*" with "***Type:*** *decease,* ***Slots:*** *agent; dead, instrument; place; time*".

To this end, this section proposes a graph-based clustering method which first groups individual event schemas into clusters, and then aggregates event types and slots in the same cluster. The main idea here is that event schemas are of the same event type if their original expressions describe the same kind of occurrence (text similarity), their predicted types are synonyms (type similarity) and they share many common semantic slots (slot set similarity). For example, in Fig. 2, "*die*" and "*decease*" are synonyms and "*agent*" and "*instrument*" are common semantic roles, therefore they are highly likely the same event type.

Based on the above idea, given instances $\mathcal{O}$ after confidence-aware schema structuralization, in which the $j$-th instance is represented as $(text^j, \hat{t^j}, SlotSet^j)$, we construct a graph to model the similarities between different individual event schemas. In the graph, each node is an event schema and the similarity between two schema nodes is computed by considering the similarity between their event expressions, the similarity between their event types, and the similarity between their slot sets:

$$
\begin{aligned}
Graph[i][j] &= Graph[j][i] \\
&= \lambda_3 * Sim(text^i, text^j) \\
&+ \lambda_4 * Sim(\hat{t}^i, \hat{t}^j) \\
&+ \lambda_5 * Sim(SlotSet^i, SlotSet^j)
\end{aligned}
\tag{5}
$$

where $\lambda_3$, $\lambda_4$, and $\lambda_5$ are hyper-parameters, and $Sim(\cdot)$ is the semantic similarity function defined in Eq. 3.

Given the schema graph, we employ the Louvain algorithm to segment and group schemas into clusters:

$$
\hat{Y} = \{\hat{y}^1, \hat{y}^2, ..., \hat{y}^{|\mathcal{O}|}\} = Louvain(Graph) \tag{6}
$$

where $\hat{y}^j \in \mathcal{Y} = \{y_1, y_2, ..., y_N\}$ indicates that the $j$-th schema is assigned to the $\hat{y}^j$-th event cluster and each cluster representing a distinct event type.

Finally, we aggregate all individual event schemas in the same cluster to obtain a complete schema. Given a cluster $y$ which can be represented as as a tuple (***Types***, ***Slots***), with ***Types*** $= \{\hat{t}^1, \hat{t}^2, ...\}$ and ***Slots*** $= \{s_1^t, s_2^t, ...\} = SlotSet^1 \cup SlotSet^2 \cup ...$ are the predicted event types/slots by summarizing event types/$SlotSets$ from all individual schemas. An example of such a cluster in Fig. 2 is "*(**Types**: {die, decease}; **Slots**: {agent; attacker; dead; instrument; place; time; victim})*".

The final event type name of this cluster is normalized by selecting the most salient prediction from $\{\hat{t}^1, \hat{t}^2, ...\}$, e.g., "*die*". For event slots, there may be synonymous slot names in ***Slots*** stand for the same semantic role, e.g., {*dead, victim*} is the synonymous set in the above example. Thus, we utilize the Louvain Algorithm again to identify synonymous event slots and then select the most salient slot to represent its synonyms, e.g., "*victim*" is chosen as the representative slot name of the synonymous set {*dead, victim*}. The final slot names of this cluster are normalized by these selected slots, e.g., the aggregated complete event schemas of the above example is "***Type**: die, **Slots**: agent; attacker; instrument; place; time; victim*" (as shown in Fig. 2).

***Summary*** – By conceptualizing diverse event expressions, structuralizing schemas by selecting and associating event types and their slots, and aggregating dispersed event knowledge across different schemas, our knowledge harvesting method can effectively address the open, diversity, and sparsity challenges, and induce conceptual, structural, and formal event schemas from PLMs.

# 3 Experiments

## 3.1 Experimental Settings

**Datasets.** We use ERE-EN [12] as our primary dataset because its event schemas are manually annotated. Furthermore, to assess event schema induction performance on different domains and languages, we further conduct experiments on various datasets including finance (ChFinAnn [14]), pandemic (Cov-19

**ERE-EN**

| ESHer | Experts | ESHer | Experts | ESHer | Experts |
|---|---|---|---|---|---|
| **Type: Transfer-Money** | **Type: Transfer-Money** | **Type:** Accusation | **Type:** Indict | **Type: Convict** | **Type: Convict** |
| **giver** | **giver** | **defendant** | **defendant** | **defendant** | **defendant** |
| **recipient** | **recipient** | **prosecutor** | **prosecutor** | **place** | **place** |
| **place** | **place** | **place** | **place** | adjudicator | adjudicator |
| currency | beneficiary | | adjudicator | sanction | |
| date | | | | person | |

**ChFinAnn**

| ESHer | Experts | ESHer | Experts |
|---|---|---|---|
| **Type: 质押 (Equity Pledge)** | **Type: Equity Pledge** | **Type: 减持 (Equity Underweight)** | **Type: Equity Underweight** |
| **出质方 (pledger)** | **pledger** | **减持方 (equityHolder)** | **equityHolder** |
| **质押方 (pledgee)** | **pledgee** | **减持数量 (tradedShares)** | **tradedShares** |
| 出质股数 (pledgedShares) | pledgedShares | 持股调整后持股比例 (laterHoldingShares) | laterHoldingShares |
| 持股数 (totalHoldingShares) | totalHoldingShares | 价格 (averagePrice) | averagePrice |
| 持股比例 (totalHoldingRatio) | totalHoldingRatio | 时间 (date) | startDate; endDate |
| 总质押股数 (totalPledgedShares) | totalPledgedShares | | |
| 时间 (date) | startDate; endDate; releasedDate | | |

**Fig. 3.** Schemas induced by ESHer and annotated by experts, in which **bold black** denotes the directly matched event types/slots; black denotes recalled ground truths; teal denotes the unmatched but reasonable ones; orange denotes the missing references; red denotes the wrong predictions. (Color figure online)

**Table 1.** Schema Coverage Comparison on ERE-EN and ChFinAnn.

| Model | ERE-EN | | | | | | | | |
|---|---|---|---|---|---|---|---|---|---|
| | # of Event Types | | | | # of Event Slots | | | | |
| | Human | Discover | Overlap | Acceptable | Human | Discover | Overlap | Acceptable | Recall |
| ESHer | 38 | 71 | 21.05% | 85.92% | 115 | 198 | 11.30% | 44.95% | 35.21% |
| ESHer (upper bound) | 38 | 100 | 21.05% | 93.00% | 115 | 371 | 19.13% | 59.30% | 49.00% |
| **Model** | ChFinAnn | | | | | | | | |
| | # of Event Types | | | | # of Event Slots | | | | |
| | Human | Discover | Overlap | Acceptable | Human | Discover | Overlap | Acceptable | Recall |
| ESHer | 5 | 44 | 100.00% | 72.73% | 35 | 231 | 37.14% | 59.31% | 15.91% |
| ESHer (upper bound) | 5 | 100 | 100.00% | 96.00% | 35 | 458 | 71.43% | 85.81% | 22.00% |

and Pandemic [11]) and daily news (New York Time[1] and People's Daily 1946–2001[2]).

**Implementation.** We use BLOOM [10] in our experiments, which is a GPT-3 like large-scale PLMs but is open-accessed. For text conceptualization, we sample in-context demonstrations from ACE [4] and DuEE [8] for both English and Chinese datasets, respectively.

## 3.2 Results of Event Schema Induction

This section assesses the event schemas induced by our method. Following [11], we use the event type/slot matching task and both qualitative and quantitative results show the effectiveness of the proposed ESHer.

1 https://catalog.ldc.upenn.edu/LDC2008T19.
2 http://en.people.cn.

**Table 2.** Event mention clustering results on ERE-EN. All values are in percentage. We run each method 10 times and report its averaged result for each metric. Note that for ESHer and its variants, due to the huge computing cost, we only run them once.

| Methods | ARI | NMI | BCubed-F1 |
|---|---|---|---|
| Kmeans | 12.51 | 37.65 | 31.01 |
| AggClus | 13.11 | 39.16 | 31.20 |
| JCSC | 17.69 | 43.40 | 37.64 |
| Triframes-CW | 5.79 | 25.73 | 33.61 |
| Triframes-Watset | 7.53 | 47.43 | 24.04 |
| ETypeClus | 10.18 | 36.17 | 28.99 |
| ESHer | **56.59** | **67.72** | **62.43** |
| - Salience | 32.84 | 57.91 | 52.51 |
| - Reliability | 52.54 | 63.86 | 61.47 |
| - Consistency | 37.51 | 66.12 | 50.69 |
| only Salience | 38.75 | 66.34 | 50.69 |
| only Reliability | 33.98 | 62.43 | 51.09 |

For qualitative results, Fig. 3 shows several schemas induced by ESHer from the ERE-EN and ChFinAnn datasets, and the comparison with the results of human experts is also shown. We can see that ESHer can induce high-quality event schemas: 1) most of the induced event types directly match the ones annotated by experts; 2) there is a large overlap between the automatically induced slots and the manually annotated ones; 3) some unmatched slots are also reasonable through our manual checking. This also shows that it is very hard to obtain high-coverage schemas only relying on experts. 4) we found some missing golden slots have been generated in text conceptualization but dropped in the confidence-aware structuralization step, therefore we believe the performance can be further improved by introducing human-in-the-loop. 5) with appropriate in-context demonstrations, ESHer can easily extend to different languages, e.g., English for ERE-EN and Chinese for ChFinAnn.

For quantitative results, we show the performances in Table 1. We can see that: for event type discovery, ESHer recover 21.05% out of 38 event types in ERE-EN and almost all (85.92%) discovered event types are acceptable. For event slot induction, ESHer recovers 11.30% out of 115 slots, 44.95% of discovered slots can be directly accepted, and 35.21% of slots can be selected from candidates. This shows that event schema is a challenging task due to the diversity and sparsity of event slots. On ChFinAnn, a typical dataset in the finance domain, we can see that ESHer is more effective and not only recover all event types but also discover lots of reasonable new types (100% Overlap and 72.73% Acceptable). This shows that domain-specific event schemas can be better induced, we believe this may be because domain-specific events are more salient in domain corpus. To assess the performance of graph-based schema aggregation, we manually check 100 individual schemas and cluster them, and its performance is shown as ESHer (upper bound) which can be regarded as the upper bound of graph-based schema aggregation. We can see that the performance gap between ESHer and ESHer (upper bound) is not large, which verifies the effectiveness of our graph-based schema aggregation component.

| Pandemic | | New York Time | | People's Daily | |
|---|---|---|---|---|---|
| **Type: Occur** | **Type: Administer** | **Type: Transport** | **Type: Charge** | **Type: 号召 (Call)** | **Type: 部署 (Deploy)** |
| disease | vaccination | origin | perpetrator | 号召者 (caller) | 部署者 (deployer) |
| agent | inoculator | destination | plaintiff | 号召对象 (callee) | 部署内容 (content of deploy) |
| instrument | administrator | place | crime | 号召内容 (content of call) | 部署任务 (mission) |
| doctor | instigator | vehicle | evidence | 地点 (place) | 部署范围 (scope of deploy) |
| patient | instrumentation | agent | penalty | 时间 (time) | 部署时间 (time) |
| person | benefit | person | place | | |
| place | place | entity | | | |

**Fig. 4.** Event schemas are induced on varying domains, in which black denotes the reasonable event types/slots; red denotes the rejected predictions.

### 3.3 Results on Event Mention Clustering

We also evaluate the effectiveness of ESHer via the event mention clustering task. Following [11], we select 15 event types with the most mentions and cluster all candidates into several groups for ERE-EN.

**Evaluation Metrics.** To evaluate whether clusters (Eq. 6) align well with the original types, we choose several standard metrics: 1) **ARI** measures the similarity; 2) **NMI** measures the normalized mutual information; 3) **BCubed-F1** measures the aggregated precision and recall. For all the above metrics, the higher the values, the better the model performance.

**Baselines.** We compare ESHer with the following methods: *Kmeans*, *AggClus*, *JCSC* [5], *Triframes-CW*, *Triframes-Watset* and *ETypeClus* [11]. We set all hyper-parameters of these clusters using the default settings of [11].

**Experimental Results.** Table 2 shows the overall results. For our approach, we use the full ESHer and its five ablated settings: ESHer-Consistency, ESHer-Salience, ESHer-Reliability, ESHer only Reliability and ESHer only Salience, where Salience, Reliability and Consistency denote different estimations described in confidence-aware schema structuralization. We can see that:

**1. ESHer outperforms all baselines on ERE-EN on all metrics.** ESHer achieves state-of-the-art performance. We believe this is because ESHer fully leverages the in-context learning ability of PLMs, therefore the diverse and sparse challenges can be effectively resolved.

**2. The proposed salience, reliability and consistency estimations are all useful and complementary to each other.** Compared with the full ESHer model, all five variants show declined performance in different degrees.

ESHer outperforms ESHer-Salience 23.75 ARI, 9.81 NMI and 9.92 BCubed-F1, and this result verifies the effectiveness of the salience score for identifying good slots. ESHer outperforms ESHer-Consistency 19.08 ARI, 1.60 NMI and 11.74 BCubed-F1, this shows that consistency estimation is also indispensable.

### 3.4 Results on Different Domains

This section assesses event schemas induced on different domains such as pandemic and daily news. Figure 4 shows the result schemas and we can see that ESHer is robust in different domains and can be generalized in different settings. Furthermore, the results also present challenges: 1) **the granularity alignment problem**: the slots in the same schema may have different granularities, e.g., the "*person*" and "*doctor, patient*" in Schema 1 on the pandemic domain; 2) **the polysemy problem**: event type "*administer*" in Schema 2 on pandemic domain misdirects the slot "*administrator*"; 3) **emotional expressions**: event schema knowledge should be objective but "*instigator*" conveys negative sentiment.

## 4 Related Work

**Event Schema Induction.** Traditional event schema induction methods mostly mine event schemas from raw corpora, and two main categories of methods have been proposed: Bottom-up concept linking methods [5] discover event types/slots by parsing and linking event expressions to external schema resources such as FrameNet; Top-down clustering methods [11] cluster event expressions according to pre-defined schema templates (e.g., the 5W1H template, or templates with the predefined number of event types/slots).

There were also some other studies such as script learning [2] and event graph schema induction [6], which focus on mining event relations and narrative schemas. This paper doesn't address these issues and leaves them as future works.

**Harvesting Knowledge from Large-Scale Language Models.** Large-scale pre-trained language models such as GPT-3 [1] and BLOOM [10] have been verified containing massive knowledge such as linguistic knowledge, factual knowledge, commonsense knowledge and reasoning knowledge. Furthermore, PLMs also have shown many emergent abilities such as in-context learning, chain-of-thought reasoning, etc. In recent years, researchers start to learn how to harvest resources from PLMs, such as knowledge graphs [13] and explanation datasets [7]. In this paper, we study how to harvest open-domain, high-quality and high-coverage event schemas from PLMs by leveraging the abilities of PLMs.

## 5 Conclusions

In this paper, we propose a new paradigm for event schema induction – knowledge harvesting from pre-trained language models (PLMs), which can effectively resolve the open, diversity and sparsity challenges by discovering, conceptualizing and structuralizing event schemas from PLMs. And an **E**vent **S**chema **H**arvester (**ESHer**) is designed to automatically induce high-quality event schemas. Empirical results show that ESHer can induce high-quality and high-coverage event schemas on different domains. Event schemas are valuable resources, we want to harvest and release an open, large-scale event schema repository to research communities.

## References


1. Brown, T., et al.: Language models are few-shot learners. In: Advances In Neural Information Processing Systems (2020)
2. Chambers, N., Jurafsky, D.: Unsupervised learning of narrative event chains. In: Proceedings of ACL 2008: HLT, pp. 789–797. Association for Computational Linguistics, Columbus, Ohio (Jun 2008). https://aclanthology.org/P08-1090
3. De Saussure, F.: Course in General Linguistics. Columbia University Press (2011)
4. Doddington, G., Mitchell, A., Przybocki, M., Ramshaw, L., Strassel, S., Weischedel, R.: The automatic content extraction (ACE) program - tasks, data, and evaluation. In: Proceedings of the Fourth International Conference on Language Resources and Evaluation (LREC 2004). European Language Resources Association (ELRA), Lisbon, Portugal (May 2004). https://www.lrec-conf.org/proceedings/lrec2004/pdf/5.pdf
5. Huang, L., et al.: Liberal event extraction and event schema induction. In: Proceedings of the 54th Annual Meeting of the Association for Computational Linguistics (Volume 1: Long Papers), pp. 258–268. Association for Computational Linguistics, Berlin, Germany (Aug 2016). https://doi.org/10.18653/v1/P16-1025, https://aclanthology.org/P16-1025
6. Li, M., et al.: The future is not one-dimensional: Complex event schema induction by graph modeling for event prediction. In: Proceedings of the 2021 Conference on Empirical Methods in Natural Language Processing, pp. 5203–5215. Association for Computational Linguistics, Online and Punta Cana, Dominican Republic (Nov 2021). https://doi.org/10.18653/v1/2021.emnlp-main.422, https://aclanthology.org/2021.emnlp-main.422
7. Li, S., et al.: Explanations from large language models make small reasoners better. arXiv preprint arXiv:2210.06726 (2022)
8. Li, X., et al.: Duee: a large-scale dataset for chinese event extraction in real-world scenarios. In: CCF International Conference on Natural Language Processing and Chinese Computing (2020)
9. Lin, Y., Ji, H., Huang, F., Wu, L.: A joint neural model for information extraction with global features. In: Proceedings of the 58th Annual Meeting of the Association for Computational Linguistics, pp. 7999–8009. Association for Computational Linguistics, Online (Jul 2020). https://doi.org/10.18653/v1/2020.acl-main.713, https://aclanthology.org/2020.acl-main.713
10. Scao, T.L., et al.: What language model to train if you have one million gpu hours? arXiv preprint arXiv:2210.15424 (2022)
11. Shen, J., Zhang, Y., Ji, H., Han, J.: Corpus-based open-domain event type induction. In: Proceedings of the 2021 Conference on Empirical Methods in Natural Language Processing, pp. 5427–5440. Association for Computational Linguistics, Online and Punta Cana, Dominican Republic (Nov 2021). https://doi.org/10.18653/v1/2021.emnlp-main.441, https://aclanthology.org/2021.emnlp-main.441

12. Song, Z., et al.: From light to rich ERE: annotation of entities, relations, and events. In: Proceedings of the The 3rd Workshop on EVENTS: Definition, Detection, Coreference, and Representation, pp. 89–98. Association for Computational Linguistics, Denver, Colorado (Jun 2015). https://doi.org/10.3115/v1/W15-0812,https://aclanthology.org/W15-0812
13. West, P., et al.: Symbolic knowledge distillation: from general language models to commonsense models. In: Proceedings of the 2022 Conference of the North American Chapter of the Association for Computational Linguistics: Human Language Technologies, pp. 4602–4625. Association for Computational Linguistics, Seattle, United States (Jul 2022). https://doi.org/10.18653/v1/2022.naacl-main.341, https://aclanthology.org/2022.naacl-main.341
14. Zheng, S., Cao, W., Xu, W., Bian, J.: Doc2EDAG: an end-to-end document-level framework for Chinese financial event extraction. In: Proceedings of the 2019 Conference on Empirical Methods in Natural Language Processing and the 9th International Joint Conference on Natural Language Processing (EMNLP-IJCNLP), pp. 337–346. Association for Computational Linguistics, Hong Kong, China (Nov 2019). https://doi.org/10.18653/v1/D19-1032, https://aclanthology.org/D19-1032

# NTDA: Noise-Tolerant Data Augmentation for Document-Level Event Argument Extraction

Liang Chen, Liu Jian, and Xu Jinan(✉)

Beijing Jiaotong University, School of Computer and Information Technology, Beijing, China
{21120367,jianliu,jaxu}@bjtu.edu.cn

**Abstract.** Event argument extraction (EAE), aiming at identifying event arguments over multiple sentences, mainly faces data sparsity problem. Cross-domain data augmentation can leverage annotated data to augment training data, but always encounters the issue of noise. The noise mainly consists of two aspects: boundary annotation differences and domain knowledge discrepancy, which may significantly impact the effectiveness of data augmentation. In this paper, we propose a new framework NTDA (**N**oise-**T**olerant **D**ata **A**ugmentation) to solve the above two issues. For annotation differences, we introduce region-based loss function to mitigate the model's sensitivity towards entity boundaries. To address the knowledge discrepancy problem, we propose a dynamic data selection strategy. Additionally, we further combine the two denoising techniques. Through conducting comprehensive experiments on three datasets, we have demonstrated the superior effectiveness of our framework compared to previous methods.

**Keywords:** event argument extraction · cross-domain data augmentation · boundary annotation difference · domain knowledge discrepancy

## 1 Introduction

Document-level event argument extraction (EAE) [1–3] aims to discover event arguments over multiple sentences, always facing data sparsity problem. Cross-domain data augmentation is a crucial way to solve data sparsity [1,4], which can effectively leverage other annotated datasets to help train the model. As the prompt structure received great attention in information extraction (IE) tasks [2,5], current models can understand the semantic meaning of labels by encoding label with input sentence, which makes the prompt-based cross-domain data augmentation get further development [6].

However, a crucial issue in cross-domain data augmentation is the impact of noise, which mainly includes two aspects: i) *Boundary annotation differences.* The annotation of entity boundaries is always ambiguous. As shown in the


H. Wang et al. (Eds.): CCKS 2023, CCIS 1923, pp. 70–82, 2023.
https://doi.org/10.1007/978-981-99-7224-1_6

left of Fig. 1, for the given sentence, the label 'place' can have four reasonable annotations, which will result in bad influence when training the model due to current neural network models easily encounter over-confidence problem [7]. This noise is also inherent within a dataset itself. ii) *Domain knowledge discrepancy*, which is the intrinsic differences between augmented data (out-of-domain dataset) and gold data (target-domain dataset). Considering not all training samples are equally important, we only need to select a subset of augmented data to train the model, achieving transferring common and helpful knowledge from out-of-domain to target domain to improve model performance.

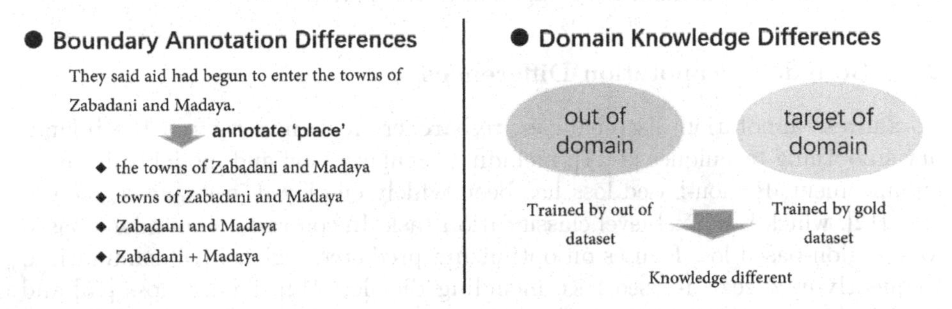


**Fig. 1.** Examples of two types of noise. Boundary annotation differences (left) mean there are several reasonable annotations for a label within a given sentence. Domain knowledge discrepancy (right) indicates that different datasets contain varying domains of knowledge.

In this paper, we propose a novel framework *NTDA* to solve the above issues. For the annotation difference problem, we introduce region-based loss to optimize the training process, which can mitigate the sensitivity of the model to the entity boundary. For domain knowledge discrepancy problem, we propose a new dynamic data selection strategy to alleviate domain discrepancy, which completes the process by which the model gradually approaches the target domain from out-of-domain. Specifically, we iteratively convert the divergence between domains into specific scores at each training epoch and select appropriate examples at that point to train the model. What's more, our method does not need to set thresholds when selecting training samples. Finally, we combine the above two techniques to denoise when augmenting training data.

Our major contributions include: i) we design a novel framework NTDA to solve noise problems encountered during cross-domain data augmentation. ii) the method of resolving boundary annotation differences can be extended to a self-denoise strategy, and it is easy to be incorporated into other model training process. iii) we conduct extensive experiments and the results have justified the effectiveness of our approach.

# 2 Related Work

## 2.1 Cross-Domain Data Augmentation

A large number of training data is usually essential for model performance [8]. Cross-domain data augmentation has been widely used to solve data sparsity in many tasks, such as event extraction (EE) and named entity recognition (NER). [1] propose implicit knowledge transfer and explication data augmentation via machine reading comprehension (MRC) structure. [5] points out that question-answering (QA) formalization can come with more generalization capability and further boost cross-domain data augmentation in NER.

## 2.2 Boundary Annotation Differences

To address annotation discrepancies, researchers have investigated the boundary smoothing techniques [9–11], including regularization and model calibration enhancement. Region-based loss has been widely employed in image segmentation [12], which is a pixel-level classification task. In contrast to boundary-based loss, region-based loss focuses on optimizing predicted regions. [13] summarized frequently-used region-based loss, including dice loss [14], tversky loss [15] and log-cosh dice loss [13]. In this paper, we introduce region-based loss to mitigate boundary annotation discrepancies.

## 2.3 Denoising for Domain Discrepancy

Previous works on denoising are based on the view that not all out-of-domain data are equally important when training model. One of the research lines is data selection, aiming at selecting examples similar to the target domain distribution. Traditional data selection methods are typically static and always rely on similarity measures [16,17], whereas current approaches tend to be dynamic and incorporate weighting strategies, such as curriculum learning [18–20]. Another line of research is leveraging all available data during training, including domain adaptation strategy [21,22] and noise robust regularization method [23].

# 3 Methodology

In this section, we provide a detailed introduction to our method *NTDA*. We first introduce our data augmentation framework (Sect. 3.1) and then detail our denoising framework, which performs two aspects of denoising: i) boundary annotation differences denoising (Sect. 3.2), ii) domain knowledge discrepancy denoising (Sect. 3.3). In Sect. 3.4, we will describe how to combine two denoising methods. The presentation of our method is also illustrated in Fig. 2.

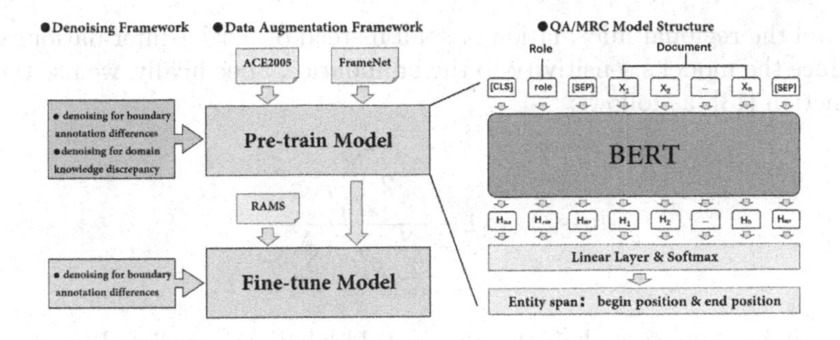


**Fig. 2.** The overview of our method NTDA.

## 3.1 Data Augmentation Framework

Following [1,6], we construct the model based on question answer (QA)/machine reading comprehension (MRC) structure, which has achieved superior performance in EAE tasks. Specifically, we append the role to the document head to form a series of new sequences as training inputs, adding one role at a time. All training inputs are organized by:

$$\text{[CLS] role [SEP] document [SEP]} \tag{1}$$

where *document* is the original input document, *role* denotes a type of role, [CLS] and [SEP] are special tokens used in BERT [24]. Next, we feed those sequences into BERT to jointly encode the role and the document, and extract the last hidden layer of BERT $H_{RP}$ as the final embedding representations of inputs:

$$H_{RP} = \text{BERT}(\text{role} + \text{document}) \tag{2}$$

Then we identify start position $P_{begin}$ and end position $P_{end}$ of the entity:

$$p_{begin} = \text{Softmax}(H_{RP}W_{begin} + b_{begin}) \tag{3}$$

$$p_{end} = \text{Softmax}(H_{RP}W_{end} + b_{end}) \tag{4}$$

where $W_{begin}$, $b_{begin}$ and $W_{end}$, $b_{end}$ are the weight matrix and bias vector for two linear layers. The index of the largest value in $p_{begin}$ and $p_{end}$ will be regarded as the start position $P_{begin}$ and the end position $P_{end}$. What's more, we notice that the document does not necessarily contain the argument for all given labels. We take the first special token [SEP] as *fake argument* under this condition.

For training the model, we first introduce out-of-domain datasets to pre-train the model, and then fine-tune the model on the in-domain dataset.

## 3.2 Denoising for Boundary Annotation Differences

To address the issue of boundary annotation differences, we propose to introduce region-based loss for adjusting the training objective. Region-based loss function focuses on the regional information of span instead of border information, which can reduce the model's sensitivity to the boundaries. Specifically, we use the dice loss function [13], as follows:

$$loss_{dice} = 1 - \frac{2\sum_{i=1}^{N} p_i q_i}{\sum_{i=1}^{N} p_i^2 + \sum_{i=1}^{N} q_i^2} \tag{5}$$

where $p_i$ and $q_i$ represent the $i$-th value in gold label and predicted label respectively, $N$ is the length of input sentence.

Our model predicts the probability of each token being in gold span as follows:

$$p_{region} = (p_{start} + p_{end})/2 \tag{6}$$

where the calculation of $p_{start}$ and $p_{end}$ have been shown in Eq. 3 and Eq. 4.

In the training process, we dynamically adjust the weight of the original boundary-based loss $loss_{bound}$ (generally is cross-entropy loss) and region-based loss $loss_{reg}$ (we use dice loss $loss_{dice}$), gradually increasing the weight of the former and decreasing the weight of the latter:

$$loss = (\alpha_{reg} + \beta_{reg} t) loss_{reg} + (\alpha_{bound} - \beta_{bound} t) loss_{bound} \tag{7}$$

where $\alpha_{reg}$, $\beta_{reg}$, $\alpha_{bound}$ and $\beta_{bound}$ are the parameters to tune the weighs and $t$ denotes the training time.

### 3.3 Denoising for Domain Knowledge Discrepancy

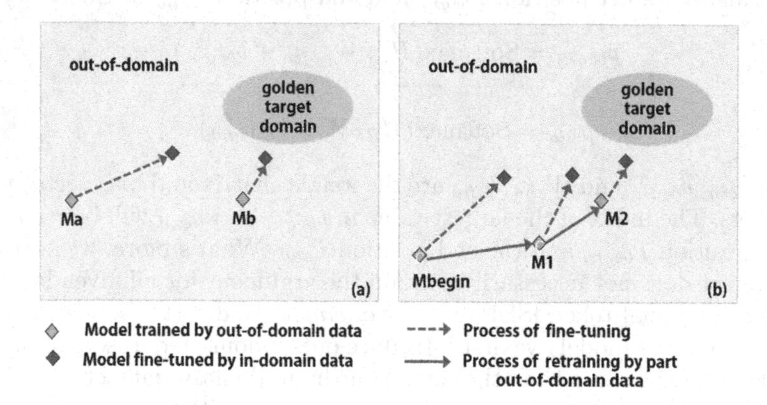


**Fig. 3.** In Figure (a), $M_b$ is more suitable than $M_a$ for fine-tuning. Figure (b) shows the iterative process of gradually approaching the target domain: $M_i$ along with fine-tuned $M_i$ compute score together to select part of the out-of-domain data to retrain $M_i$ and get $M_{i+1}$, and we expect $M_{i+1}$ is closer to the target domain than $M_i$.

Based on the two-stage data augmentation (DA) framework described in Sect. 3.1, there is such a hypothesis: when fine-tuning model with the same in-domain data, the closer model trained on out-of-domain data is to the target domain in the first stage, the better model performance we get by fine-tuning in the second stage. As shown in Fig. 3 (a), $M_b$ is more suitable than $M_a$ for fine-tuning. Therefore, we provide a novel dynamic data selection strategy to optimize DA framework by obtaining the model closer to target domain at the first pre-training stage.

In this section, we first define a new score as the criterion for data selection and then detail our strategy for addressing domain discrepancy, which contains two stages: 1) warm-up stage: in this stage, we expect to get a beginning model as close to the target domain as possible through a simple way, which is trained on part of the out-of-domain data and will be used at the beginning of the iterative stage. 2) iteration stage: we devise an iterative approach to accomplish the process that model gradually approaches the target domain by dynamic data selection, as presented in Fig. 3 (b). To better understand our method, we summarize the training procedure in Algorithm 1.

**Measuring Discrepancy.** For each input, we can compute its probability vectors $p_{begin}$ and $p_{end}$ according to Eq. 3 and Eq. 4, representing the possibility of each token as the starting position and ending position separately. We define $Score$ as follows:

$$Score = P_{golden-begin} + P_{golden-end} \tag{8}$$

where $P_{golden-begin}$ and $P_{golden-end}$ respectively denote the value in $p_{begin}$ and $p_{end}$ corresponding to the start position and the end position of the golden argument (note that it is not necessarily the maximum value in probability vectors). Then we further define $\Delta Score$ to measure the discrepancy of each input between out-of-domain model and target domain model:

$$\Delta Score = Score_{in} - Score_{out} \tag{9}$$

where $Score_{in}$ and $Score_{out}$ are computed on different models. $\Delta Score$ and $Score$ will all dynamically change with the model at the training process.

**Warm-Up Stage.** We first train a model with all in-domain data and use it to calculate $Score$ of out-of-domain data via Eq. 8 (**Lines** 3–4). Sentences in out-of-domain dataset with the highest scores will be selected for training $M_{begin}$ (**Line** 5).

**Iteration Stage.** We denote the model trained/retrained on (part of) out-of-domain data $S_{out}$ as $M_{out}$ and denote the model fine-tuned on in-domain data $D_{in}$ as $M_{in}$. For each iteration, our target is to select samples from out-of-domain dataset to retrain $M_{out}$ and let it approach target domain. Specifically, for the

**Algorithm 1.** Addressing Domain Knowledge Discrepancy

**Input**: in-domain data $D_{in}$, out-of-domain data $S_{out}$,
**Output**: Model $\theta$

1: convert $D_{in}, S_{out}$ into MRC structure
2: /* stage 1: warm-up stage */
3: Train a model on $D_{in}$
4: Calculate *score* of $S_{out}$ on model
5: Choose a subset to train $M_{begin}$ $(M_{out}^{i=0})$
6: /* stage 2: iteration stage */
7: **while** not convergence **do**
8: Fine-tune $M_{out}^{i}$ and get $M_{in}^{i}$
9: Calculate $\Delta Score$ of $S_{out}$ on $M_{in}^{i}$ , $M_{out}^{i}$
10: Select subset with $\Delta Score > 0$ to retrain $M_{out}^{i}$ and get $M_{out}^{i+1}$
11: **if** iteration number $> 4$ **then**
12: break
13: **end if**
14: Update $\theta$ by $M_{out}^{i}$
15: **end while**
16: **return** $\theta$

i-th iteration, we first fine-tune the model $M_{out}^{i}$ (obtained from iteration $i-1$, $M_{begin}$ is used as $M_{out}^{0}$ for the first iterative process) on all in-domain data $D_{in}$ and get $M_{in}^{i}$ (**Line** 8). Then $Score_{in}^{i}$ and $Score_{out}^{i}$ on $S_{out}$ can be computed by $M_{in}^{i}$ and $M_{out}^{i}$ based on Eq. 8. We focus on the differences between $Score_{in}^{i}$ and $Score_{out}^{i}$:

$$\Delta Score^{i} = Score_{in}^{i} - Score_{out}^{i}, i = 0, 1... \tag{10}$$

We select the data in $S_{out}$ with $\Delta Score^{i} > 0$ to retrain $M_{out}^{i}$ and get the model $M_{out}^{i+1}$, which is closer to the target domain than $M_{out}^{i}$ (**Line** 10). We can do multiple iterations to make $M_{out}$ gradually approach the target domain.

### 3.4 Combining Two Denoising Methods

It is easy to combine the above two denoising methods: we use the loss function in Eq. 7 to train the model when addressing domain knowledge discrepancy.

## 4 Experiments

### 4.1 Experiments Settings

**Datasets.** We evaluate our method on three common datasets, which can be classified into two groups: target domain dataset (RAMS [25]) and out-of-domain dataset (ACE 2005 [26] and FrameNet [27]). RAMS is a well-established document-level dataset in Event Argument Extraction (EAE), providing 3,194 documents and 17,026 arguments. For out-of-domain datasets, we introduce two datasets with different tasks: i) ACE 2005, a sentence-level dataset in EAE, contains 599 sentences and 8,100 arguments. ii) FrameNet, a sentence-level dataset in semantic role labeling (SRL) task, is annotated with 19,391 sentences and 34,219 arguments. The detailed statistics are tabulated in Table 1.

**Table 1.** Summary statistics of datasets.

| Dataset | RAMS | | | ACE05 | FrameNet |
|---|---|---|---|---|---|
| | Train | Dev | Test | Train | Train |
| #Sentence | 3,194 | 399 | 400 | 599 | 19,391 |
| #Event | 7,329 | 924 | 871 | 4,859 | 19,391 |
| #Argument | 17,026 | 2,188 | 2,03 | 8,100 | 34,219 |
| #Role | 65 | 60 | 60 | 22 | – |

**Implementation.** In our method, we adopt $\text{BERT}_{\text{cased}}$ [24] as the pre-trained language model to conduct experiments. The model is trained for 10 epochs with the batch size of 24. The learning rate is selected in [1e−5, 3e−5, 5e−5] and the weights for different loss are chosen from [0.1, 0.3, 0.5, 0.7, 1]. We use Adam optimizer [28] to optimize all parameters.

**Baselines.** We compare our denoising approach with the following methods: **CrossE** [29] computes the cross-entropy difference between two language models for input sentences and trains the model only on data that is close to the target domain. **BertSim** [18] choose training samples according to BERT representation similarities. **CurriCE** [20] is a type of curriculum learning method that dynamically selects sentences. **DomainBp** [30] is an effective domain adaptation approach.

**Table 2.** Performance of different strategies for addressing noise issue. Role-Prompt and Role-Prompt$_{DA}$ are the baseline model without denoising strategy. The results of our approach are shown at the bottom of the table. Here 'A' and 'K' indicate denoising for boundary annotation differences and denoising for domain knowledge discrepancy. 'AK' means combining the above two methods.

| Method | RAMS | | |
|---|---|---|---|
| | P | R | F1 |
| Role-Prompt | 43.91 | 42.14 | 42.47 |
| Role-Prompt$_{DA}$ | 46.41 | 40.40 | 43.20 |
| CrossE | 47.43 | 41.00 | 43.98 |
| BertSim | 43.32 | 43.60 | 43.46 |
| CurriCE | 44.84 | 42.40 | 43.59 |
| DomainBp | 42.49 | 41.90 | 42.20 |
| NTDA(A) | 49.49 | 40.85 | 44.75 |
| NTDA(K) | 46.05 | 41.70 | 43.77 |
| NTDA(AK) | 47.42 | 43.15 | 45.18 |

### 4.2 Main Result

Table 2 demonstrates the main results of our denoising framework compared to baselines. The metrics are precision (P), recall (R), and f1 score (F1). As we can see, our method outperforms previous approaches by about 1.5% in terms of f1 score. NTDA(A) and NTDA(K) achieve notable gains in f1, obtaining 1.55% and 0.57% improvement respectively. Such results have verified the effectiveness of our framework.

For our approach can beat the baselines, we speculate the main reasons are: 1) commonly used data selection methods (CrossE, BertSim, CurriCE) may be unreasonable because some useful examples can also be filtered out. 2) previous methods are for ordinary sentences, not designed for the form of question answer (QA)/machine reading comprehension (MRC). But our approach is well-designed for sequences containing the role and document. We also observe that DomainBp leads to about 1% performance degradation, which means that the domain adaption strategy may not be suitable for noise reduction in cross-domain data augmentation.

## 5 Discussion

We conduct a series of studies to further justify the effectiveness of our model, including the effect of dice loss (Sect. 5.1) when denoising for boundary annotation differences, the effect of iteration number (Sect. 5.2) when denoising for domain knowledge discrepancy, and the denoising approach extending to a self-denoise strategy (Sect. 5.3).

**Table 3.** The effect of dice loss in different settings. '+ loss' denotes the training model with dice loss and the value in '( )' indicates the utilization stage of dice loss.

| Setting | model | P | R | F1 |
|---|---|---|---|---|
| In-domain | Role-Prompt | 43.91 | 42.14 | 42.47 |
| | + Loss | 43.21 | 42.95 | 43.08 |
| DA | Role-Prompt$_{DA}$ | 46.41 | 40.40 | 43.20 |
| | + Loss (1-th) | 46.30 | 43.20 | 44.70 |
| | + Loss (2-th) | 45.45 | 42.75 | 44.06 |
| | + Loss (1&2) | 47.42 | 43.15 | 45.18 |
| Zero-shot | Role-Prompt | 27.46 | 18.51 | 22.11 |
| | + Loss | 19.50 | 29.00 | 23.32 |

### 5.1 Effect of Dice Loss

We conduct a thorough ablation study to explore the effect of dice loss on cross-domain data augmentation. Table 3 shows the results. Our experiments contain three settings: 1) **In-domain**, training model only with the target domain dataset RAMS. 2) **DA**, performing cross-domain data augmentation. 3) **Zero-shot**, training model only with the out of domain datasets ACE2005 and FrameNet. From the results in the settings 1) and 3), it can be observed that training with dice loss can yield 0.5%-1% f1 improvement. Such results suggest that our method is capable of self-denoising and we further explore it in Sect. 5.3. For setting 2), we see that the effectiveness of the dice loss is both in the 1-th stage and 2-th stage. Due to the larger annotation discrepancies between out of domain dataset and target domain dataset, the use of dice loss in both stages achieves higher performance gains.

### 5.2 Effect of Iteration Number

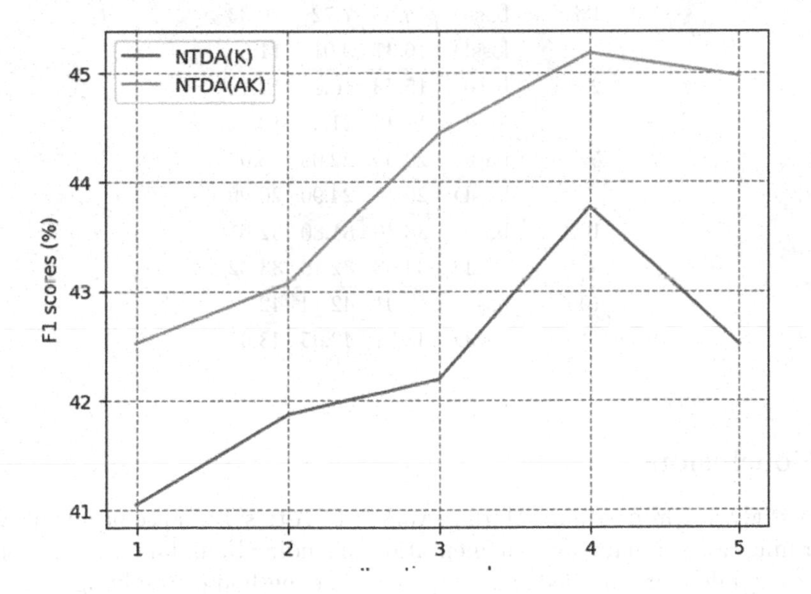


**Fig. 4.** Effect of iteration number. NTDA(K) is the model training only with boundary loss and NTDA(AK) is combining boundary loss and region loss.

We conduct experiments with different iteration numbers for addressing domain knowledge discrepancy and illustrate the results in Fig. 4. We can observe that f1 reaches a maximum when the iteration number is set to 4 in most cases. By training model with boundary loss and region loss together, model performance gets further improved. By analyzing the trend, the f1 score shows a progressive increase, indicating that the model completes the process by which gradually approaching the target domain from out-of-domain.

### 5.3 Self-Denoising

Due to the issue of boundary annotation differences also inherent within a dataset itself, we can regard training with dice loss as a self-denoise strategy. We conduct experiments in low-resource environments to justify it. Specifically, we randomly select subsets for RAMS with the ratio 1%, 2%, 5%, 10%, 100% according to the event type. We conducted 10 experiments under each setting and calculated the average as the final results. Results are provided in Table 4. It can be observed that training with dice loss (lossD) performs better in all low-resource settings, which indicates that this training strategy can be extended to be a self-denoise approach. What's more, training with dice loss can produce more stable results due to the reduced sensitivity of the model to different boundary annotations.

**Table 4.** The results of model training without dice loss (base) and with dice loss (lossD) in different low-resource scenarios.

| Setting | model | P | R | F1 |
|---|---|---|---|---|
| 1% | base | 15.58 | 7.72 | 10.33 |
| | lossD | 16.37 | 9.01 | 11.63 |
| 2% | base | 15.33 | 10.00 | 12.10 |
| | lossD | 17.42 | 11.05 | 13.52 |
| 5% | base | 29.09 | 22.05 | 25.09 |
| | lossD | 29.47 | 24.90 | 26.99 |
| 10% | base | 33.46 | 31.80 | 32.61 |
| | lossD | 34.63 | 32.10 | 33.32 |
| 100% | base | 43.91 | 42.14 | 42.47 |
| | lossD | 43.21 | 42.95 | 43.08 |

## 6 Conclusion

In this study, we propose a novel framework NTDA to solve noise problems when performing cross-domain data augmentation, including boundary annotation differences and domain knowledge discrepancy. The method of resolving boundary annotation differences can be extended to a self-denoise strategy and can be easily integrated into any QA/MRC structure model. In addition, experimental results show that our framework outperforms previous methods. We hope that our work can stimulate the advancement of data augmentation research.

**Acknowledgments.** The research work described in this paper has been supported by Fundamental Research Funds for the Central Universities (No. 2023JBMC058), the National Key R&D Program of China (2020AAA0108001), and the National Nature Science Foundation of China (No. 61976015, 61976016, 61876198 and 61370130). The authors would like to thank the anonymous reviewers for their valuable comments and suggestions to improve this paper.

## References

1. Liu, J., Chen, Y., Xu, J.: Machine reading comprehension as data augmentation: a case study on implicit event argument extraction. In: Proceedings of the 2021 Conference on EMNLP, pp. 2716–2725 (2021)
2. Du, X., Cardie, C.: Event extraction by answering (almost) natural questions. arXiv preprint arXiv:2004.13625 (2020)
3. Liu, J., Chen, Y., Xu, J.: Event extraction as machine reading comprehension. In: Proceedings of EMNLP, pp. 1641–1651 (2020)
4. Li, J., Yu, J., Xia, R.: Generative cross-domain data augmentation for aspect and opinion co-extraction. In: Proceedings of the 2022 Conference of the NAACL, pp. 4219–4229 (2022)
5. Li, X., Feng, J., Meng, Y., et al.: A unified MRC framework for named entity recognition. arXiv preprint arXiv:1910.11476 (2019)
6. Liu, J., Liang, C., Xu, J.: Document-level event argument extraction with self-augmentation and a cross-domain joint training mechanism. Knowl.-Based Syst. **257**, 109904 (2022)
7. Zhu, E., Li, J.: Boundary smoothing for named entity recognition. arXiv preprint arXiv:2204.12031 (2022)
8. Ding, B., Liu, L., Bing, L., et al.: DAGA: data augmentation with a generation approach for low-resource tagging tasks. arXiv preprint arXiv:2011.01549 (2020)
9. Müller, R., Kornblith, S., Hinton, G.E.: When does label smoothing help? In: Advances in Neural Information Processing Systems, vol. 32 (2019)
10. Szegedy, C., Vanhoucke, V., Ioffe, S., et al.: Rethinking the inception architecture for computer vision. In: Proceedings of the IEEE Conference on Computer Vision and Pattern Recognition, pp. 2818–2826 (2016)
11. Zhu, Y.: LPS@ LT-EDI-ACL2022: an ensemble approach about hope speech detection. In: Proceedings of ACL, pp. 183–189 (2022)
12. Masood, S., Sharif, M., Masood, A., et al.: A survey on medical image segmentation. Curr. Med. Imaging **11**(1), 3–14 (2015)
13. Jadon, S.: A survey of loss functions for semantic segmentation. In: Proceedings of CIBCB, pp. 1–7. IEEE (2020)
14. Sorensen, T.A.: A method of establishing groups of equal amplitude in plant sociology based on similarity of species content and its application to analyses of the vegetation on Danish commons. Biol. Skar. **5**, 1–34 (1948)
15. Tversky, A.: Features of similarity. Psychol. Rev. **84**(4), 327 (1977)
16. Moore, R.C., Lewis, W.: Intelligent selection of language model training data. In: Proceedings of ACL, pp. 220–224 (2010)
17. Kirchhoff, K., Bilmes, J.: Submodularity for data selection in machine translation. In: Proceedings of EMNLP, pp. 131–141 (2014)
18. Dou, Z.Y., Anastasopoulos, A., Neubig, G.: Dynamic data selection and weighting for iterative back-translation. arXiv preprint arXiv:2004.03672 (2020)
19. Wang, W., Caswell, I., Chelba, C.: Dynamically composing domain-data selection with clean-data selection by "co-curricular learning" for neural machine translation. arXiv preprint arXiv:1906.01130 (2019)
20. Platanios, E.A., Stretcu, O., Neubig, G., et al.: Competence-based curriculum learning for neural machine translation. arXiv preprint arXiv:1903.09848 (2019)
21. Jia, C., Liang, X., Zhang, Y.: Cross-domain NER using cross-domain language modeling. In: Proceedings of ACL, pp. 2464–2474 (2019)

22. Liu, Z., Winata, G.I., Fung, P.: Zero-resource cross-domain named entity recognition. arXiv preprint arXiv:2002.05923 (2020)
23. Shi, H., Tang, S., Gu, X., et al.: Alleviate dataset shift problem in fine-grained entity typing with virtual adversarial training. In: proceedings of IJCAI, pp. 3898–3904 (2021)
24. Devlin, J., Chang, M.W., Lee, K., et al.: Bert: pre-training of deep bidirectional transformers for language understanding. arXiv preprint arXiv:1810.04805 (2018)
25. Ebner, S., Xia, P., Culkin, R., et al.: Multi-sentence argument linking. arXiv preprint arXiv:1911.03766 (2019)
26. Strassel, S., Cole, A.W.: Corpus development and publication. In: Proceedings of LREC, Genoa, Italy. http://papers.ldc.upenn.edu/LREC2006/CorpusDevelopmentAndPublication.pdf (2006)
27. Baker, C.F., Fillmore, C.J., Cronin, B.: The structure of the FrameNet database. Int. J. Lexicogr. **16**(3), 281–296 (2003)
28. Kingma, D.P., Ba, J.: Adam: a method for stochastic optimization. arXiv preprint arXiv:1412.6980 (2014)
29. Moore, R.C., Lewis, W.: Intelligent selection of language model training data. In: Proceedings of ACL, pp. 220–224 (2010)
30. Xu, M., Zhang, J., Ni, B., et al.: Adversarial domain adaptation with domain mixup. AAAI **34**(04), 6502–6509 (2020)

# Event-Centric Opinion Mining via In-Context Learning with ChatGPT

Lifang Wu[1], Yan Chen[1], Jie Yang[1], Ge Shi[1(✉)], Xinmu Qi[2], and Sinuo Deng[1]

[1] Faculty of Information Technology, Beijing University of Technology, Beijing, China
shige@bjut.edu.cn

[2] Stony Brook Institute, Anhui University, Anhui, China

**Abstract.** Events represent fundamental constituents of the world, and investigating expressions of opinion centered around events can enhance our comprehension of events themselves, encompassing their underlying causes, effects, and consequences. This helps us to understand and explain social phenomena in a more comprehensive way. In this regard, we introduce ChatGPT-opinion mining as a framework that transforms event-centric opinion mining tasks into question-answering (QA) utilizing large language model. We employ this approach in the context of the event-centric opinion mining task that utilizes an event-argument structure. In our study, we primarily leverage in-context learning methods to construct demonstrations. Through comprehensive comparative experiments, we illustrate that the model achieves superior results within the ChatGPT-opinion mining framework when the demonstrations exhibit diversity and possess higher semantic similarity, comparable to those of supervised models. Moreover, ChatGPT-opinion mining surpasses the supervised model, particularly when there is limited availability of the same data.

**Keywords:** Large Language Model · Opining Mining · ChatGPT

## 1 Introduction

Events serve as fundamental building blocks of the world [1]. In our daily lives, we articulate, exchange, and disseminate our viewpoints on events based on personal comprehension, emotions, and attitudes. Karamebeck et al. [2] highlighted that examining diverse viewpoints on events has the potential to trigger prejudiced emotional responses and convictions concerning societal concerns. Thus, the exploration of event-centric opinions bears significant social and personal implications and warrants ample attention. Nonetheless, current research predominantly concentrates on entity-centric opinion mining, which underscores individuals' sentiments towards entities. In contrast, event-centric opinion mining centers on the events themselves or specific facets thereof. When expressing their opinions, users may implicitly allude to the details of events. Therefore, in event-centric opinion mining, it becomes imperative to extract not only users'


H. Wang et al. (Eds.): CCKS 2023, CCIS 1923, pp. 83–94, 2023.
https://doi.org/10.1007/978-981-99-7224-1_7

opinions on the events but also their opinions on specific event-related aspects. To our best knowledge, EcoV1 [3] introduces and expounds upon the task of event-centric opinion mining, drawing on event argument structure and expression classification theory. It suggests that users can articulate their opinions concerning three types of event arguments: (1) event, (2) sub-event, and (3) entity. Given the prevalence of large language models, we posit that these models can extract precise event arguments by leveraging the content of users' opinions, while minimizing resource consumption.

Due to the recent advancements, large language models (LLMs) have exhibited a remarkable capacity to in-context learning frameworks. LLMs can generate results for novel test inputs without requiring parameter tuning and by utilizing only a limited number of task-specific demonstrations. Within the in-context learning framework, LLMs has demonstrated favorable performance across diverse NLP tasks, including machine translation [4,5] and question answering [6,7]. Drawing on these indications, we redirect our focus to ChatGPT and present ChatGPT-opinion mining as a solution for event-centric opinion mining. We transform event-centric opinion mining into a question-answering (QA) process and steer ChatGPT towards generating more precise responses by constructing in-context prompts. While constructing the prompts, we devised multiple approaches, including random selection, demonstration selection based on k-nearest neighbors ($k$NN), demonstration selection based on part-of-speech (POS) tagging. Moreover, we design a task description, with the intention of mitigating model illusions. Through these methodologies, we have conducted an extensive array of experiments, revealing that LLMs attains optimal outcomes when prompts exhibit diversity and higher semantic similarity, yielding results comparable to those of supervised models.

Generally, the contributions of this paper are:

- We introduce ChatGPT-opinion mining, which, to the best of our knowledge, represents the pioneering endeavor of utilizing ChatGPT for opinion mining and attains results comparable to those of supervised models.
- Within ChatGPT-opinion mining, we devised multiple in-context techniques, such as $k$NN selection demonstration and Pos Tagging selection demonstration, and conducted an extensive range of experiments.
- The experimental results reveal some key findings concerning in-context learning: firstly, ChatGPT-opinion mining primarily acquires knowledge regarding the data structure rather than the annotation content, and secondly, the demonstration should possess a diverse and semantically coherent space. Moreover, ChatGPT-opinion mining outperforms the supervised model when in the case of small amount of data.

## 2 Related Work

### 2.1 Opinion Mining

In general, opinions can be expressed on anything, and current research on opinion mining has focused on entity-centric [8]. The entity-centric opinion mining task primarily categorizing the document holder's sentiments towards entities and their attributes at the document level [9,10], sentence level [11,12], aspect level [13,14]. There are also some studies on event-related [15,16]. However, these studies usually treat events as a special entity and do not take into account the unique event-centric connotations. Therefore, in response to the difference between entities and events, event-centric opinion mining is needed to better understand the perception of events. To the best of our knowledge, most of the publicly available datasets for opinion mining are based on the SemEval Challenge [17,18]. According to the public benchmark in the event-centric opinion mining task, [3] helps us to perform method evaluation.

### 2.2 Large Language Models and In-Context Learning

The creation of Large Language Models (LLMs) [19,20] has attracted a great deal of attention. LLMs are capable of solving a wide range of natural language processing tasks and have achieved excellent performance [21,22]. Methods for downstream tasks using LLMs can be divided into two categories: fine-tuning, which performs model parameter tuning, and prompt engineering, which does not tune parameters. The fine-tuning uses supervised learning methods to train on a specified dataset. Select the appropriate pre-training model and add an additional layer at the top to adjust the model to the output of the downstream task [23–25].

Large Language Models (LLMs) demonstrate an in-context learning (ICL) ability, that is, learning from a few examples in the context [26]. Unlike fine-tune's approach, ICL does not need to adjust the parameters of the model to achieve the desired performance. Dong et al. conducted a detailed analysis and summary of the progress of ICL and found that the demonstrated samples can have a great impact on the output of LLMs [26]. Gonen et al. found that mutual information is a useful selection metric. So they designed a method to construct a prompt, which selected the example with lower perplexity as the demonstration, and the performance was significantly improved [27]. Rubin et al. scored a demonstration through a model and selected the one with the highest score by using the method of comparative learning [28].

## 3 Method

In this work, we propose ChatGPT-opinion mining framework. We focus on Opinion Target Extraction (OTE) in event-centric opinion mining. According to EcoV1 [3], given the event title $E$ and the opinion expression sentence $S$, OTE aims to recognize a span in $E$ corresponding to the target argument of $S$. We convert Opinion Target Extraction to QA processes, and which refer to solving such opinion mining tasks by constructing in-context prompts with large language models (ChatGPT) for QA. Our work follows the general paradigm of in-context learning. The whole process is divided into three steps: (1) construction of prompt: given the opinion expression sentence $S$ with the title $E$, we need to construct *prompt(S, T)*. (2) Input the constructed prompt into the large language model to get the text sequence $W = \{w_1, ..., w_n\}$ returned by the model. (3) Transform the text sequence $W$ into the corresponding labels.

### 3.1 Prompt Construction

Figure 1 is an example of prompt that includes two parts: Task Description and Few-shot Demonstration

**Task Description.** The Task Description is of great help to the large language model to understand the output task and enables the model to output the content more accurately. It contains two parts (1) the first sentence is the definition of the task: *The task is to [Task Description]*, for OTE, [Task Description] is to identify the corresponding argument in the given opinion expression sentence. It is worth noting that is that, if there is no specific task description in the prompt and only follows the general paradigm of in-context learning, the large language model will produce more wrong answers and will be illusory. (2)The second sentence: *'Here are some examples'*, which means the end of the description, while telling the model to perform in-context learning. From the comparison experiments, we can see that the presence of task description helps the model to better understand the event-centric opinion mining task, and it works better.

**Few-Shot Demonstration.** We use the Few-shot as a demonstration, combined with the questions and sentences to form the in-context. There are several advantages to this approach. The first is that the output of the large language model can be more standardized. By providing a few-shot demonstration, it ensures that ChatGPT is consistent across responses to similar questions. Demonstration construct can be used to define the expected response to a particular question, thus reducing ambiguity and inconsistency in responses. The second is the ability to allow large language models to quickly adapt to new domains, and the models use examples of Few-shot to quickly learn the event-centric opinion mining domain and improve performance capabilities. Demonstration in OTE format is:

$$Title : [The\ input\ title]$$

$$Sentence : [The\ input\ sentence]\ //\ [The\ input\ argument]$$

### 3.2 In-Context Demonstrations Selection

This section describes the method of in-context demonstrations selection

**Random Selection of In-Context Demonstrations.** One of the simplest methods is to randomly pick k demonstrations from the training set. In this work, we randomly select multiple demonstrations. Unlike selecting only a single demonstration, this approach randomly selects multiple demonstrations from the sample set as inputs to the model. By introducing multiple demonstrations, the model can obtain more diverse input information, expand its knowledge and linguistic expression, and provide more comprehensive responses. In the

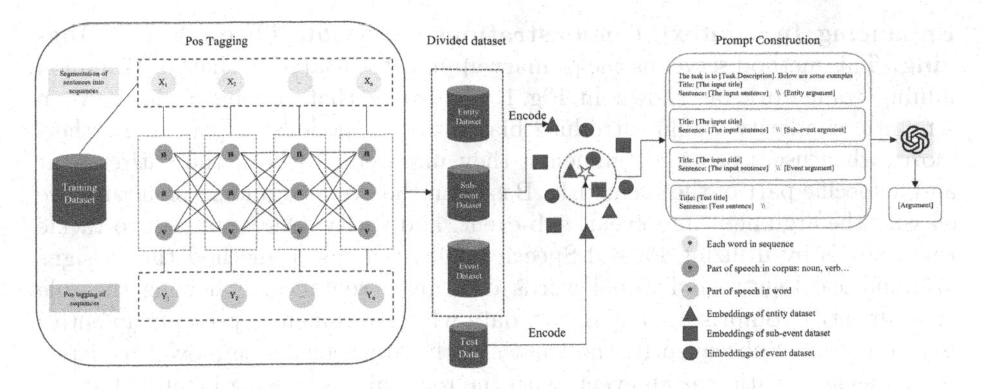


**Fig. 1.** ChatGPT-opinion mining Framework

process of random selection, we have designed 5 types of methods: (1) Random selection of demonstrations from the overall dataset. (2) Random selection of demonstrations from the dataset labeled as entity type. (3) Random selection of demonstrations from the dataset labeled as sub-event type. (4) Random selection of demonstrations from the data set labeled as event type. (5) Random selection of demonstrations from the data set labeled as entity, sub-event, and event type. However, this approach has a notable limitation: there is no assurance that the chosen demonstrations are semantically similar to the input.

**In-Context Demonstrations Selection Through $k$NN-Augmentation.** To address the problem of semantic relatedness mentioned in above section, one solution is to retrieve the k nearest neighbors ($k$NN) of the input sequence from the training set. This can be achieved by calculating representations for all training examples and then identifying the k training examples that are most similar to the input test sequence. Also in this work, we investigate the effect of the distance between contextual examples and test samples on ChatGPT performance. Specifically, to find the k closest samples in the training set, We used a pre-trained language model (Bert [23]) of [CLS] embedding as a representation of the sentence, and use Euclidean distance to measure the similarity of the two sentences.

$$dist(X, Y) = \sqrt{\sum_{i=1}^{n}(x_i - y_i)}, \quad where\ i = 1, 2 \ldots n \tag{1}$$

$dist(X, Y)$ refers to the distance from the test set data to the training set, and n is set to 768. It is worth noting that for each test sample, we find the k most similar examples in the training set as the demonstration, aiming to make the model output more accurate.

**Enhancing In-Context Demonstrations Selection Through Pos Tagging.** This method serves as the primary approach within the ChatGPT-opinion mining framework, as shown in Fig. 1. We argue that relying exclusively on semantic similarity would introduce bias into the model's responses. Furthermore, when users provide comments, they may refer to both the entire event and a specific part of the event title. Based on the event-argument structure, we classify the argument into event, sub-event, and entity. One approach to tackle this issue is by utilizing Part-of-Speech (POS) tagging, a method that assigns grammatical tags to individual words in a given sentence. Following the rule that an entity comprises solely nouns, data whose argument represents an entity is filtered out. Subsequently, the length of the argument is employed to determine its classification as an event, with the remaining set being identified as the sub-event. We employ a Conditional Random Fields(CRF)-based algorithm for lexical annotation. Let $P(y|x)$ be a conditional random field and the conditional probability of a random scalar Y taking a value of y, conditional on the variable X taking a value of x, be of the following form:

$$P(y|x) = \frac{1}{Z(x)} exp(\sum_{k=1}^{m} \sum_{i=1}^{n} \lambda_k f_k(s, i, l_i, l_{i-1})) \tag{2}$$

where $Z(x) = \sum_{l'} exp(\sum_{k=1}^{m} \sum_{i=1}^{n} \lambda_k f_k(s, i, l_i^{'}, l_{i-1}^{'})$ is normalization factor,$f_k$ is feature functions,$\lambda_k$ is feature weights, $i$ represents the position of the word in the sentence, $l_i$ represents the label of the word at the current position. The length of sentence $s$ is $n$, the number of feature functions is $m$. Given a training set X and a corresponding sequence of labels Y, the model parameters $\lambda_k$ and conditional probabilities to be learned by the feature function. By utilizing this lexical annotation, the dataset can be partitioned effectively, enabling us to enhance the demonstrations selection process and promote greater diversity in the demonstrations. Subsequently, the $k$NN method is applied to select a demonstration from each dataset, including event, subevent, and entity, serving as an in-context prompt. Combining each type into a demonstration mitigates the risk of over-representing specific sentence types or inadvertently reinforcing biases present in the in-context data.

# 4 Experiments

In this chapter, we will describe in detail the experiments and the analysis of the results. We use ChatGPT(GPT-3.5) as the Large Language Model for all experiment. For GPT-3.5 parameters, we set temperature: 0.5, presence_penalty:0, frequency_penalty:0. It is worth noting that in the few-shot demonstration, we use all 3-shot.

## 4.1 Dataset

ECO Bank [3] is a new corpus of event-centric opinion mining in both Chinese and English. The Chinese data was collected from the social media network

(WeChat Top Topics), which contains 14,710 opinion expressions and 821 event titles. The English data were collected from the W2E dataset [29] and manually shortened to meaningful text when necessary in the event title collection, which contains 11,775 opinion expression sentences and 167 event titles.

### 4.2 Evaluation Metrics

To evaluate the work of opinion target extraction, we use the dataset with a golden annotated opinion snippet. Split the snippet into sentences, and use accuracy as well as overlap-F1 evaluation metrics at the sentence level. Specifically, (1) Accuracy refers to whether the extracted argument can exactly match the annotated opinion sentence. (2) Overlap-F1, which is the F1 value of the overlap between the predicted argument and the golden argument.

$$Overlap - F_1 = \frac{2 * precision * recall}{precision + recall} \tag{3}$$

Precision is the sum of the overlapping characters of the predicted argument and its corresponding arguments in the golden set divided by the total number of characters of the predicted argument. Recall refers to the sum of the overlapping characters of the predicted argument and its corresponding argument in the golden set divided by the total number of characters in the golden argument.

### 4.3 Main Results

This chapter focuses on the experimental comparison of the methods presented in Sect. 3. In which we have different baselines in different comparison experiments.

Initially, following the random selection method proposed in the article, we conducted experiments on the complete Chinese dataset due to the ChatGPT API quota limitation. The baseline for comparison was set as random selection. Random selection involves randomly choosing data from the training set. Random selection (entity) specifically pertains to randomly selecting data from the training set where the argument represents the entity.

Analysis of Table 1 reveals that (1) within the random selection in-context demonstration, different argument types exert a significant influence on LLM. For instance, when the demonstration includes two entities and a sub-event, LLM exhibits a preference for the overall event as the argument, resulting in a recall rate of 93.82. Hence, we assert that the inclusion of diverse argument types in the in-context demonstration is crucial for the ChatGPT-opinion mining task. (2) The best performance is attained when there is one representative argument for each type. We contend that, in the context of ChatGPT-opinion mining, enhancing the diversity of in-context data will yield improved results. (3) Furthermore, we observe that LLM exhibits sensitivity to variations in content performance under identical arguments. For instance, in the case of entities, the standard deviation of Overlap-F1 reaches 13.113, indicating a significant disparity.

**Table 1.** Main results of in-context demonstrations

| Method | ZH | | | | EN | | | |
|---|---|---|---|---|---|---|---|---|
| | Accuracy | Overlap-F1 | | | Accuracy | Overlap-F1 | | |
| | | P | R | F1 | | P | R | F1 |
| Supervised Model | | | | | | | | |
| SpanR | 49.21 | – | – | 77.83 | 26.50 | – | - | 53.31 |
| MRC | 64.89 | – | – | 84.89 | 54.29 | – | – | 76.98 |
| ChatGPT-opinion mining | | | | | | | | |
| Random Selection | 42.78 ± 2.097 | 68.94 ± 3.955 | 93.82 ± 1.905 | 79.34 ± 2.122 | – | – | - | - |
| Random Selection (entity) | 29.21 ± 4.370 | 64.79 ± 10.108 | 62.11 ± 16.002 | 63.32 ± 13.113 | – | – | – | – |
| Random Selection (sub-event) | 34.74 ± 9.917 | 71.27 ± 0.925 | 85.37 ± 8.440 | 77.57 ± 3.586 | – | – | – | – |
| Random Selection (event) | 35.87 ± 0.021 | 63.34 ± 0.035 | 99.90 ± 0.007 | 77.52 ± 0.028 | – | – | – | – |
| Random Selection (all type) | 40.10 ± 4.405 | 67.24 ± 2.451 | 94.61 ± 1.954 | 78.59 ± 1.969 | – | – | – | – |
| Knn Augmentation (title+sentence) | 45.35 | 72.76 | 88.02 | 79.67 | 31.97 | 60.92 | 76.13 | 67.68 |
| Knn Augmentation (title+sentence+argument) | 27.62 | 73.94 | 56.11 | 63.80 | 25.80 | 63.39 | 60.18 | 61.75 |
| Pos Tagging (ChatGPT-opinion mining) | 45.25 | 71.02 | 96.39 | 81.78 | 32.45 | 59.23 | 82.69 | 69.02 |

Secondly, we conducted additional experiments on both the Chinese and English datasets using the proposed $k$NN method. In this case, the baseline was set as the $k$NN method incorporating sentence+title+argument. More specifically, we directly employ the pre-trained BERT language model for sentence feature extraction. The training set extracts features for sentence+title+argument, while the test set focuses on sentence+title features. The results demonstrate that the $k$NN method using the same sentence+title combination yields the best performance across both the Chinese and English datasets. Thus, we contend that LLM leverages the in-context setting to acquire knowledge about the underlying data structure, moving beyond a singular focus on the relationship between input and annotation.

Finally, we employ the proposed POS tagging method to conduct experiments on both the Chinese and English datasets, with all other methods serving as the baseline. The training dataset is partitioned into entities, sub-events, and events based on the argument, followed by the identification of their respective k-nearest neighbors. This approach not only captures the diversity within the demonstrations but also ensures semantic similarity with the test set. Analysis of Table 1 reveals that incorporating POS tagging in the English and Chinese datasets results in a two-percentage-point improvement over the standalone $k$NN method. This finding substantiates the significance of demonstration diversity in the task at hand.

In summary, in the event-centric opinion mining task, ChatGPT-opinion mining is sensitive to the choice of demonstrations. Experimental findings demonstrate that the model performs optimally when the demonstrations exhibit diversity and possess higher semantic similarity. Surprisingly, the results of ChatGPT-opinion mining with Pos tagging method are comparable to the supervised model results. In contrast, ChatGPT-opinion mining operates without the need for fine-tuning or parameter updates, significantly reducing computational and time requirements.

### 4.4 Task Description Results

Concurrently, we investigate the impact of task description on ChatGPT-opinion mining, and the corresponding results are presented in Table 2. It can be observed that the inclusion of the task description leads to improvements, to some extent, in various types of demonstrations. In this context, we contend that the task description plays a crucial role in enhancing the understanding of the task by LLM and facilitating the generation of more precise answers. Importantly, the task description should exhibit clarity, specificity, and comprehensiveness to ensure the accurate comprehension of the task's scope and objectives by the large language model, resulting in the generation of appropriate responses.

**Table 2.** Main results of task description

| Method | With Task Description | | | | Without Task Description | | | |
|---|---|---|---|---|---|---|---|---|
| | Accuracy | Overlap-F1 | | | Accuracy | Overlap-F1 | | |
| | | P | R | F1 | | P | R | F1 |
| Random Selection | 42.78 ± 2.097 | 68.94 ± 3.955 | 93.82 ± 1.905 | 79.34 ± 2.122 | 26.66 ± 15.804 | 67.69 ± 0.656 | 88.36 ± 7.750 | 76.60 ± 3.359 |
| Random Selection (entity) | 29.21 ± 4.370 | 64.79 ± 10.108 | 62.11 ± 16.002 | 63.32 ± 13.113 | 19.01 ± 1.758 | 57.18 ± 12.881 | 58.54 ± 21.467 | 57.57 ± 16.717 |
| Random Selection (sub-event) | 34.74 ± 9.917 | 71.27 ± 0.925 | 85.37 ± 8.44 | 77.57 ± 3.586 | 22.86 ± 7.944 | 63.59 ± 0.832 | 92.32 ± 5.647 | 75.25 ± 1.609 |
| Random Selection (event) | 35.87 ± 0.021 | 63.34 ± 0.035 | 99.90 ± 0.007 | 77.52 ± 0.028 | 35.95 ± 0.042 | 63.22 ± 0.098 | 99.86 ± 0.049 | 77.42 ± 0.085 |
| Random Selection (all type) | 40.10 ± 4.405 | 67.24 ± 2.451 | 94.61 ± 1.954 | 78.59 ± 1.969 | 33.91 ± 3.393 | 61.96 ± 2.764 | 98.78 ± 0.668 | 76.13 ± 2.287 |

### 4.5 K-Shot Demonstration Results

We conducted experiments to evaluate the impact of different numbers of demonstrations. The experiments were conducted using the English dataset. Due to the computational constraints and token limitations of ChatGPT, we were able to use a maximum of 9-shots. The results are shown in Fig. 2. The improvement of the supervised models' MRC is not significant as the number of shots increases. This is because the amount of data in the k-shot setting is small, making it challenging to adjust the parameters during supervised model training. However, the results of ChatGPT-opinion mining exhibit an upward trend, suggesting that performance can be improved further by incorporating additional demonstrations.

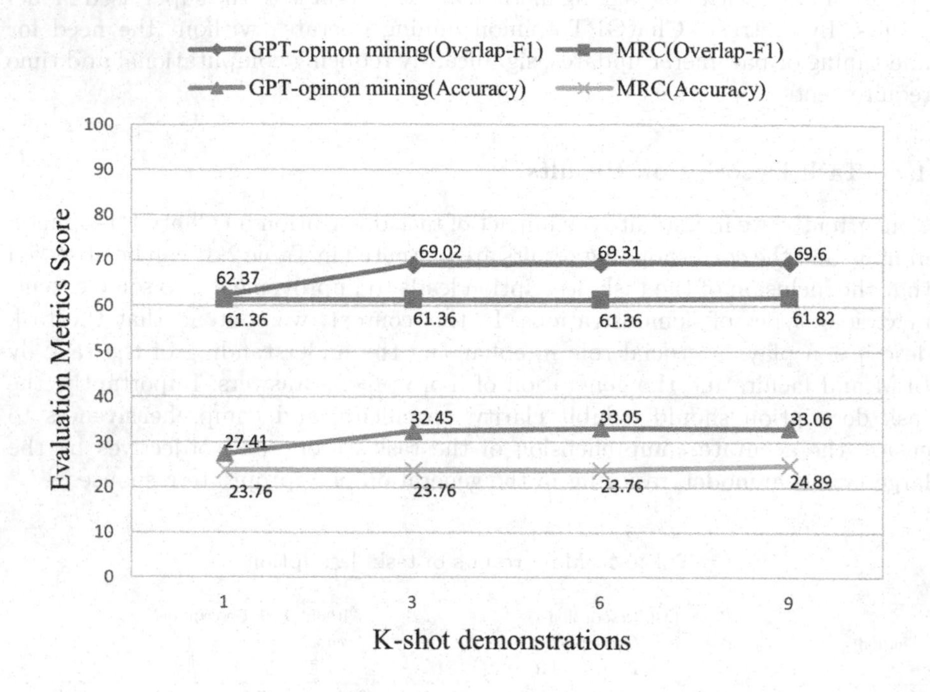


**Fig. 2.** Comparisons by varying k-shot demonstrations

## 5 Conclusion

This paper presents ChatGPT-opinion mining, a framework that converts event-centric opinion mining tasks into question-answering (QA) using the powerful language model ChatGPT. We hypothesize that the large language model can accurately extract event arguments by leveraging users' expressed opinions, thus achieving the mining task with reduced resource consumption. Subsequently, we develop a task description to mitigate potential model hallucination and design demonstrations for experimental purposes using the in-context learning approach. In the selection of demonstrations, we employ both part-of-speech (POS) tagging and k-nearest neighbors ($k$NN) methods. The experimental results indicate that the model achieves optimal performance, comparable to supervised models, when the demonstrations are diverse and have high semantic similarity. However, a limitation of this framework is the lack of a fine-tuned model for extracting [cls] features to compute k-nearest neighbors, which may introduce biases in the distribution of the semantic space. Considering the limitations of ChatGPT's input token limit, we conducted experiments with a maximum of 9 shots. We hold the belief that augmenting the number of demonstrations will lead to further improvements in the results.

**Acknowledgement.** This work was supported in part by the National Natural Science Foundation of China under Grant NO. 62106010, 62236010 and Beijing Postdoctoral Research Foundation under Grant NO. Q6042001202101.

## References

1. Russell, B.: The Analysis of Matter. Taylor & Francis (2022)
2. Karamibekr, M., Ghorbani, A.A.: Sentence subjectivity analysis in social domains. In: 2013 IEEE/WIC/ACM International Joint Conferences on Web Intelligence (WI) and Intelligent Agent Technologies (IAT). vol. 1, pp. 268–275. IEEE (2013)
3. Xu, R., et al.: ECO v1: towards event-centric opinion mining. In: Findings of the Association for Computational Linguistics: ACL 2022, pp. 2743–2753 (2022)
4. Vilar, D., Freitag, M., Cherry, C., Luo, J., Ratnakar, V., Foster, G.: Prompting palm for translation: assessing strategies and performance. arXiv preprint arXiv:2211.09102 (2022)
5. Vidal, B., Llorens, A., Alonso, J.: Automatic post-editing of MT output using large language models. In: Proceedings of the 15th Biennial Conference of the Association for Machine Translation in the Americas (Volume 2: Users and Providers Track and Government Track), pp. 84–106 (2022)
6. Robinson, J., Rytting, C.M., Wingate, D.: Leveraging large language models for multiple choice question answering. arXiv preprint arXiv:2210.12353 (2022)
7. Lazaridou, A., Gribovskaya, E., Stokowiec, W., Grigorev, N.: Internet-augmented language models through few-shot prompting for open-domain question answering. arXiv preprint arXiv:2203.05115 (2022)
8. Liu, B., et al.: Web data mining: exploring hyperlinks, contents, and usage data, vol. 1. Springer (2011). https://doi.org/10.1007/978-3-642-19460-3
9. Rhanoui, M., Mikram, M., Yousfi, S., Barzali, S.: A CNN-BiLSTM model for document-level sentiment analysis. Mach. Learn. Knowl. Extr. **1**(3), 832–847 (2019)
10. Rao, G., Huang, W., Feng, Z., Cong, Q.: Lstm with sentence representations for document-level sentiment classification. Neurocomputing **308**, 49–57 (2018)
11. Zhang, Y., Zhang, Z., Miao, D., Wang, J.: Three-way enhanced convolutional neural networks for sentence-level sentiment classification. Inf. Sci. **477**, 55–64 (2019)
12. Peng, H., Cambria, E., Zou, X.: Radical-based hierarchical embeddings for Chinese sentiment analysis at sentence level. In: The Thirtieth International Flairs Conference (2017)
13. Chen, Z., Qian, T.: Transfer capsule network for aspect level sentiment classification. In: Proceedings of the 57th Annual Meeting of the Association for Computational Linguistics, pp. 547–556 (2019)
14. Huang, B., Carley, K.M.: Syntax-aware aspect level sentiment classification with graph attention networks. arXiv preprint arXiv:1909.02606 (2019)
15. Paltoglou, G.: Sentiment-based event detection in t witter. J. Am. Soc. Inf. Sci. **67**(7), 1576–1587 (2016)

16. Maynard, D., Roberts, I., Greenwood, M.A., Rout, D., Bontcheva, K.: A framework for real-time semantic social media analysis. J. Web Semant. **44**, 75–88 (2017)
17. Pontiki, M., et al.: Semeval-2016 task 5: aspect based sentiment analysis. In: ProWorkshop on Semantic Evaluation (SemEval-2016), pp. 19–30. Association for Computational Linguistics (2016)
18. Hercig, T., Brychcín, T., Svoboda, L., Konkol, M.: UWB at SemEval-2016 task 5: aspect based sentiment analysis. In: Proceedings of the 10th International Workshop on Semantic Evaluation (SemEval-2016), pp. 342–349 (2016)
19. Du, N., et al.: Glam: efficient scaling of language models with mixture-of-experts. In: International Conference on Machine Learning, pp. 5547–5569. PMLR (2022)
20. Thoppilan, R., et al.: LaMDA: language models for dialog applications. arXiv preprint arXiv:2201.08239 (2022)
21. Frantar, E., Alistarh, D.: SparseGPT: massive language models can be accurately pruned in one-shot. In: Proceedings of the 40th International Conference on Machine Learning, pp. 10323–10337 (2023)
22. Wei, J., Wang, X., Schuurmans, D., Bosma, M., Chi, E., Le, Q., Zhou, D.: Chain of thought prompting elicits reasoning in large language models. arXiv preprint arXiv:2201.11903 (2022)
23. Devlin, J., Chang, M.W., Lee, K., Toutanova, K.: Bert: pre-training of deep bidirectional transformers for language understanding. arXiv preprint arXiv:1810.04805 (2018)
24. Wei, J., et al.: Finetuned language models are zero-shot learners. arXiv preprint arXiv:2109.01652 (2021)
25. Radford, A., Wu, J., Child, R., Luan, D., Amodei, D., Sutskever, I., et al.: Language models are unsupervised multitask learners. OpenAI Blog **1**(8), 9 (2019)
26. Dong, Q., et al.: A survey for in-context learning. arXiv preprint arXiv:2301.00234 (2022)
27. Gonen, H., Iyer, S., Blevins, T., Smith, N.A., Zettlemoyer, L.: Demystifying prompts in language models via perplexity estimation. arXiv preprint arXiv:2212.04037 (2022)
28. Rubin, O., Herzig, J., Berant, J.: Learning to retrieve prompts for in-context learning. arXiv preprint arXiv:2112.08633 (2021)
29. Hoang, T.A., Vo, K.D., Nejdl, W.: W2E: a worldwide-event benchmark dataset for topic detection and tracking. In: Proceedings of the 27th ACM International Conference on Information and Knowledge Management, pp. 1847–1850 (2018)

# Relation Repository Based Adaptive Clustering for Open Relation Extraction

Ke Chang and Ping Jian(✉)

Beijing Institute of Technology, Beijing, China
{changke,pjian}@bit.edu.cn

**Abstract.** Clustering-based relation discovery is one of the important methods in the field of open relation extraction (OpenRE). However, samples residing in semantically overlapping regions often remain indistinguishable. In this work, we propose an adaptive clustering method based on a relation repository to explicitly model the semantic differences between clusters to mitigate the relational semantic overlap in unlabeled data. Specifically, we construct difficult samples and use bidirectional margin loss to constrain the differences of each sample and apply self-supervised contrastive learning to labeled data. Combined with contrastive learning of unlabeled data, we construct a relation repository to explicitly model the semantic differences between clusters. Meanwhile, we place greater emphasis on the difficult samples located on the boundary, enabling the model to adaptively adjust the decision boundary, which lead to generate cluster-friendly relation representations to improve the effect of open relation extraction. Experiments on two public datasets show that our method can effectively improve the performance of open relation extraction.

**Keywords:** open relation extraction · contrastive learning · adaptive clustering

## 1 Introduction

The goal of Open Relation Extraction (OpenRE) is to mine structured information from unstructured text without being restricted by the set of predefined relations in the original text. Methods for dealing with open relation extraction can be roughly divided into two categories. One is Open Information Extraction (OpenIE), which extracts relational phrases of different relational types from sentences. However, this approach is limited by the redundancy of different relation phrases. The other category is unsupervised relation discovery, which focuses on unsupervised relation clustering. Furthermore, the self-supervised signal provides an optimization direction for relation clustering. Hu et al. [6] proposed a relation-oriented clustering method to predict both predefined relations and novel relations.

In current methods, the encoder is guided to update relation representations using pseudo-labels generated through clustering. However, these methods still


H. Wang et al. (Eds.): CCKS 2023, CCIS 1923, pp. 95–106, 2023.
https://doi.org/10.1007/978-981-99-7224-1_8

face challenges when dealing with difficult samples that are classified incorrectly due to semantic overlap between clusters. Specifically, instances with highly similar contexts but different relation types tend to lie at the boundary of two clusters in the semantic space. As a result, during training, blurred decision boundaries lead to the generation of incorrect guidance signals, causing these instances to oscillate between the two clusters. This phenomenon significantly impedes the accurate semantic description of relations and the appropriate categorization of relation types.

By integrating the instance and class perspectives, we propose a novel approach that leverages a relational repository to store relation representations in clusters after each epoch. This allows us to address the limitation of optimizing instances and clusters simultaneously under a single perspective. We utilize cluster representations to capture and model the semantic distinctions between clusters, enabling the model to effectively learn and optimize the decision boundary. In addition, the introduction of the sample attention mechanism on the decision boundary during the training process can improve the classification of difficult samples from the perspective of clustering.

The major contributions of our work are as follows: (1) For predefined relations, bidirectional margin loss is used to distinguish difficult samples, and instance-level self-supervised contrastive learning is enhanced for knowledge transfer. (2) For novel relations, cluster semantics are aligned with relational semantics on the basis of constructing a relation repository, and weights are used to emphasize difficult samples in training. (3) Experiment results and analyses on two public datasets demonstrate the effectiveness of our proposed method.

## 2 Related Work

Open relation extraction is used for extracting new relation types. The Open Information Extraction (OpenIE) regards the relation phrases within the sentence as individual relation types, but the same relation often has multiple surface forms, resulting in redundant relation facts.

Unsupervised relation clustering methods focus on relation types. Recently, Hu et al. [6] is an adaptive clustering model to iteratively get pseudo-labels on the BERT-encoded relation representations, and then used the pseudo-labels as self-supervised signals to train relation classifier and optimize the encoder. Zhao et al. [16] followed SelofORE's iterative generation pseudo-label scheme as part of unsupervised training. In order to obtain the relation information from the predefined data, they learned low-dimensional relation representations oriented to clustering constraints with the help of labeled data. This method does not need to design complex clustering algorithms to complete the identification of relational representations. Different from them, we proposed a method based on relation repository to explicitly model the difference in cluster semantics.

# 3 Method

The training data set $D$ includes predefined relation data $D^l = \{(\boldsymbol{s}_i^l, y_i^l)\}_{i=1}^N$ and novel relation data set $D^u = \{\boldsymbol{s}_i^u\}_{i=1}^M$, $N$ and $M$ represent the number of relation instances in each data set, $\boldsymbol{s}_i^l$ in $D^l$ and $\boldsymbol{s}_i^u$ in $D^u$ are all relation instances, including the sentence, as well as the head entity and tail entity in the text. And the $y_i^l \in \mathcal{Y}^l = \{1, ..., C^l\}$ is the relation label corresponding to the instance $\boldsymbol{s}_i^l$, the label is visible to the model during training, and the one-hot vector corresponding to $y_i^l$ is represented as $\boldsymbol{y}_i^l$. $C^u$ is provided as prior knowledge to the model.

Our goal is to automatically cluster relation instances in all unlabeled datasets into $C^u$ categories, in particular, $C^l \cap C^u = \emptyset$. Considering that the data to be predicted in real-world scenarios does not only come from unlabeled data, we use labeled and unlabeled data to evaluate the discriminative ability of the model during testing.

## 3.1 Relation Representations

Given a sentence $\boldsymbol{x} = (x_1, \ldots, x_T)$, where $T$ is the number of tokens in the sentence, $e_h$ and $e_t$ are two entities in the sentence and marked with their start and end positions. The combination of them forms a relation instance $\boldsymbol{s} = (\boldsymbol{x}, e_h, e_t)$.

For the sentence $\boldsymbol{x}$ of the relation instance $\boldsymbol{s}$, each token is encoded as $h \in R^d$ by the encoder $\boldsymbol{f}$, where $d$ represents the output dimension. The $\boldsymbol{f}$ here is the pre-trained language model BERT [2]. We use the maximum pooling of the token hidden layer vectors related to the head entity and the tail entity to obtain the hidden layer vectors of the two entities:

$$\begin{aligned} h_1, \ldots, h_T &= \text{BERT}(x_1, \ldots, x_T) \\ h_{ent} &= \text{MAXPOOL}([h_s, \ldots, h_e]) \end{aligned} \tag{1}$$

where $h_{ent} \in R^d$ represents the entity representation, $s$ and $e$ represent the start and end positions of an entity, respectively. The concatenation of the head entity representation $h_{head}$ and the tail entity representation $h_{tail}$ is regarded as a relation representation, [,] represents the concatenation operation:

$$\boldsymbol{z}_i = [h_{head}, h_{tail}] \tag{2}$$

where the relation representation $\boldsymbol{z}_i \in R^{2\times d}$.

## 3.2 Bidirectional Margin Loss

To create a sample with the same relation type but different contexts from the original, we randomly substitute the head entity and tail entity with other words of the same entity type, and the representation of new sample is recorded as $\boldsymbol{z}_i^+$. Furthermore, we randomly choose an instance of a different relation type from the original instance and replace its head entity and tail entity with synonyms found in the original instance. This allows us to construct a sample $z_i^-$ with a similar context but a distinct relation type.

In order to measure the difference between two difficult samples in the labeled data in the same semantic space, the loss $L^H$ is used to limit the difference between the cosine similarity between the original sample and the two difficult samples to the range of $[-m_2, -m_1]$:

$$\begin{aligned}\mathcal{L}^H &= max(0, sim(\boldsymbol{z}_i, \boldsymbol{z}_i^-) - sim(\boldsymbol{z}_i, \boldsymbol{z}_i^+) + m_1) \\ &+ max(0, -sim(\boldsymbol{z}_i, \boldsymbol{z}_i^-) + sim(\boldsymbol{z}_i, \boldsymbol{z}_i^+) - m_2)\end{aligned} \tag{3}$$

where $sim(,)$ is calculated by cosine similarity, the negative of $m_1$ and the negative of $-m_2$ represent the upper and lower bounds of semantic differences, and $m_1$ is set to 0.1 and $m_2$ is 0.2 during training.

### 3.3 Knowledge Transfer

The objective of knowledge transfer is to obtain information pertaining to relation representations from labeled data and learn relation representations that can be used to cluster unknown categories. In this paper, contrastive learning is used for joint training on mixed datasets to transfer relational knowledge from labeled data to unlabeled data. First we use the positive samples in Sect. 3.2 to construct a positive sample set.

In each batch, for relation instance $\boldsymbol{s}_i$ in dataset $D$, where $i \in \mathcal{N} = \{1, \dots, N\}$ is the sample number in the same batch, after obtaining the relation representation $\boldsymbol{z}_i$ through relation encoding, follow the traditional contrastive learning strategy, using NCE [4] as the contrastive loss function between instances:

$$\mathcal{L}_i^{NCE-I} = -\log \frac{\exp\left(cos(\boldsymbol{z}_i, \hat{\boldsymbol{z}}_i)/\tau\right)}{\sum_n \mathbb{1}_{[n \neq i]} \exp\left(cos(\boldsymbol{z}_i, \hat{\boldsymbol{z}}_n)/\tau\right)} \tag{4}$$

where $\hat{\boldsymbol{z}}_i$ represent a positive example of $\boldsymbol{z}_i$, $\tau$ is the temperature coefficient, $\mathbb{1}_{[n \neq i]}$ means that the expression value is 1 if and only if $n$ is not equal to $i$, otherwise it is 0.

Unlike traditional self-supervised contrastive learning tasks, there are labeled data in each batch, in order to fully learn the relational knowledge of these labeled data, we use an additional loss. Except for the constructed positive samples, all instances consistent with the current instance label are regarded as more positive samples, while other class instances of the same batch are negative samples. Since the instances of the same category are in the same positive sample set, it indirectly constrains the distribution consistency within the class, and the loss function is as below:

$$\mathcal{L}_i^{NCE-L} = -\frac{1}{|P(i)|} \sum_{p \in P(i)} \log \frac{\exp\left(cos(\boldsymbol{z}_i, \boldsymbol{z}_p)/\tau\right)}{\sum_n \mathbb{1}_{[n \neq i]} \exp\left(cos(\boldsymbol{z}_i, \boldsymbol{z}_n)/\tau\right)} \tag{5}$$

where $P(i) = \{p \in \mathcal{N} \backslash i : y_p = y_i\}$ represents the set of sample numbers with the same label with the $i$th instance $\boldsymbol{s}_i$ in a batch. For unlabeled datasets, $P(i) = \emptyset$, $\mathcal{L}_i^{NCE-L} = 0$. We construct the contrastive learning loss:

$$\mathcal{L}^{CL} = \frac{1}{N}\sum_{i}^{N}((1-\lambda)\mathcal{L}_i^{NCE-I} + \lambda\mathcal{L}_i^{NCE-L}) \tag{6}$$

where $\mathcal{L}^{NCE-I}$ only has a pair of positive samples, $\mathcal{L}^{NCE-L}$ use samples of the same relational type as the positive sample set, and constrain the encoder to learn representations that are sensitive to the semantic features of relations.$\lambda$ is used to balance $\mathcal{L}_i^{NCE-I}$ and $\mathcal{L}_i^{NCE-L}$, avoiding the overfitting of predefined relation.

### 3.4 Adaptive Clustering

Adaptively adjusting the clustering boundary method is used for unlabeled data clustering, after each training epoch, each sample's pseudo-label is modified to the label set $\mathcal{Y} = \{\hat{y}_1, \ldots, \hat{y}_{BN}\}$, $\hat{y}_i \in [1, C^u]$, where $B$ is the batch number of unlabeled data sets.

In order to facilitate the measurement of the association of cross-category instances with different categories, we use a repository set of size $BN/(C^u - 1)$ $\mathcal{M} = \{M_1, \ldots, M_{C^u}\}$ to store the enhanced instance of each category. For the positive sample representation $\hat{z}^u$ with the current pseudo-label $\hat{y}_i$, other positive sample data except $M_{\hat{y}_i}$ are used as comparison sets $Q_i$, $Q_i = \{\hat{z}^u | \hat{z}^u \in M_j \quad \forall j \in [1, C^u] \quad and \quad j \neq \hat{y}_i\}$. After each backpropagation, the new relation representation $\hat{z}^u$ enters the corresponding queue $M_{\hat{y}_i}$, and the oldest representation added to the queue will be removed. The repository set maintains instances of each category, which can be used as a basis to realize the division of relational types. The process flow of this module for unlabeled data is shown in Fig. 1, each category corresponds to a list to store related instances.

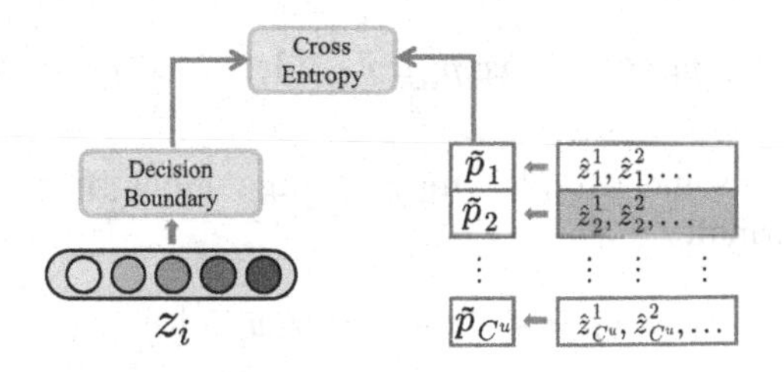


**Fig. 1.** Adaptive Clustering

In order to discover new relations using relation representations, we update decision boundaries by maximizing the intra-cluster similarity and minimizing the inter-cluster similarity and then updating the representations according to the relation repository. The instance representations of each category stored independently are used to construct the cluster center. For the current instance representation $z_i^u$, we use $\tilde{p}_{i,j}$ to calculate the probability that it belongs to the category $j$:

$$\tilde{p}_{i,j} = \frac{\sum_{\hat{z}^u \in M_j} \exp\left(\cos\left(z_i^u, \hat{z}^u\right)/\tau\right)}{\sum_{j'=1}^{C^u} \sum_{\hat{z}^u \in M_{j'}} \exp\left(\cos\left(z_i^u, \hat{z}^u\right)/\tau\right)} \tag{7}$$

where $\tau$ is the temperature coefficient. This formulation measures the semantic similarity of the current instance representation to instances of all categories. The clustering decision boundary is shown below:

$$p_i = \text{Softmax}\left(W^\top z_i^u + b\right) \in \mathcal{R}^{C^u} \tag{8}$$

where $W$ and $b$ are the parameters of the decision boundary, and $z_i^u$ is mapped to a $C^u$ dimensional vector, each dimension represents the probability $p_{i,j}$ of the corresponding category.

To align class semantics with relation categories, we minimize the cross-entropy between the cluster assignment $\tilde{p}_i$ based on the semantic similarity in the feature space and the prediction $p_i$ generated based on the decision boundary:

$$\mathcal{L}^{CD} = -\frac{1}{N} \sum_{i=1}^{N} \sum_{j=1}^{C^u} \tilde{p}_{i,j} \log p_{i,j} \tag{9}$$

Due to the setting of relation repositories, samples are assigned to the most similar category under the constraint of loss, while according to the adaptive decision boundary, relation repositories are updated in time with the semantic features corresponding to them. Following each epoch of training, the parameters of the encoder and the decision boundary are optimized, the label of the instance is updated by maximum likelihood estimation, and the relation repository is updated according to the label:

$$\hat{y}_i = \underset{j}{argmax}\ p_{i,j}, j \in \{1, \ldots, C^u\} \tag{10}$$

During training, some samples may change label repeatedly in adjacent epochs, which is formalized as:

$$\boldsymbol{s}_i^e = \boldsymbol{s}_i^{e-1} + \mathbb{1}[\hat{y}_i^e \neq \hat{y}_i^{e-1}] \tag{11}$$

where $\boldsymbol{s}_i^e$ represents the instance $\boldsymbol{s}_i$ in the $e$th epoch of training. These samples may be the difficult samples at the decision boundary. With the help of the attention mechanism, higher weights are given to these samples so that the model can achieve the correct prediction of the difficult samples:

$$w_i^e = \frac{\boldsymbol{s}_i^e}{\sum_j^N \boldsymbol{s}_j^e} \tag{12}$$

where $w_i^e$ represents the weight of $z_i^u$ in the $e$th epoch of training.

We can update the weights in the instance discriminative loss $\mathcal{L}^{NCE-I}$, and update $\mathcal{L}^{CL}$:

$$\mathcal{L}^{NCE-I} = \sum_{i=1}^{N} w_i^e \mathcal{L}_i^{NCE-I} \tag{13}$$

$$\mathcal{L}^{CL} = (1-\lambda)\mathcal{L}^{NCE-I} + \frac{\lambda}{N}\sum_{i}^{N}(\mathcal{L}_i^{NCE-L}) \tag{14}$$

We set a cross-entropy loss in order to avoid the catastrophic forgetting phenomenon of predefined relations in the process of guiding the discovery of new relations. We use the softmax layer $\sigma$ to map the relation representation $z_i^l \in \mathbb{R}^{C^l}$ to a posterior distribution $p_c = \sigma(z_i^l)$ with dimension $C^l$. The loss function is defined as follows:

$$\mathcal{L}^{CE} = -\sum_{c=1}^{C_l} y_c \log(p_c) \tag{15}$$

The total loss is:

$$\mathcal{L} = \alpha\mathcal{L}^{H} + \mathcal{L}^{CL} + \mathcal{L}^{CD} + \beta\mathcal{L}^{CE} \tag{16}$$

where $\alpha$ and $\beta$ are hyperparameters used to balance the overall loss.

## 4 Experiments

### 4.1 Datasets

To assess the performance of our method, we conduct experiments on two relation extraction datasets. **FewRel** [5] consists of texts from Wikipedia that are automatically annotated with Wikidata triple alignments in a far-supervised manner followed by manual inspection. It contains 80 relation types, there are 700 instances in each type. **TACRED** [15] is a large-scale human-annotated relation extraction dataset, including 41 relation types.

For FewRel, 64 types of relation in the original training set will be used as labeled data, and the 16 types of relation in the original verification set will be used as unlabeled data sets to discover new relations. Each type of data is divided into the training set and the test set according to 9:1. For TACRED, after removing instances labeled "No Relation", the remaining 21,773 instances are used for training and evaluation. Afterward, the 0–30 relation types are regarded as labeled datasets, and the 31–40 relation types are regarded as unlabeled datasets.In each dataset, 1/7 of the data is randomly selected as the test set, and the rest of the data is divided into the train set.

We use $B^3$ [1], $V-measure$ [11] and $ARI$ [7] to evaluate the performance of the model, they are used to measure the accuracy and recall of clustering, the uniformity and completeness of clusters, and the consistency between clusters and the true distribution.

### 4.2 Baselines

We select these OpenRE baselines for comparison:

**Discrete-state Variational Autoencoder (VAE)** [10]. VAE exploits the reconstruction of entities and predicted relations to achieve open-domain relation extraction.

**HAC with Re-weighted Word Embeddings (RW-HAC)** [3]. RW-HAC utilizes entity type and word embedding weights as relational features for clustering.

**Entity Based URE (Etype+)** [12]. Etype+ relies on entity types and uses a link predictor and two additional regularizers on top of VAE.

**Relational Siamese Network (RSN)** [13]. RSN learns the similarity of predefined relation representations from labeled data and transfers relation knowledge to unlabeled data to identify new relations.

**RSN with BERT Embedding (RSN-BERT)** [13]. This method is based on the RSN model and uses word embeddings encoded by BERT instead of standard word vectors.

**Self-supervised Feature Learning for OpenRE (SelfORE)** [6]. SelfORE uses a large-scale pre-trained language model and self-supervised signals to achieve adaptive clustering of contextual features.

**Relation-Oriented Open Relation Extraction (RoCORE)** [16]. RoCORE learns relation-oriented representations from labeled data with predefined relations and uses iterative joint training to reduce the bias caused by labeled data.

The unsupervised benchmark models include VAE, RE-HAC, EType+, the self-supervised benchmark model is SelfORE, and the supervised benchmark models include RSN, RSN-BERT, and RoCORE.

### 4.3 Implementation Details

Referring to the settings of the baseline model, we use BERT-Base-uncased to initialize the word embedding. At the same time, in order to avoid overfitting, we refer to the settings of Zhao et al. [16] and only fine-tune the parameters of Layer 8. We use Adam [8] as the optimizer, 5e−4 as learning rate, and the batch size is 100. $\alpha$ is 5e−4, 1e−3 on the FewRel and TACRED, $\beta$ is set to 0.8, $\lambda$ is set to 0.35 on the two datasets, this parameter depends on the importance of hard samples in predefined relations on different datasets. We use the "merge and split" method [14] when updating pseudo-labels to avoid cluster degradation caused by unbalanced label distribution. All experiments are trained on GeForce RTX A6000 with 48 GB memory.

### 4.4 Main Results

The main results are shown in Table 1. The method proposed in this paper exceeds the strong baseline model RoCORE on three main evaluation indicators

**Table 1.** Experimental results produced by baselines and proposed model on FewRel and TACRED in terms of $B^3$, V-measure, ARI. The horizontal line divides unsupervised and supervised methods.

| Dataset | Method | $B^3$ | | | $V-measure$ | | | $ARI$ |
|---|---|---|---|---|---|---|---|---|
| | | Prec. | Rec. | $F_1$ | Hom. | Comp. | $F_1$ | |
| FewRel | VAE | 30.9 | 44.6 | 36.5 | 44.8 | 50.0 | 47.3 | 29.1 |
| | RW-HAC | 25.6 | 49.2 | 33.7 | 39.1 | 48.5 | 43.3 | 25.0 |
| | EType+ | 23.8 | 48.5 | 31.9 | 36.4 | 46.3 | 40.8 | 24.9 |
| | SelfORE | 67.2 | 68.5 | 67.8 | 77.9 | 78.8 | 78.3 | 64.7 |
| | RSN | 48.6 | 74.2 | 58.9 | 64.4 | 78.7 | 70.8 | 45.3 |
| | RSN-BERT | 58.5 | **89.9** | 70.9 | 69.6 | **88.9** | 78.1 | 53.2 |
| | RoCORE | **75.2** | 84.6 | 79.6 | 83.8 | 88.3 | 86.0 | 70.9 |
| | **Ours** | 78.5 | 82.6 | **80.5** | 85.6 | 88.7 | **87.1** | **72.4** |
| TACRED | VAE | 24.7 | 56.4 | 34.3 | 20.8 | 36.2 | 26.4 | 15.9 |
| | RW-HAC | 42.6 | 63.3 | 50.9 | 46.9 | 59.7 | 52.6 | 28.1 |
| | EType+ | 30.2 | 80.3 | 43.9 | 26.0 | 60.7 | 36.4 | 14.3 |
| | SelfORE | 57.6 | 51.0 | 54.1 | 63.0 | 60.8 | 61.9 | 44.7 |
| | RSN | 62.8 | 63.4 | 63.1 | 62.4 | 66.3 | 64.3 | 45.9 |
| | RSN-BERT | 79.5 | **87.8** | 83.4 | 84.9 | 87 | 85.9 | 75.6 |
| | RoCORE | 87.1 | 84.9 | 86.0 | **89.5** | 88.1 | 88.8 | 82.1 |
| | **Ours** | **85.9** | 87.3 | **86.6** | 89.1 | **89.3** | **89.2** | **82.6** |

$B^3F_1$, $V-measureF_1$ and $ARI$ on all datasets, bringing 0.9%/0.6%, 1.1%/0.4% and 1.5%/0.5% growth respectively. Utilizing RoCORE and conducting paired t-tests on key performance indicators through multiple experiments, the one-tailed p-values on the two datasets are as follows: 0.002/0.024, 0.011/0.019, and 0.004/0.005, all of which are less than 0.05 indicates that our method exhibits significant differences from the RoCORE method in terms of the aforementioned indicators. It reveals that the method in this paper can effectively use the relation repository sets to model the semantic differences of different relations compared with other models. The encoder is then encouraged to generate cluster-oriented deep relation representations.

### 4.5 Ablation Analysis

In order to deeply analyze the influence of each key module on the performance of the model, we construct some ablation experiments, and the experiment results are the average results of multiple experiments (Table 2).

**Bidirectional Margin Loss.** Bidirectional margin loss can handle difficult samples better. Comparative analysis reveals that the model's performance on both datasets deteriorates after removing the margin loss, with a more pronounced decline observed in TACRED. This suggests that difficult samples within predefined relations have varying effects on different datasets.

**Knowledge Transfer.** Knowledge transfer of predefined relations greatly facilitates the discovery of new relations. Notably, the impact of knowledge transfer

**Table 2.** Abalation study of our method.

| Dataset | Method | $B^3$ | | | $V-measure$ | | | $ARI$ |
|---|---|---|---|---|---|---|---|---|
| | | Prec. | Rec. | $F_1$ | Hom. | Comp. | $F_1$ | |
| FewRel | **Ours** | 78.5 | 82.6 | **80.5** | 85.6 | 88.7 | **87.1** | **72.4** |
| | w/o margin loss | 78.3 | 82.4 | 80.3 | 85.5 | 88.1 | 86.8 | 72.2 |
| | w/o knowledge transfer | 77.1 | 73.8 | 75.4 | 83.3 | 84.7 | 84.0 | 68.7 |
| | w/o ID training | 74.6 | 76.6 | 75.6 | 81.6 | 85.3 | 83.4 | 69.8 |
| | w/o weight $w_i^e$ | 74.6 | 82.4 | 78.3 | 82.3 | 87.2 | 84.7 | 69.3 |
| TACRED | **Ours** | 85.9 | 87.3 | **86.6** | 89.1 | 89.3 | **89.2** | **82.6** |
| | w/o margin loss | 86.4 | 86.0 | 86.2 | 89.2 | 88.6 | 88.9 | 82.1 |
| | w/o knowledge transfer | 83.9 | 84.7 | 84.3 | 87.2 | 87.0 | 87.1 | 79.1 |
| | w/o ID training | 85.3 | 79.5 | 82.3 | 85.6 | 87.0 | 86.3 | 78.2 |
| | w/o weight $w_i^e$ | 85.6 | 81.9 | 83.7 | 88.9 | 86.1 | 87.5 | 78.6 |

on the FewRel dataset, in the absence of supervised contrastive loss for predefined relations, is more substantial than on TACRED. This underscores the beneficial role of knowledge transfer in enabling the encoder to learn relation representations.

**Adaptive Clustering.** Adaptive clustering holds equal importance in conjunction with knowledge transfer of predefined relations. Despite employing the knowledge within the relation repository to update pseudo-labels as a substitute, its effectiveness remains inferior to the cluster assignment guided by the clustering boundary. This highlights the efficacy of iteratively updating the decision boundary for the clustering of new relations.

**Sample Attention Mechanism.** Incorporating the difficult sample attention mechanism enhances the model's ability to discriminate between classes. The removal of the weighting strategy significantly diminishes the clustering effect on different datasets, underscoring the importance of emphasizing difficult samples with ambiguous semantics to improve the model's class discrimination ability.

### 4.6 Visualization Analysis

In order to show intuitively how our method helps refine the relation representation space, t-SNE [9] is used to visualize each relation representation in the semantic space. We randomly select 8 categories from the training set of FewRel, with a total of 800 relation representations, and reduce the dimension of each representation from $2 \times 768$ to 2 dimensions. The change of the relational semantic space during the training process is shown in Fig. 2, after training for 10, 30, and 52 epochs, the representation in the cluster is more compact than before, and the boundary between each cluster is more clear, and the clusters of each relation category have been aligned with the semantics.

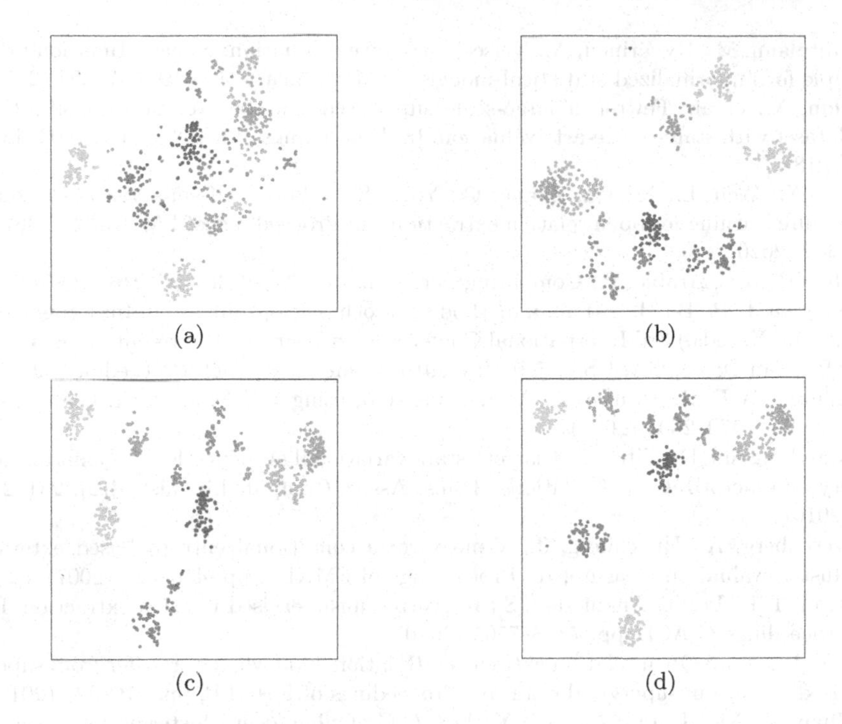


**Fig. 2.** Visualization of the relation representations

## 5 Conclusion

In this paper, we propose a relation repository-based adaptive clustering for open relation extraction. Our main contribution is to enhance the model's capability to classify difficult samples. The proposed method leverages bidirectional margin loss and adaptive clustering to enhance the prediction performance for both predefined and novel relations. Experiments and analysis demonstrate the effectiveness of our method.

**Acknowledgements.** We thank the anonymous reviewers. This work is supported by the Open Project Program of the National Defense Key Laboratory of Electronic Information Equipment System Research under Grant DXZT-JCZZ-2017-009.

## References

1. Bagga, A., Baldwin, B.: Entity-based cross-document coreferencing using the vector space model. In: Proceeding of ACL, pp. 79–85 (1998)
2. Devlin, J., Chang, M., Lee, K., Toutanova, K.: BERT: pre-training of deep bidirectional transformers for language understanding. In: Proceedings of NAACL, pp. 4171–4186 (2019)
3. ElSahar, H., Demidova, E., Gottschalk, S., Gravier, C., Laforest, F.: Unsupervised open relation extraction. CoRR abs/1801.07174 (2018)

4. Gutmann, M., Hyvärinen, A.: Noise-contrastive estimation: a new estimation principle for unnormalized statistical models. J. Mach. Learn. Res. **9**, 297–304 (2010)
5. Han, X., et al.: Fewrel: a large-scale supervised few-shot relation classification dataset with state-of-the-art evaluation. In: Proceedings of EMNLP, pp. 4803–4809 (2018)
6. Hu, X., Wen, L., Xu, Y., Zhang, C., Yu, P.S.: Selfore: self-supervised relational feature learning for open relation extraction. In: Proceedings of EMNLP, pp. 3673–3682 (2020)
7. Hubert, L.J., Arabie, P.: Comparing partitions. J. Classif. **2**, 193–218 (1985)
8. Kingma, D.P., Ba, J.: Adam: a method for stochastic optimization. In: Bengio, Y., LeCun, Y. (eds.) 3rd International Conference on Learning Representations, ICLR 2015, San Diego, CA, USA, 7–9 May 2015, Conference Track Proceedings (2015)
9. Laurens, V.D.M., Hinton, G.: Visualizing data using t-SNE. J. Mach. Learn. Res. **9**(2605), 2579–2605 (2008)
10. Marcheggiani, D., Titov, I.: Discrete-state variational autoencoders for joint discovery and factorization of relations. Trans. Assoc. Comput. Linguist. **4**(2), 231–244 (2016)
11. Rosenberg, A., Hirschberg, J.: V-measure: a conditional entropy-based external cluster evaluation measure. In: Proceedings of EMNLP, pp. 410–420 (2007)
12. Tran, T.T., Le, P., Ananiadou, S.: Revisiting unsupervised relation extraction. In: Proceedings of ACL, pp. 7498–7505 (2020)
13. Wu, R., et al.: Open relation extraction: Relational knowledge transfer from supervised data to unsupervised data. In: Proceedings of EMNLP, pp. 219–228 (2019)
14. Zhan, X., Xie, J., Liu, Z., Ong, Y., Loy, C.C.: Online deep clustering for unsupervised representation learning. In: 2020 IEEE/CVF Conference on Computer Vision and Pattern Recognition, pp. 6687–6696 (2020)
15. Zhang, Y., Zhong, V., Chen, D., Angeli, G., Manning, C.D.: Position-aware attention and supervised data improve slot filling. In: Proceedings of EMNLP, pp. 35–45 (2017)
16. Zhao, J., Gui, T., Zhang, Q., Zhou, Y.: A relation-oriented clustering method for open relation extraction. In: Proceedings of EMNLP, pp. 9707–9718 (2021)

# Knowledge Integration and Knowledge Graph Management

# LNFGP: Local Node Fusion-Based Graph Partition by Greedy Clustering

Chao Tian[1], Tian Wang[2], Ding Zhan[1], Yubiao Chang[1], Xingyu Wu[2], Cui Chen[2], Xingjuan Cai[1(✉)], Endong Tong[2], and Wenjia Niu[2]

[1] School of Computer Science and Technology, Taiyuan University of Science and Technology, Taiyuan 030024, China
{1521236687,zd334488,changyubiao}@stu.tyust.edu.cn, xingjuancai@163.com
[2] Beijing Key Laboratory of Security and Privacy in Intelligent Transportation, Beijing Jiaotong University, Beijing 100044, China

**Abstract.** Graph partitioning manages large RDF datasets in various applications such as file systems, databases and distributed computing frameworks. Research on graph partitioning can be generally categorized into two types: vertex partitioning and edge partitioning. Due to the independent nature of vertex partitioning, which facilitates easier management and maintenance, vertex partitioning methods have become more practical and popular. However, most existing vertex partitioning methods primarily focus on operating partitions in the original dimensional space, overlooking the control of partition locality from different dimensions. This oversight can adversely affect query efficiency. Therefore, we propose a graph partitioning method based on local node fusion. Based on constructing the co-occurrence matrix to calculate property weight values, we utilize a greedy clustering algorithm to achieve weight-sensitive node fusion. By constructing abstract super-nodes, we achieve a multi-granularity RDF graph that combines regular nodes with super-nodes. By setting a cost threshold, we iteratively apply an edge-cut selection mechanism, ultimately achieving vertex-based graph partitioning and super-node de-fusion. Extensive experiments are conducted on synthetic and real RDF datasets, validating the effectiveness of our proposed method.

**Keywords:** Graph Partitioning · RDF Query · Propetry Clustering · Distributed Knowledge Graph

## 1 Introduction

Knowledge graph is a graphical model used to represent knowledge and information. It consists of various elements, such as entities, relationships, and properties, and can describe and represent various types of knowledge and information. The storage of knowledge graphs is based on the RDF (Resource Description Framework) model, which serves as the standard model for publishing and


H. Wang et al. (Eds.): CCKS 2023, CCIS 1923, pp. 109–120, 2023.
https://doi.org/10.1007/978-981-99-7224-1_9

exchanging data on the web. RDF has been widely adopted in various applications for its ability to represent data in a standardized format. An RDF dataset can naturally be seen as a graph, where the subjects and objects are the graph's vertices, and the predicates are the edges connecting the vertices. SPARQL is a query language for querying and retrieving data from the RDF data model [1]. Answering a SPARQL query is equivalent to finding subgraph matches of the query graph on the RDF graph.

With the development and application of the Semantic Web, RDF datasets have been continuously growing, leading to performance issues in managing and querying RDF data on a single machine [2]. To address this issue, distributed solutions have emerged as well. This article focuses on optimizing a distributed RDF system designed explicitly for evaluating SPARQL queries [3]. In this system, the RDF graph is divided into several subgraphs, denoted as $\{F_1, ...., F_n\}$, referred to as partitions distributed across multiple nodes. One critical challenge is reducing the communication costs between different partitions during the distributed query evaluation process and enhancing partition locality to improve query efficiency.

Most graph partitioning techniques consider dividing the data in the original dimensional space. They only focus on achieving load balancing to reduce storage costs or minimizing cut edges or vertices to reduce communication costs while ignoring the control of partitioning locality in different dimensions. Only a few methods achieve partitioning based on workload, which can assign nodes with similar workloads to the same partition, thereby increasing locality. However, this approach will change as query demands change and workloads evolve. On the other hand, we propose a graph partitioning method based on local node fusion, which leverages the similarity of property weights. The idea is to store all nodes connected by edges with similar property weights in the same partition. In a distributed engine, the nodes being queried should ideally reside in the same partition. This helps to reduce the overall query performance overhead associated with accessing nodes across different partitions.

Our contribution is introducing a novel graph partitioning approach based on local node fusion. This method utilizes local node fusion as a critical technique for partitioning the graph, resulting in a new graph partitioning scheme. We calculate property weights by constructing a property co-occurrence matrix to achieve node fusion based on property similarity. Furthermore, we propose a property clustering algorithm incorporating weight sensitivity for node fusion. By constructing super nodes, we facilitate the creation of a multi-granularity RDF graph. We employ an iterative edge-cut selection mechanism by setting a cost threshold to determine which edges need to be cut. This enables the automatic partitioning of the graph and the resolution of super-node fusion. To evaluate the effectiveness of our proposed technique, we conducted experiments using synthetic and real RDF datasets. We employed various performance metrics to assess the performance of our approach and validate its efficacy.

The remaining sections of this paper are organized as follows. In Sect. 2, we provide a review of related work on graph partitioning techniques. Section 3 introduces the property co-occurrence matrix and the multi-granularity RDF graph, along with our proposed greedy clustering algorithm. In Sect. 4, we present extensive experiments conducted on synthetic and real RDF datasets to validate the effectiveness of our approach. Finally, Sect. 5 concludes the paper with a summary of our contributions.

## 2 Related Work

The most crucial aspect in distributed systems is partitioning an RDF graph into multiple subgraphs. Typically, graph partitioning techniques can be categorized based on how they divide the graph data: vertex partitioning, edge partitioning, and other approaches.

**Vertex Partitioning.** In distributed systems, most graph partitioning approaches are based on vertex partitioning. This method assigns the graph's nodes to individual partitions, and to ensure data integrity, each partition replicates the nodes at its boundaries. In vertex partitioning, nodes are the fundamental units of the graph data, and each node has certain properties or connections to other nodes. Virtuoso [4] and SHAPE [5] are examples of hash-based partitioning methods. In this approach, the elements of the triples are mapped to hash functions and then subjected to modulo operations. The number of nodes assigned to each partition is calculated using the equation $f(v) = hash(v)\ mod\ n$. This partitioning technique can also be extended to edge partitioning by assigning IDs to each edge and performing modulo operations for partitioning.

**Edge Partitioning.** This method assigns nodes of a graph to different subgraphs, which may result in some edges between nodes being cut off to maintain the integrity of the subgraphs. Edge partitioning has been widely used in many cloud-based distributed RDF systems, such as S2RDF [6], where edges with different properties are stored in separate tables. SPARQLGX [7] is based on vertical partitioning, where each triple element is distributed to different partitions or tables, typically storing two out of the three elements of a triple instead of the complete triple.

**Other Partitioning.** With the continuous improvement of partitioning techniques, several other partitioning methods have emerged that consider additional information in addition to the graph data itself [8]. For example, the DiploCloud [9] method requires defining templates as partitioning units before partitioning the graph data. Partout [10] utilizes query workloads to distribute RDF triples among the desired partitions, defining the query workload as the partition. Yars2 [11] is based on range partitioning, where RDF triples are distributed based on certain range values of the partition key.

Our approach belongs to vertex partitioning, primarily aimed at improving partition locality. However, unlike other methods employed in workload, we utilize node fusion to merge nodes with similar weights and construct super nodes.

We can effectively reduce the number of edges cut between partitions by transforming the data from a common-dimensional space into a multi-dimensional RDF graph. We aim to optimize partition locality across dimensions while considering the connectivity between partitions.

# 3 Method

We will describe our method in three steps. Firstly, we introduce the overall structure of the method. Secondly, we assign values to the dataset using a property co-occurrence matrix and achieve node fusion through a greedy algorithm. Lastly, we set a cost threshold and implement automatic partitioning through an edge selection mechanism.

## 3.1 Model Overview

The overall architecture of LNFGP is shown in Fig. 1. It consists of several steps: initializing RDF data, constructing a multi-granularity RDF graph, and performing data auto-partitioning.

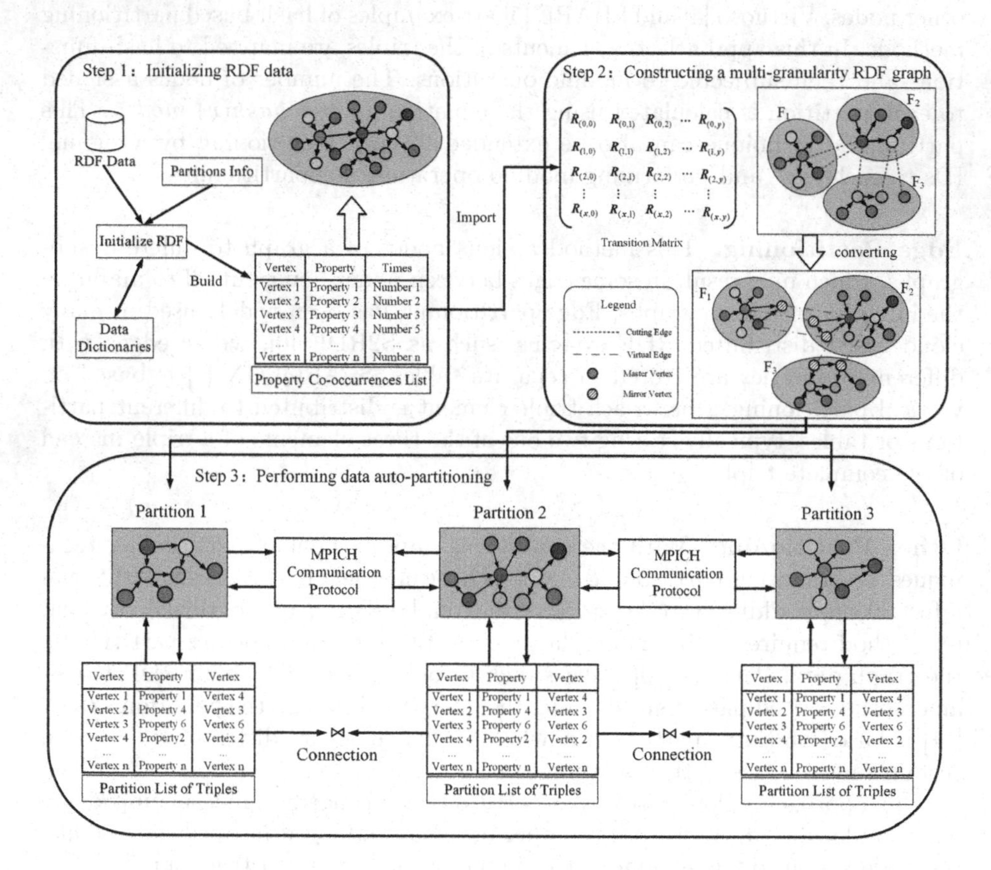


Fig. 1. The overall architecture diagram of LNFGP for partitioning.

RDF data and partitioning information are first obtained in the initialization phase of RDF data. Then, this information is used to initialize the data. Subsequently, the initialized RDF data is used to build an attribute co-occurrence list, which organizes and records the co-occurrence relationships between node attributes. Additionally, a data dictionary enables bidirectional conversion between the data and the dictionary, facilitating subsequent data processing and querying.

In constructing a multi-granularity RDF graph, the previously built attribute co-occurrence list is first transformed into an attribute co-occurrence matrix and then normalized. Next, a greedy clustering algorithm merges weight-sensitive nodes, forming super nodes. This process transforms the original RDF graph into a multi-granularity RDF graph. In the data auto-partitioning phase, we set a cost threshold and use an iterative edge selection mechanism to determine the edges that need cut. This enables automatic data partitioning and the resolution of super-nodes. Once the partitioning is completed, we store the information of each partition in triple lists and utilize MPICH for bidirectional communication between the partitions.

### 3.2 Property Co-Occurrence Matrix Generation

During the initialization stage of graph data, we create a property-based list $L = \{L(1), ..., L(n)\}$, where $L(i)$ is composed of a triplet $< v, e, x >$. Here, $v$ represents the subject in the RDF dataset, $e$ represents the property in the

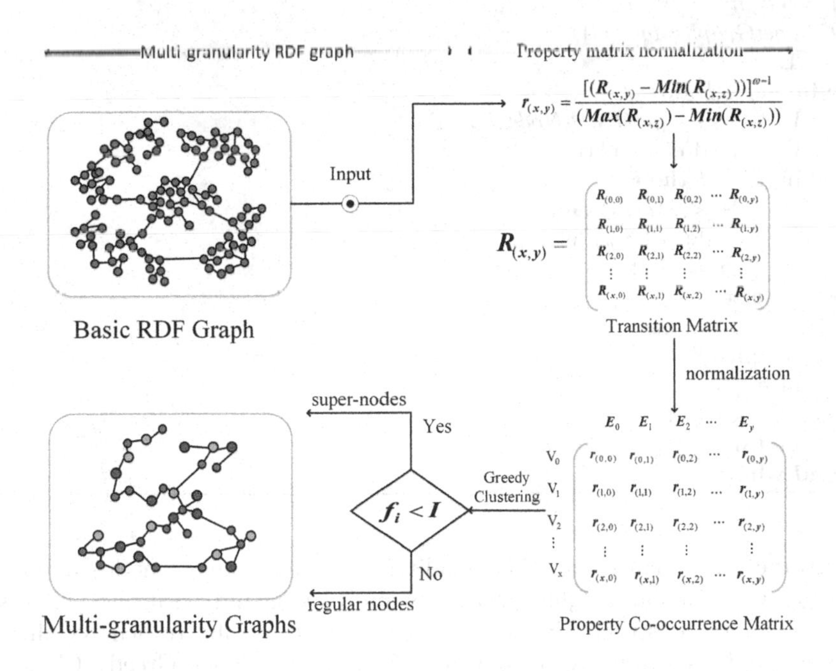


**Fig. 2.** Construct a multi-granularity RDF graph.

RDF dataset, and $x$ represents the count of occurrences when the subject and property are the same.

As shown in Fig. 2, in the matrix, $V_x$ represents nodes, $E_y$ represents properties, and $R_{(x,y)}$ represents the number of times the node and property co-occur in the triplets. The graph consists of a multi-granularity RDF graph and property matrix normalization. To transform the base RDF graph into a multi-granularity graph, we must convert the property co-occurrence list into a property co-occurrence matrix. During the matrix normalization process, the transformation of $R_{(x,y)}$ to $r_{(x,y)}$ is accomplished using Eq.(1):

$$r_{(x,y)} = \frac{\left[(R_{(x,y)} - Min(R_{(x,z)}))\right]^{\omega-1}}{(Max(R_{(x,z)}) - Min(R_{(x,z)}))} \tag{1}$$

where $R_{(x,y)}$ represents a specific element in the matrix, and we obtain the maximum and minimum values of the row where that element belongs. The variable $\omega$ represents the range for matrix normalization. By normalizing the matrix, we bring all the elements to a standardized scale, eliminating differences in dimensions among the data. The normalized values of the matrix elements are then assigned as weights to the property edges.

**Algorithm 1.** Node greedy clustering algorithm

**Input:** Property list $L$, Graph $G = (V, E, L)$, Mumber of clusters $K$
**Output:** Required clusters $F = \{f_1, ..., f_k\}$

1: $I = |G| / n$;
2: $E = getGraphEdges(G)$;
3: $i = 1$;
4: **while** $e_i \in E$ **do**
5: $V(v_n, v_m) = getGraphNodes(G, e_i)$;
6: $W = getWeights(L(v_n, e_i))$;
7: **if** $|f_i| < I$ **then**
8: **if** $W < L(e_i)$ **then**
9: Add $(v_n, v_m)$ into $f_i$;
10: $i = i + 1$;
11: **else**
12: continue;
13: **end if**
14: **else**
15: $i = i + 1$;
16: **end if**
17: **end while**

In order to enhance partition locality, we aim to merge nodes connected by edges with similar weights into the same cluster and treat the clusters as super nodes to construct a multi-granularity RDF graph. To achieve this, we have adapted the greedy algorithm and proposed the Node Greedy Clustering Algorithm.

Algorithm 1 presents the Node Greedy Clustering Algorithm, which iteratively assigns nodes to clusters. It requires input parameters such as the property list $L$, the graph $G$, and the desired number of clusters $K$. The algorithm starts by initially estimating the number of nodes needed for each cluster in the first line. Then, it obtains all property edges from the graph $G$ and iterates through them. Based on the 5th and 6th lines of Algorithm 1, retrieve the two vertices $v_n$and $v_m$connected by the edge and the weights of their edges. If the number of nodes $f_i$ in a particular cluster is less than the threshold $I$, it is considered a super node. This super node can then be used to construct a multi-granularity RDF graph based on the nodes provided by the algorithm.

### 3.3 Edge Selection Mechanism

In this section, we will select the edges to be cut and determine the final partitions. Using Algorithm 1, we obtained the multi-granularity RDF graph, which includes super and regular nodes. However, there are different edges between these super and regular nodes. We establish a cost threshold equation to minimize the number of cut edges. Given a set of properties $L' \subseteq L$, the cost of selecting $L'$ as the cut edge is defined as:

$$cost(L') = \sum_{L' \subseteq L} \omega \cdot \frac{1}{K} |f_i| + \max_{L' \subseteq L} |e| \tag{2}$$

The cost threshold is composed of two parts: $\sum_{L' \subseteq L} \omega \cdot \frac{1}{K} |f_i|$ represents load balancing, and $\max_{L' \subseteq L} |e|$ represents the maximum number of cut edges. The parameter $\omega$ is used for normalization, $K$ represents the size of the partitions, and $|e|$ represents the number of cut edges.

**Algorithm 2.** Edge selection algorithm

**Input:** Clusters $F = \{f_1, ..., f_k\}$, Graph $G = (V, E, L)$, Number of partitions $X$
**Output:** Cut edge list $L_c$

1: $L_c \leftarrow \phi$;
2: $T = (1 + \alpha) |V| / X$;
3: $L_i = getEdge(F)$;
4: $j = 1$;
5: **while** $L_i \neq \phi$ **do**
6:     **if** $cost(L_j) < T$ **then**
7:         $L_c \leftarrow L_j$;
8:     **else**
9:         $j = j + 1$;
10:     **end if**
11: **end while**

Algorithm 2 demonstrates how to perform edge cuts. First, we input the cluster list and graph $G$ from the previous section. Following lines 2 and 3 of

Algorithm 2, we obtain the edges $L_i$ between each pair of clusters and the cost threshold $T$. Here, $V$ represents the number of nodes in graph $G$, and $\alpha$ controls the balance rate. When $L_i$ is not empty, we compare the cost calculated by the cost function with $T$ to minimize the number of cuts. We use line 6 of Algorithm 2 to determine whether the cost is below the threshold. If it is below the threshold, the edge is stored in the cut list; otherwise, we move on to the next edge for evaluation.

# 4 Experiments

## 4.1 Setup

We conducted experiments on both virtual and real datasets to validate our method. All experiments were executed in the environment of CentOS 7. The partitioned datasets were stored in eight virtual machines, each equipped with a CPU with 8 cores running at 2.62 GHz, 32 GB of RAM, and 1 TB disk.

*1) Datasets and Partitioning Environments.* The experiments used two types of datasets: virtual and real datasets. Virtual datasets included WatDiv [12] and LUBM [13], while the real datasets included DBpedia [14] and YAGO4 [15]. Table 1 presents each dataset's number of triples, entity count, property count, and dataset size. We employed two different partitioning environments to evaluate the partitioning techniques: distributed RDF storage environment and a pure federated environment. Koral [16] and gStoreD [17] were used for the distributed RDF storage environment, while FedX [18] was utilized for the pure federated environment.

**Table 1.** Statistics of two types of datasets.

| Categories | Dataset | Triples | Entities | Properties | Size |
|---|---|---|---|---|---|
| Virtual | WatDiv | 110,006,886 | 5,212,743 | 86 | 15.7 GB |
| | LUBM | 106,827,145 | 65,512,368 | 18 | 14.6 GB |
| Real | DBpedia | 445,728,227 | 51,541,671 | 1721 | 63.5 GB |
| | YAGO4 | 329,524,333 | 15,636,745 | 1628 | 47.6 GB |

*2) Partitioning Techniques and Benchmark Queries.* We compared three partitioning techniques: Subject-Based, Horizontal, and Min-Edgecut, and the LNFGP method proposed in this paper. All of these partitioning techniques fulfill the requirements of being configurable, applicable to RDF datasets, and scalable to medium to large-scale datasets. We employed two query benchmarks on the dataset. The first benchmark consists of individual BGP queries that do not involve other SPARQL features such as OPTIONAL, UNION, and FILTER. The second benchmark includes complex queries that involve multiple BGPs or aggregations.

### 4.2 Evaluation Metric

For the evaluation of partitioning, we will consider five aspects: partitioning execution time, query time, partition imbalance, scalability and partition ranking. Four partitioning methods, eight partitions, and five benchmarks are included in the evaluation. The formulas for partition imbalance and partition ranking calculation are as follows.

**Partitioning Imbalance.** Assuming there are a total of $n$ partitions, represented as $\{F_1, F_2, ..., F_n\}$. As the number of triples increases, the imbalance within the partition, ranging from 0 to 1, can be represented as follows:

$$b := \frac{2\sum_{i=1}^{n}(i \times |F_i|)}{(n-1) \times \sum_{j=1}^{n}|F_i|} - \frac{n+1}{n-1} \tag{3}$$

**Partition Ranking.** Let t be the total number of partitioning techniques and b be the total number of benchmark executions used in the evaluation. Let $1 \leq r \leq t$ denote the rank number and $O_p(r)$ denote the occurrences of a partitioning technique $p$ placed at rank $r$. The sorting score range of partitioning technique $p$ is defined between 0 and 1 as follows:

$$s := \sum_{r=1}^{t} \frac{O_p(r) \times (t-r)}{b(t-1)} \tag{4}$$

### 4.3 Evaluation Results

**Partitioning Execution Time:** Table 2 compares the total time spent generating 8 partitions using 4 datasets in the evaluation. The subject-based approach has the lowest time cost, followed by LNFGP. The optimal results are highlighted in bold, while the second-best results are underlined. Topic-based queries only require scanning the dataset once to place the triples directly. In contrast, the advantages of LNFGP in query performance can compensate for the shortcomings in partitioning time, which is acceptable.

**Query Time:** One of the most critical evaluation results for partitioning is achieved by assessing the runtime performance of queries using each selected partitioning technique. We performed complete benchmark queries on data partitions created by all partitioning techniques and measured the time taken to execute these benchmark queries. We randomly selected three queries for each dataset in the benchmark, where $*$ denotes complex queries, and the rest are individual BGP queries. The results, as shown in Table 3, demonstrate that the execution time of $WQ_3^*$ in the LNFGP partition is 30% to 40% faster than Subject-Based, thereby proving that the proposed technique outperforms other partitioning methods in the majority of benchmark queries.

**Table 2.** Execution time of partitioning methods on datasets (in seconds).

| Dataset | Subject-Based | Horizontal | Min-Edgecut | LNFGP |
|---|---|---|---|---|
| WatDiv | **1773.62** | 1886.29 | 1949.11 | 1776.89 |
| LUBM | **1365.79** | 1463.87 | 2672.41 | 1381.08 |
| DBpedia | **21147.51** | 31476.85 | 71558.36 | 21237.09 |
| YAGO4 | **32654.31** | 43072.19 | 63765.62 | 38671.82 |

**Partition Imbalance:** Figure 3a, displays the partitioning imbalance generated by the selected partitioning technique. The partitioning imbalance is calculated using Eq. (3), and it shows that horizontal partitioning minimizes the partitioning imbalance across all datasets, followed by LNFGP, Subject-Based, and Min-Edgecut (Fig. 4).

**Table 3.** Execution time of different query statements (in milliseconds).

| Queries | WatDiv | | | LUBM | | | YAGO4 | | |
|---|---|---|---|---|---|---|---|---|---|
| | $WQ_1$ | $WQ_2$ | $WQ_3^*$ | $LQ_1$ | $LQ_2$ | $LQ_3^*$ | $YQ_1$ | $YQ_2$ | $YQ_3^*$ |
| SB | 259 | 1,043 | 47,569 | 692 | 763 | 85,147 | 1,842 | 7,409 | 79,150 |
| Ho | 677 | 840 | 53,792 | 366 | 428 | 78,953 | 817 | 17,374 | 186,737 |
| ME | 374 | 946 | 67,284 | 1,382 | 670 | 94,153 | 1,341 | 25,471 | 97,639 |
| LNFGP | **139** | **253** | **13,657** | **118** | **275** | **42,564** | **459** | **3,048** | **61,724** |

SB=Subject-Based, Ho=Horizontal, ME=Min-Edgecut, *=complex queries.

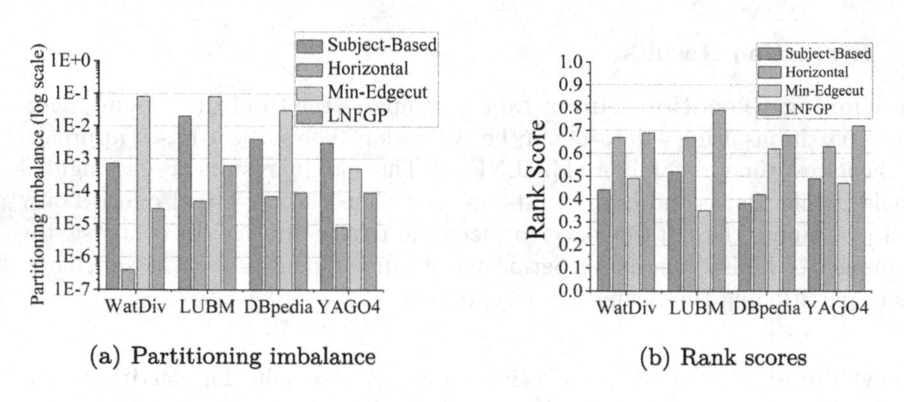


(a) Partitioning imbalance (b) Rank scores

**Fig. 3.** Rank scores and partitioning imbalance of the partitioning techniques.

**Partition Ranking:** Ranking scores indicate the overall ranking of specific methods relative to other selected methods in the completed benchmark executions. It is a value between 0 and 1, where 1 represents the highest ranking. Figure 3b represents the ranking scores of each partitioning technique calculated using Eq. (4). The results show that LNFGP has the highest ranking score, followed by horizontal, Min-Edgecut, and Subject-Based.

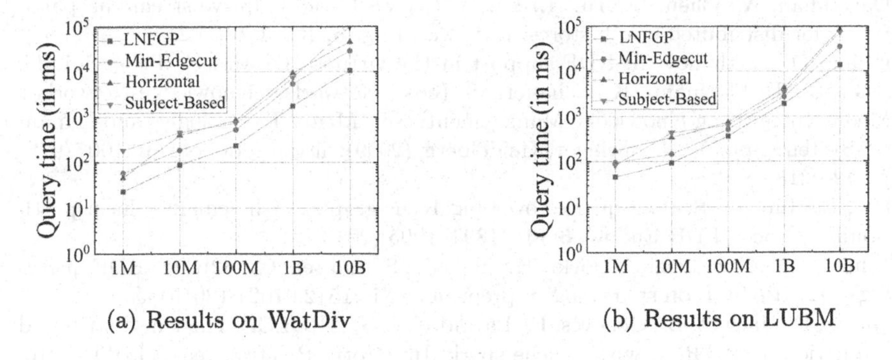


(a) Results on WatDiv (b) Results on LUBM

**Fig. 4.** Scalability of different methods on two datasets.

**Scalability:** We conducted scalability experiments on two virtual datasets by continuously increasing the scale of the datasets while keeping the number of partitions and query statements fixed. We partitioned the data using four partitioning techniques and measured the query time. The results indicate that LNFGP outperforms other methods regarding query time as the dataset scales up.

## 5 Conclusion

We propose a graph partitioning method based on local node fusion to achieve this. Firstly, we calculate attribute weights by constructing an attribute co-occurrence matrix and normalizing the weights to constrain their range. We introduce a greedy clustering algorithm to merge nodes into super nodes. Finally, automatic partitioning is achieved by setting a cost threshold and employing an iterative edge-cut selection mechanism. Overall results demonstrate that the proposed technique outperforms previous methods regarding query runtime performance, overall ranking scores, and scalability. We aim to further enhance LNFGP's time cost during partitioning by leveraging techniques such as index tables to expedite data loading and partitioning. Additionally, we can explore the application of LNFGP in different attribute graphs for further research.

**Acknowledgements.** The work was supported by the program under Grant No. 315087705.

## References


1. Saleem, M., Khan, Y., Hasnain, A., Ermilov, I., Ngonga Ngomo, A.C.: A fine-grained evaluation of SPARQL endpoint federation systems. Semant. Web **7**(5), 493–518 (2016)
2. Özsu, M.T.: A survey of RDF data management systems. Front. Comp. Sci. **10**, 418–432 (2016)

3. Davoudian, A., Chen, L., Tu, H., Liu, M.: A workload-adaptive streaming partitioner for distributed graph stores. Data Sci. Eng. **6**, 163–179 (2021)
4. Erling, O., Mikhailov, I.: RDF support in the virtuoso DBMS. In: Pellegrini, T., Auer, S., Tochtermann, K., Schaffert, S. (eds.) Networked Knowledge-Networked Media: Integrating Knowledge Management, New Media Technologies and Semantic Systems, pp. 7–24. Springer, Heidelberg (2009). https://doi.org/10.1007/978-3-642-02184-8_2
5. Lee, K., Liu, L.: Scaling queries over big RDF graphs with semantic hash partitioning. Proc. VLDB Endow. **6**(14), 1894–1905 (2013)
6. Schätzle, A., Przyjaciel-Zablocki, M., Skilevic, S., Lausen, G.: S2RDF: RDF querying with SPARQL on spark. arXiv preprint arXiv:1512.07021 (2015)
7. Graux, D., Jachiet, L., Genevès, P., Layaïda, N.: SPARQLGX: efficient distributed evaluation of SPARQL with apache spark. In: Groth, P., et al. (eds.) ISWC 2016. LNCS, vol. 9982, pp. 80–87. Springer, Cham (2016). https://doi.org/10.1007/978-3-319-46547-0_9
8. Abadi, D.J., Marcus, A., Madden, S.R., Hollenbach, K.: Scalable semantic web data management using vertical partitioning. In: Proceedings of the 33rd International Conference on Very Large Data Bases, pp. 411–422 (2007)
9. Wylot, M., Cudré-Mauroux, P.: Diplocloud: efficient and scalable management of RDF data in the cloud. IEEE Trans. Knowl. Data Eng. **28**(3), 659–674 (2015)
10. Galárraga, L., Hose, K., Schenkel, R.: Partout: a distributed engine for efficient RDF processing. In: Proceedings of the 23rd International Conference on World Wide Web, pp. 267–268 (2014)
11. Harth, A., Umbrich, J., Hogan, A., Decker, S., et al.: Yars2: a federated repository for querying graph structured data from the web. ISWC/ASWC **4825**, 211–224 (2007)
12. Aluç, G., Hartig, O., Özsu, M.T., Daudjee, K.: Diversified stress testing of RDF data management systems. In: Mika, P., et al. (eds.) ISWC 2014. LNCS, vol. 8796, pp. 197–212. Springer, Cham (2014). https://doi.org/10.1007/978-3-319-11964-9_13
13. Guo, Y., Pan, Z., Heflin, J.: LUBM: a benchmark for owl knowledge base systems. J. Web Semant. **3**(2–3), 158–182 (2005)
14. Lehmann, J., et al.: Dbpedia-a large-scale, multilingual knowledge base extracted from wikipedia. Semant. web **6**(2), 167–195 (2015)
15. Pellissier Tanon, T., Weikum, G., Suchanek, F.: YAGO 4: a reason-able knowledge base. In: Harth, A., et al. (eds.) ESWC 2020. LNCS, vol. 12123, pp. 583–596. Springer, Cham (2020). https://doi.org/10.1007/978-3-030-49461-2_34
16. Janke, D., Staab, S., Thimm, M.: Koral: a glass box profiling system for individual components of distributed RDF stores. In: CEUR Workshop Proceedings (2017)
17. Peng, P., Zou, L., Özsu, M.T., Chen, L., Zhao, D.: Processing SPARQL queries over distributed RDF graphs. VLDB J. **25**, 243–268 (2016)
18. Schwarte, A., Haase, P., Hose, K., Schenkel, R., Schmidt, M.: FedX: optimization techniques for federated query processing on linked data. In: Aroyo, L., et al. (eds.) ISWC 2011. LNCS, vol. 7031, pp. 601–616. Springer, Heidelberg (2011). https://doi.org/10.1007/978-3-642-25073-6_38

# Natural Language Understanding and Semantic Computing

# Multi-Perspective Frame Element Representation for Machine Reading Comprehension

Shaoru Guo[1], Yong Guan[1], and Ru Li[1,2](✉)

[1] School of Computer and Information Technology, Shanxi University, Taiyuan, China
liru@sxu.edu.cn

[2] Key Laboratory of Computational Intelligence and Chinese Information Processing of Ministry of Education, Shanxi University, Taiyuan, China

**Abstract.** Semantics understanding is a critical aspect of Machine Reading Comprehension (MRC). In recent years, researchers have started leveraging the semantic knowledge provided by FrameNet to enhance the performance of MRC systems. While significant efforts have been dedicated to Frame representation, there is a noticeable lack of research on Frame Element (FE) representation, which is equally crucial for MRC. We propose a groundbreaking approach called the Multi-Perspective Frame Elements Representation (MPFER) method. It aims to comprehensively model FEs from three distinct perspectives: FE definition, Frame (semantic scenario), and FE relation. By considering these multiple perspectives, our proposed model significantly improves the representation and understanding of FEs in MRC tasks. To validate the effectiveness of the MPFER method, we conducted extensive experiments. The results clearly demonstrate that our proposed model outperforms existing state-of-the-art methods. The superiority of our approach highlights its potential for advancing the field of MRC and showcasing the importance of properly modeling FEs for better semantic understanding.

**Keywords:** FrameNet · Frame element representation · Machine reading comprehension

## 1 Introduction

Machine Reading Comprehension (MRC) involves the ability of machines to read and understand a text passage *semantically*, and accurately answer questions related to it. FrameNet [1,4], as a widely adopted knowledge base, provides *schematic* scenario representation that can potentially be utilized to enhance text understanding.

In particular, *Frame* (F) is defined as a composition of *Lexical Units* (LUs) and a set of *Frame Elements* (FEs). FEs are basic semantic units of a Frame,


H. Wang et al. (Eds.): CCKS 2023, CCIS 1923, pp. 123–134, 2023.
https://doi.org/10.1007/978-981-99-7224-1_10

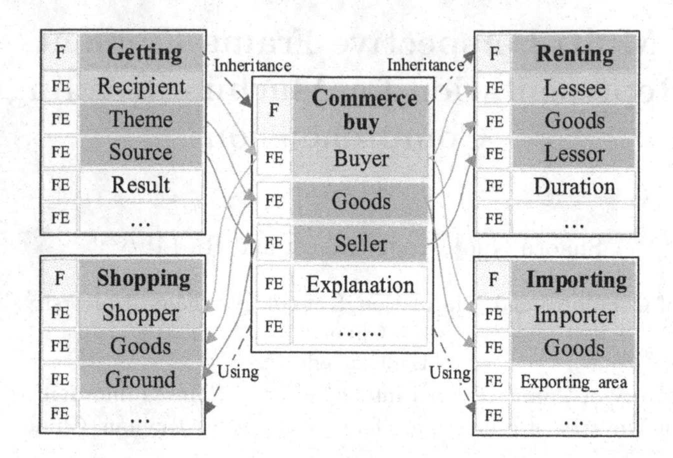


**Fig. 1.** An Example of F-to-F and FE-to-FE. Dash lines represent F-to-F. Solid lines represent FE-to-FE.

which are more semantically rich than traditional semantic role labels. In addition, FrameNet arranges relevant Frames into a network by defining Frame-to-Frame (F-to-F) relations. As each Frame contains one or multiple FEs, every F-to-F in FrameNet can thus correspond to one or more underlying FE-to-FE relations, and the relations between FEs are the same as the corresponding relations between Frames. For instance, in Fig. 1, FE ***Goods*** inherits from FE ***Theme*** (solid green line, FE-to-FE) as Frame **Commerce_buy** inherits from Frame **Getting** (dash red line, F-to-F).

Note both Frame and Frame Elements are critical for MRC. For example, in dataset MCTest, given a passage *She walked into the kitchen ... start baking when she saw she was all out of* ***flour****. She would have to go to the store to* ***get*** *some*, and a question ***What*** did Jill need to **buy** to make her pie? machines should have semantic knowledge that 1) **Get** has relation with **Buy**, and 2) **Flour** that Jill has to get represents **what** Jill need to buy. Fortunately, Frame provides rich semantics relation, e.g. **Get** and **Buy** in the given passage/question evoke Frame Getting and Commerce_buy respectively, and they are semantically connected as F-to-F in FrameNet. Meanwhile, FEs provide additional information to the semantic structure of a sentence. For example, FrameNet would mark **Flour** as a semantic unit *Theme* FE and **what** as a semantic unit *Goods* FE, and they are semantically related as FE-to-FE (green line), as shown in Fig. 1.

Recently, there has been a growing trend among researchers to leverage Frame semantic knowledge in order to develop models with enhanced semantic inference capabilities [5–7,19]. These studies highlight the importance of incorporating Frame information into modeling approaches. However, we observe that these models only pay attention to Frame and ignore critical Frame Elements (FEs) information, indicating that the current models still suffer from insufficient semantic information and relations and can be further enhanced by incorporating FEs knowledge.

In this paper, we propose a novel Multi-Perspective Frame Elements Representation (MPFER) method, which leverages three semantic perspectives to better model FEs. In addition, we further integrate the FE representation to MRC architecture to improve its performance. The key contributions of this work are summarized as:

1. To our best knowledge, our work is the first attempt to encode Frame Elements into their distributed representations.
2. We propose a novel *Multi-Perspective Frame Element Representation* (MPFER) method, which models FEs from three perspectives: FE definition, Frame (Semantic scenario), and FE-to-FE relations, to acquire richer and deeper semantic understanding of sentences.
3. Extensive experimental results on MCTest dataset demonstrate our proposed model is significantly better than ten state-of-the-art models for machine reading comprehension.

## 2 Multi-Perspective Frame Element Representation Model (MPFER)

Figure 2 depicts the comprehensive framework of our MPFER model, which comprises four essential components:

(1) **FE Definition Representation**: This component encodes each Frame Element (FE) by utilizing its corresponding definitions, resulting in the nominal representation $FE_m^d$. By incorporating the FE definitions, we capture the essence and characteristics of each FE.
(2) **Frame Representation**: The goal of this component is to obtain the representation of the Frame scenario, denoted as $F$. Building on the Frame representation approach (FRA) [7], we leverage existing techniques to capture the semantic context and scenario associated with the Frame
(3) **FE Relation Representation**: Leveraging FE-to-FE relations, this component models the semantic relationships between FEs using an attention schema, denoted as $FE_m^r$. By considering the connections and dependencies between FEs, we enhance the understanding of the overall semantic structure.
(4) **Multi-Perspective Fusion**: This component integrates the representations from $FE_m^d$, $F$ and $FE_m^r$, combining them in a multi-perspective fusion process. This integration generates a more comprehensive and meaningful representation of the FEs, denoted as $FE_m^*$. By combining information from different perspectives, we enhance the semantic inference ability and capture a richer understanding of the text.

### 2.1 FE Definition Representation

FE Definition Representation (FED) module aims to represent each FE *from its definition perspective.* To achieve this, we utilize the definition provided for

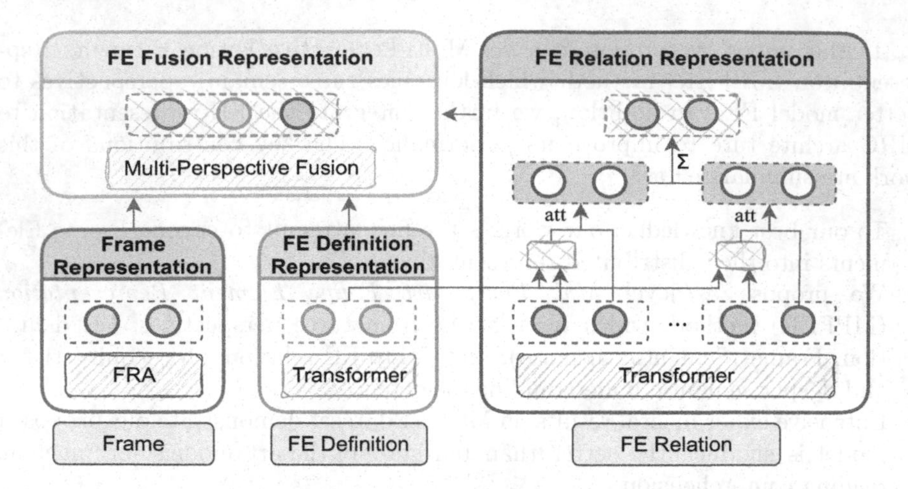


**Fig. 2.** The Architecture of our Proposed MPFER Method.

each FE in resources like FrameNet. These definitions capture the essence and characteristics of the FE. For example, in FrameNet, FE **Goods** is defined as *anything (including labor or time, for example) which is exchanged for Money in a transaction.* For given $FE_m$, we feed its definition $D = \{d_1, d_2, \ldots, d_i, \ldots, d_N\}$ into transformer-based encoder [16] to generate $FE_m^d$. This representation captures the essential information conveyed by the FE definitions and serves as a valuable input for further analysis and understanding in the MRC task.

$$FE_m^d = Transformer(d_i)(i = 1, \ldots, N) \tag{1}$$

where $d_i$ represents the $i$-th word in definition, and $N$ is the total number of words in $D$.

## 2.2 Frame Representation

Each FE is defined within the context of a specific Frame. Therefore, to capture *the semantic scenario perspective* of each FE, we model them based on their associated Frames. The Frame Representation (FR) module is designed to encode each Frame in FrameNet into a vector representation, denoted as $F$.

To construct this vector representation $F$, we utilize the Frame Relation Attention (FRA) model [7]. The FRA model leverages the relationships between Frames, known as Frame-to-Frame (F-to-F) relations, to model the Frames using an attention schema. By incorporating the attention mechanism, the FRA model captures the salient features and connections between Frames, enabling a more comprehensive understanding of the semantic scenarios they represent.

The Frame Representation module, through the use of the FRA model, encodes each Frame into a vector representation $F$. This representation incorporates information from the associated Frames and their relationships, thereby enhancing the semantic inference abilities of the model in the context of MRC tasks.

### 2.3 FE Relation Representation

The FE Relation Representation (FER) module serves as the core component of our MPFER method. It aims to enhance the representation of Frame Elements (FEs) by incorporating semantic connections from the *FE-to-FE relation perspective*

Given Frame Element $FE_m$, $FE_m^+ = \{FE_{m,1}, \ldots, FE_{m,w}, \ldots\}$ represents its expanded FEs, including all FEs linking to $FE_m$ through FE-to-FE relations. This expanded set includes all FEs that are linked to $FE_m$ through FE-to-FE relations. By including these linked FEs, we capture the semantic relationships and connections that contribute to a more comprehensive understanding of $FE_m$.

To emphasize the relevant FEs and bridge any semantic gaps, attention schemes are employed. These attention schemes are designed to assign higher weights to the relevant FEs, ensuring that they receive more focus and influence in the representation. This approach helps mitigate the influence of less relevant but linked FEs, enabling the model to concentrate on the most pertinent information for the given task.

$$FE_m^r = \sum_{w=1}^{W} att(FE_{m,w}) \cdot FE_{m,w} \tag{2}$$

$$att(FE_{m,w}) = \frac{exp(FE_m^d \cdot FE_{m,w})}{\sum_{k=1}^{W} exp(FE_m^d \cdot FE_{m,k})} \tag{3}$$

Here, $FE_m^r$ is FE relation representation, and $W$ stands for the total number of FEs in $FE_m^+$.

### 2.4 Multi-Perspective Fusion

Clearly, FED, FR and FER modules represent FEs from three different semantic perspectives. To effectively capture and integrate these perspectives, we introduce the Multi-Perspective Fusion (MPF) module, which strengthens the FE representation.

Denote the different meaningful perspectives as $H^p = [h_1^p, h_2^p, \ldots; h_i^p, \ldots, h_M^p]$. $h_i^p$ is the $i$-th perspective of FE, and $M$ is the number of perspectives. To model the interactions between these perspectives and generate a more comprehensive representation, we adopt source2token self-attention mechanism [13]. This mechanism allows us to highlight the most important perspectives for a specific FE and emphasize their contributions. The importance value $t_i$ for $h_i^p$ is calculated by:

$$t_i = W^T \sigma(W h_i^p + b) \tag{4}$$

where $\sigma(\cdot)$ is an activation function, $W$ is weight matrices, and $b$ is the bias term.

Let $S$ represents the computation scores for all perspectives in $H^p$. $S$ is computed as follows:

$$S = softmax(t) \tag{5}$$

Noted that $\sum_{i=1}^{N} S_i = 1$, where $S_i$ is the weight of $h_i^p$ when computing the representation.

The output of the attention mechanism is a weighted sum of the embedding for all perspectives in $H^p$.

$$FE_m^* = \sum_{i=1}^{N} S_i \cdot h_i^p \tag{6}$$

$FE_m^*$ is the final representation of FE that combine with multiple perspective information.

By considering the attention weights and aggregating the perspectives accordingly, the MPF module generates a better representation that combines the most significant and relevant aspects of the FEs from multiple semantic perspectives. This integrated representation helps strengthen the overall understanding of the FEs and enhances the performance of the model in MRC tasks.

### 2.5 Final Model for MRC

For all text including passage, question and option, we use SEMAFOR [2,3] to automatically process sentences with Frame semantic annotations [9]. This allows us to incorporate frame semantics into the analysis of the text, including the passage, question, and answer options.

Except for the embedding lookup layer, our model is the same as the baseline [7].

The embedding lookup layer maps F and FE to embeddings $e(x)$. For the FE embeddings, we employ our MPFER method to obtain more comprehensive and meaningful representations.

Following the embedding lookup layer, a neural network are utilized to encode sequence $f(x)$. This encoding step captures the contextual information and prepares it for further analysis.

Finally, the Softmax Linear layer is applied to select the best answer $a^*$ from 4 available options.

$$a_i = softmax(f(x))(i = 1, \ldots, 4) \tag{7}$$

$$a^* = argmax(a_i) \tag{8}$$

By incorporating Frame semantics and utilizing our MPFER method for FE embeddings, our model extends the baseline architecture and enhances its ability to understand and answer questions in the MRC task.

## 3 Experiments

This section presents a detailed description of our experimental setup, including the datasets used, the results obtained, and an in-depth analysis of those results.

**Table 1.** The Performance Comparison of 11 Different Models on Two MCTest Datasets.

| Method | MCTest-160 (%) | MCTest-500 (%) |
|---|---|---|
| Richardson et al. [12] | 69.16 | 63.33 |
| Wang et al. [17] | 75.27 | 69.94 |
| Li et al. [10] | 74.58 | 72.67 |
| Attentive Reader [8] | 46.3 | 41.9 |
| Neural Reasoner [11] | 47.6 | 45.6 |
| Parallel-Hierarchical [15] | 74.58 | 71.00 |
| Reading Strategies [14] | 81.7 | 82.0 |
| Bert [18] | 73.8 | 80.4 |
| BERT+DCMN+ [18] | 85.0 | 86.5 |
| FSR [7] | 86.1 | 84.2 |
| MPFER | 87.5 | 85.5 |

### 3.1 Datasets for MRC

For evaluating the performance of our model on the multiple-choice machine comprehension task, we utilize the MCTest dataset [12]. The MCTest dataset comprises two subsets, namely MCTest-160 and MCTest-500.

MCTest-160: This subset of the MCTest dataset contains 160 passages accompanied by multiple-choice questions. Each passage is followed by four answer options, among which only one option is correct.

MCTest-500: The MCTest-500 subset expands upon the MCTest-160 dataset and consists of 500 passages accompanied by multiple-choice questions. Similar to MCTest-160, each passage is associated with four answer options, of which one is correct. The questions in this subset also assess the understanding of the passage through various types of inquiries.

By utilizing the MCTest dataset, we evaluate our model's ability to comprehend passages and accurately select the correct answer from the provided options. The dataset's inclusion of diverse question types helps assess the model's capability to handle different comprehension challenges, including inference, and logical reasoning.

### 3.2 Experiment Results

Table 1 shows the remarkable performance of our MPFER model in various evaluation metrics. Specifically, our model achieves an outstanding accuracy on the MCTest-160 dataset, surpassing all ten state-of-the-art methods by a significant margin. This exceptional result demonstrates the effectiveness and superiority of our approach. Moreover, our model exhibits impressive competitiveness on the MCTest-500 dataset, outperforming nine existing methods and performing on

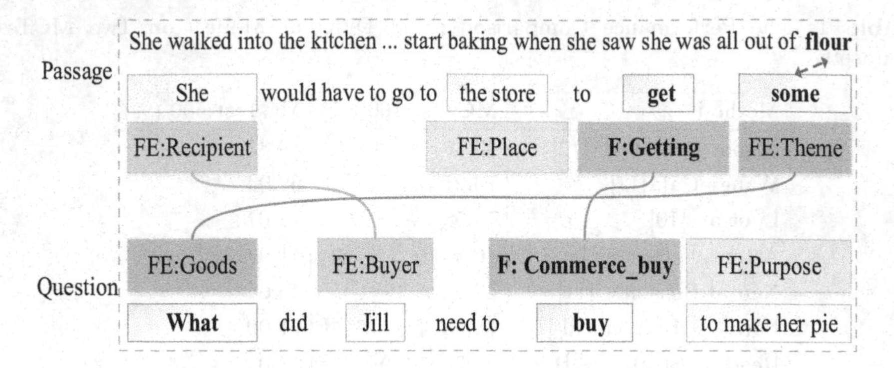


**Fig. 3.** A Case Study Example.

par with the BERT+DCMN+ model. This achievement underscores the robustness and versatility of our MPFER model across different evaluation scenarios.

It is worth emphasizing the notable performance boost our model provides over the Frame-based FSR model [7]. By integrating frame element semantic knowledge, our MPFER model surpasses the Frame-based FSR model by 1.4% and 1.3% in terms of accuracy. This result highlights the significant advantages and benefits of incorporating FE semantic knowledge into the model's architecture. The successful integration of FE semantic knowledge enhances the model's understanding and reasoning abilities, leading to improved performance in machine comprehension tasks.

Overall, the findings from our experiments not only demonstrate the superior performance of our MPFER model compared to state-of-the-art methods, but also highlight the potential of integrating FE semantic knowledge to achieve remarkable advancements in the field of machine comprehension. Our research contributes to the ongoing efforts to enhance the capabilities of natural language processing models.

### 3.3 Ablation Study

Recall in Sect. 2, we proposed three different methods, namely, FED, FER, FR, for FE representation. We conduct experiments to evaluate the effectiveness of different semantic perspectives. Note FE embedding in w/o *All* are initialized randomly with a uniform distribution between [–1, 1]. Table 2 shows their detailed results:

(1) **Individual Effectiveness**: We found that eliminating any of the semantic perspectives would have a detrimental impact on the model's performance. This observation confirms the individual effectiveness of the three different perspectives in our proposed model. It highlights the significance of considering FE definitions, FE-to-FE relations, and Frame representations to achieve a comprehensive understanding of text.

**Table 2.** The Performance Comparison Between MPFER and Four FE Representation Methods.

| Method | 160 (%) | 500 (%) |
|---|---|---|
| MPFER | 87.5 | 85.5 |
| w/o FE Relation | 86.6 | 84.6 |
| w/o Frame | 87.3 | 85.2 |
| w/o FE Definition | 87.0 | 84.9 |
| w/o ALL | 86.1 | 84.2 |

(2) **Random Initialization**: We compared the performance of models using randomly initialized FE embeddings. The results indicated that the performance significantly degraded when using random initialization. This suggests that training FE vectors rather than relying on random initialization is preferable for capturing the semantic nuances of FEs.

(3) **Comparison with MPFER Model**: When comparing the four models without FE relation to our MPFER model, we observed that the absence of FE-to-FE relations had the most detrimental effect on performance. This finding emphasizes the critical role of leveraging FE-to-FE relations to enrich semantic information and bridge semantic gaps in text understanding. Incorporating FE-to-FE relations helps establish meaningful connections between FEs and enhances the model's ability to capture intricate semantic relationships.

### 3.4 Case Study

Figure 3 presents a case study example from the MCTest dataset, demonstrating the effectiveness of our MPFER model in correctly answering questions. In this example, the passage and questions involve the Frames "Getting" and "Commerce_buy", which are semantically related. Similarly, their corresponding FEs also have meaningful relations.

By leveraging the FE-to-FE relations in FrameNet, we can establish connections between specific FEs. In this particular case, we can infer that the FE "Recipient" corresponds to the FE "Buyer", and the FE "Goods" corresponds to the FE "Theme" (with some instances of "Goods" mapping to "flour"). These connections allow us to deduce that the answer to the question is "flour".

This case study exemplifies how our MPFER model can leverage FE-to-FE relations to enhance semantic understanding and facilitate accurate question answering. By leveraging the semantic connections between FEs within Frames, we can infer the correct answer by mapping relevant FEs from the passage to the corresponding FEs in the question. This highlights the capability of our model to effectively utilize semantic information encoded in Frame Elements, leading to accurate comprehension and reasoning in MRC tasks.

## 4 Conclusion and Future Work

We introduce the Multi-Perspective Frame Element Representation (MPFER) method as a novel approach for comprehensive FE representations. To the best of our knowledge, this is the first work that addresses the challenge of capturing and utilizing Frame Elements (FEs) in a comprehensive manner. Through extensive experiments, we demonstrate that MPFER outperforms the state-of-the-art (SOTA) methods on two benchmark datasets. This implies that the utilization of frame semantic knowledge enhances the model's ability to comprehend and reason about textual information, leading to improved performance on MRC tasks.

In terms of future work, there are two important directions to consider. Firstly, with the growing scale and abilities of large language models (LLMs) like ChatGPT, leveraging these LLMs to further enhance the performance of FE representation and MRC becomes increasingly crucial. Exploiting the vast knowledge and context encoded within these models can lead to more accurate and comprehensive understanding of texts. Additionally, integrating external knowledge sources has the potential to significantly augment the model's capabilities in understanding and reasoning. By incorporating factual knowledge and commonsense reasoning, the model can acquire a broader understanding of the world, enabling it to make more informed predictions and generate contextually appropriate responses. This integration of external knowledge sources can be achieved through methods such as knowledge graph embeddings, pre-training with external corpora, or leveraging external knowledge bases.

**Acknowledgements.** We thank anonymous reviewers for their insightful comments and suggestions. This work is supported by the China-Singapore International Joint Laboratory for Language Intelligence (No. 202204041101016), the 1331 Engineering Project of Shanxi Province, China. This work is also supported by the Research on Key Technology and Application for Automatic Generation of Enterprise Innovation Capability Portrait for Strategic Emerging Industries (No. 202102020101008).

## References

1. Baker, C.F., Fillmore, C.J., Lowe, J.B.: The berkeley framenet project. In: Boitet, C., Whitelock, P. (eds.) 36th Annual Meeting of the Association for Computational Linguistics and 17th International Conference on Computational Linguistics, COLING-ACL 1998, Proceedings of the Conference, Université de Montréal, Montréal, Quebec, Canada, 10–14 August 1998, pp. 86–90. Morgan Kaufmann Publishers/ACL (1998). https://doi.org/10.3115/980845.980860. https://aclanthology.org/P98-1013/
2. Das, D., Chen, D., Martins, A.F.T., Schneider, N., Smith, N.A.: Frame-semantic parsing. Comput. Linguist. **40**(1), 9–56 (2014). https://doi.org/10.1162/COLI_a_00163
3. Das, D., Schneider, N., Chen, D., Smith, N.A.: Probabilistic frame-semantic parsing. In: Human Language Technologies: Conference of the North American Chapter

of the Association of Computational Linguistics, Proceedings, Los Angeles, California, USA, 2–4 June 2010, pp. 948–956. The Association for Computational Linguistics (2010). https://aclanthology.org/N10-1138/
4. Fillmore, C.J.: Frame semantics. In: Linguistics in the Morning Calm, pp. 111–138 (1982)
5. Guan, Y., Guo, S., Li, R., Li, X., Tan, H.: Frame semantic-enhanced sentence modeling for sentence-level extractive text summarization. In: Moens, M., Huang, X., Specia, L., Yih, S.W. (eds.) Proceedings of the 2021 Conference on Empirical Methods in Natural Language Processing, EMNLP 2021, Virtual Event/Punta Cana, Dominican Republic, 7–11 November 2021, pp. 4045–4052. Association for Computational Linguistics (2021). https://doi.org/10.18653/v1/2021.emnlp-main.331
6. Guan, Y., Guo, S., Li, R., Li, X., Zhang, H.: Integrating semantic scenario and word relations for abstractive sentence summarization. In: Moens, M., Huang, X., Specia, L., Yih, S.W. (eds.) Proceedings of the 2021 Conference on Empirical Methods in Natural Language Processing, EMNLP 2021, Virtual Event / Punta Cana, Dominican Republic, 7–11 November, 2021. pp. 2522–2529. Association for Computational Linguistics (2021). https://doi.org/10.18653/v1/2021.emnlp-main.196, https://doi.org/10.18653/v1/2021.emnlp-main.196
7. Guo, S., et al.: A frame-based sentence representation for machine reading comprehension. In: Jurafsky, D., Chai, J., Schluter, N., Tetreault, J.R. (eds.) Proceedings of the 58th Annual Meeting of the Association for Computational Linguistics, ACL 2020, Online, 5–10 July 2020, pp. 891–896. Association for Computational Linguistics (2020). https://doi.org/10.18653/v1/2020.acl-main.83
8. Hermann, K.M., et al.: Teaching machines to read and comprehend. In: Cortes, C., Lawrence, N.D., Lee, D.D., Sugiyama, M., Garnett, R. (eds.) Advances in Neural Information Processing Systems 28: Annual Conference on Neural Information Processing Systems, Montreal, Quebec, Canada, 7–12 December 2015, pp. 1693–1701 (2015). https://proceedings.neurips.cc/paper/2015/hash/afdec7005cc9f14302cd0474fd0f3c96-Abstract.html
9. Kshirsagar, M., Thomson, S., Schneider, N., Carbonell, J.G., Smith, N.A., Dyer, C.: Frame-semantic role labeling with heterogeneous annotations. In: Proceedings of the 53rd Annual Meeting of the Association for Computational Linguistics and the 7th International Joint Conference on Natural Language Processing of the Asian Federation of Natural Language Processing, ACL 2015, Beijing, China, 26–31 July 2015, vol. 2: Short Papers, pp. 218–224. The Association for Computer Linguistics (2015). https://doi.org/10.3115/v1/p15-2036
10. Li, C., Wu, Y., Lan, M.: Inference on syntactic and semantic structures for machine comprehension. In: McIlraith, S.A., Weinberger, K.Q. (eds.) Proceedings of the Thirty-Second AAAI Conference on Artificial Intelligence, (AAAI-18), the 30th innovative Applications of Artificial Intelligence (IAAI-18), and the 8th AAAI Symposium on Educational Advances in Artificial Intelligence (EAAI-18), New Orleans, Louisiana, USA, 2–7 February 2018, pp. 5844–5851. AAAI Press (2018). https://www.aaai.org/ocs/index.php/AAAI/AAAI18/paper/view/16333
11. Peng, B., Lu, Z., Li, H., Wong, K.: Towards neural network-based reasoning. CoRR abs/1508.05508 (2015). http://arxiv.org/abs/1508.05508
12. Richardson, M., Burges, C.J.C., Renshaw, E.: Mctest: a challenge dataset for the open-domain machine comprehension of text. In: Proceedings of the 2013 Conference on Empirical Methods in Natural Language Processing, EMNLP 2013, A meeting of SIGDAT, a Special Interest Group of the ACL, Grand Hyatt Seattle,

Seattle, Washington, USA, 18–21 October 2013, pp. 193–203. ACL (2013). https://aclanthology.org/D13-1020/
13. Shen, T., Zhou, T., Long, G., Jiang, J., Pan, S., Zhang, C.: Disan: directional self-attention network for RNN/CNN-free language understanding. In: McIlraith, S.A., Weinberger, K.Q. (eds.) Proceedings of the Thirty-Second AAAI Conference on Artificial Intelligence, (AAAI-18), the 30th Innovative Applications of Artificial Intelligence (IAAI-18), and the 8th AAAI Symposium on Educational Advances in Artificial Intelligence (EAAI-18), New Orleans, Louisiana, USA, 2–7 February 2018, pp. 5446–5455. AAAI Press (2018). https://www.aaai.org/ocs/index.php/AAAI/AAAI18/paper/view/16126
14. Sun, K., Yu, D., Yu, D., Cardie, C.: Improving machine reading comprehension with general reading strategies. In: Burstein, J., Doran, C., Solorio, T. (eds.) Proceedings of the 2019 Conference of the North American Chapter of the Association for Computational Linguistics: Human Language Technologies, NAACL-HLT 2019, Minneapolis, MN, USA, 2–7 June 2019, vol. 1 (Long and Short Papers), pp. 2633–2643. Association for Computational Linguistics (2019). https://doi.org/10.18653/v1/n19-1270
15. Trischler, A., Ye, Z., Yuan, X., He, J., Bachman, P.: A parallel-hierarchical model for machine comprehension on sparse data. In: Proceedings of the 54th Annual Meeting of the Association for Computational Linguistics, ACL 2016, Berlin, Germany, 7–12 August 2016, vol. 1: Long Papers. The Association for Computer Linguistics (2016). https://doi.org/10.18653/v1/p16-1041
16. Vaswani, A., et al.: Attention is all you need. In: Guyon, I., et al. (eds.) Advances in Neural Information Processing Systems 30: Annual Conference on Neural Information Processing Systems 2017, Long Beach, CA, USA, 4–9 December 2017, pp. 5998–6008 (2017). https://proceedings.neurips.cc/paper/2017/hash/3f5ee243547dee91fbd053c1c4a845aa-Abstract.html
17. Wang, H., Bansal, M., Gimpel, K., McAllester, D.A.: Machine comprehension with syntax, frames, and semantics. In: Proceedings of the 53rd Annual Meeting of the Association for Computational Linguistics and the 7th International Joint Conference on Natural Language Processing of the Asian Federation of Natural Language Processing, ACL 2015, Beijing, China, 26–31 July 2015, vol. 2: Short Papers, pp. 700–706. The Association for Computer Linguistics (2015). https://doi.org/10.3115/v1/p15-2115
18. Zhang, S., Zhao, H., Wu, Y., Zhang, Z., Zhou, X., Zhou, X.: DCMN+: dual co-matching network for multi-choice reading comprehension. In: The Thirty-Fourth AAAI Conference on Artificial Intelligence, AAAI 2020, The Thirty-Second Innovative Applications of Artificial Intelligence Conference, IAAI 2020, The Tenth AAAI Symposium on Educational Advances in Artificial Intelligence, EAAI 2020, New York, NY, USA, 7–12 February 2020, pp. 9563–9570. AAAI Press (2020). https://ojs.aaai.org/index.php/AAAI/article/view/6502
19. Zhang, X., Sun, X., Wang, H.: Duplicate question identification by integrating framenet with neural networks. In: McIlraith, S.A., Weinberger, K.Q. (eds.) Proceedings of the Thirty-Second AAAI Conference on Artificial Intelligence, (AAAI-18), the 30th innovative Applications of Artificial Intelligence (IAAI-18), and the 8th AAAI Symposium on Educational Advances in Artificial Intelligence (EAAI-18), New Orleans, Louisiana, USA, 2–7 February 2018, pp. 6061–6068. AAAI Press (2018). https://www.aaai.org/ocs/index.php/AAAI/AAAI18/paper/view/16308

# A Generalized Strategy of Chinese Grammatical Error Diagnosis Based on Task Decomposition and Transformation

Haihua Xie[1(✉)], Xuefei Chen[1], Xiaoqing Lyu[2], and Lin Liu[3]

[1] Yanqi Lake Beijing Institute of Mathematical Sciences and Applications, Beijing, China
haihua.xie@bimsa.cn

[2] Wangxuan Institute of Computer Technology, Peking University, Beijing, China
lvxiaoqing@pku.edu.cn

[3] School of Software, Tsinghua University, Beijing, China
linliu@tsinghua.edu.cn

**Abstract.** The task of Chinese Grammatical Error Diagnosis (CGED) is considered challenging due to the diversity of error types and subtypes, as well as the imbalanced distribution of subtype occurrences and the emergence of new subtypes, which pose a threat to the generalization ability of CGED models. In this paper, we propose a sentence editing and character filling-based CGED strategy that conducts task decomposition and transformation based on different types of grammatical errors, and provides corresponding solutions. To improve error detection accuracy, a refined set of error types is designed to better utilize training data. The correction task is transformed into a character slot filling task, the performance of which, as well as its generalization for long-tail scenarios and the open domain, can be improved by large-scale pre-trained models. Experiments conducted on CGED evaluation datasets show that our approach outperforms comparison models in all evaluation metrics and has good generalization.

**Keywords:** Chinese Grammatical Error Diagnosis · Model Generalization · Sequence Editing based Grammatical Error Correction

## 1 Introduction

Grammatical Error Correction (GEC) tools are used for detecting and correcting grammar errors in natural language texts on sentence basis, e.g. word redundancy, missing words or word misuse [1]. Sequence editing based models are considered the state-of-the-art solution for GEC tasks with good interpretability and inference speed [2]. In sequence editing based GEC models, error types and correction operations are represented as edit tags, hence, the design of edit tags becomes a critical task in the process.

A sequence editing based GEC model, GECToR [3], was originally proposed for English, which defines a unique set of edit tags, including 1,167 token-dependent APPEND, 3,802 REPLACE and 29 token-independent g-transformations. To train a sequence tagging model with 5,000 different tag types, GECToR collected a large training data set of over 1 million parallel sentences from real-world data and 9 million parallel sentences from synthetic data.


H. Wang et al. (Eds.): CCKS 2023, CCIS 1923, pp. 135–147, 2023.
https://doi.org/10.1007/978-981-99-7224-1_11

**Table 1.** Common error types in CGED.

| Error Types | Examples |
|---|---|
| Character(s) Redundancy | 我是感到非常高兴。<br>I am feel very happy. |
| Character(s) Missing | 我感(到)非常高兴。<br>I fee(l) very happy. |
| Character(s) Misuse | 吾感到非常高兴。<br>Me feel very happy. |
| Character(s) Out-of-order | 我感到高兴非常。<br>I feel happy very. |

GECToR is often used as the benchmark model in evaluations of Chinese Grammatical Error Diagnosis (CGED) [4,5]. However, Chinese grammar has many features that are distinct from English. As shown in Table 1, grammatical errors in Chinese often occur at the character level, resulting in irregular subtype errors (e.g., missing specific characters), and leading to a significant long-tail problem. Based on our statistics of 10,000 cases of missing character(s), there are 1,922 subtypes, and 90% of them appear less than 10 times, with 67% appearing only once. It is in a similar case for other error types. Given the scarcity of annotation data for certain subtypes and the highly unbalanced distribution of occurrences, it is challenging to train a CGED model using data-driven approaches based on annotated data.

To address these challenges, we propose TECF, which utilizes multiple NLP techniques, including Tagging, Editing, and Character Filling, and decomposes and transforms the CGED problem into tasks that can effectively utilize existing resources and pre-trained models. The main contributions of this paper are outlined below.

1) The accuracy of grammatical error detection is improved by refining error subtypes to increase the utilization of training data.
2) The performance and generalization of error correction is enhanced by transforming the problem to tasks of tag based editing (TbE) and character slot filling (CSF).
3) An iterative beam search mechanism is applied in CSF to provide more diverse solutions for error correction and handle the long-tail problem of CGED.

## 2 Related Work

There are two main types of Grammatical Error Correction (GEC) models, which are Sequence Generation (SG) based GEC and Sequence Editing (SE) based GEC. SG-based GEC adopts an encoder-decoder structure, which directly generates the correction result based on the embedding of the input sentence [7,9]. A copy mechanism was applied to speed up the decoding process by reducing the computation time of text generation [6], because the input sequence and the target sequence often have much duplicated content. Because of the relatively faster decoding speed and the better interpretability and controllability of the output results, SE-based GECs have drawn much attention in recent years [10]. In 2019 PIE was proposed [2], which designs four types of edit operations, including 'copy', 'append', 'delete' and 'replacement', to indicate grammatical errors and the corresponding corrections. In 2020, GECToR was proposed by [3], in which the tag vocabulary contains 4971 basic token-level transformations and 29 token-independent g-transformations.

In existing SE-based GEC models, the operations for correcting grammatical errors are specified in edit tags. For example, if a word misuse is detected, the edit tag suggests a replacement word. There are two main drawbacks of such mode.

- The size of the tag vocabulary is too large to get sufficient training samples.
- The alternative correction options are limited to the pre-defined set of tags.

Due to the above drawbacks, the current SE-based GEC methods do not work well in languages with limited annotated data or in open-domain scenarios [12].

## 3 Methodology

TECF has two major tasks, a refined tag set based sequence tagging for grammatical error detection (RefTag) and an editing and character filling based error correction (ECFCor), as shown in Fig. 1.

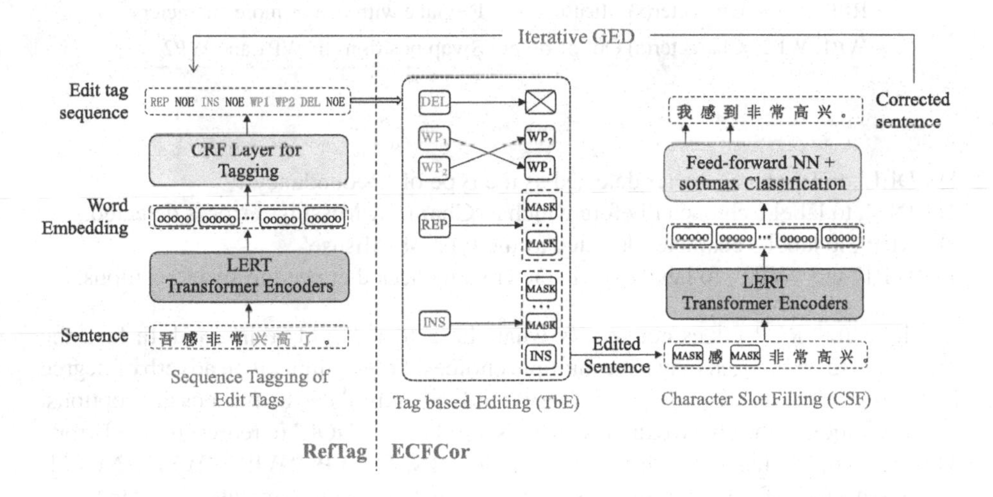


**Fig. 1.** Framework of TECF for Chinese grammatical errors detection and correction.

### 3.1 Refined Set Based Error Tagging (RefTag)

RefTag is a sequence tagging model that uses edit tags to represent the types of grammatical errors and the corresponding edit operations.

**Model Structure.** RefTag is built on LERT[1] and CRF. LERT is a pre-trained language model trained based on multiple linguistic tasks, including dependency parsing and named entity recognition, which are useful for grammatical analysis. In RefTag, LERT is the underlying model for extracting and representing linguistic features.

CRF is used in the decoding layer to produce the tagging result. CRF takes the correlation of tags into consideration [11], so it helps to eliminate ill-formed tagging sequences involving multiple types of tags.

Given a sentence $S = \{s_1, ... s_K\}$, a embedding result is generated by LERT as $U = \{u_1, ... u_K\}$, which passes through the CRF layer to produce a tag sequence $T = \{t_1, ... t_K\}$ for $S$, $t_i \in TS$, which is the tag set (see the next section).

**The Tagging Strategy.** Corresponding to the four error types in Table 1, there are four kinds of edit tags as shown in Table 2.

**Table 2.** Edit tags in RefTag of TECF.

| Tag | Error Type | Correction Operation |
|---|---|---|
| DEL | Character(s) Redundancy | Delete |
| INS | Character(s) Missing | Insert one or more characters |
| REP | Character(s) Misuse | Replace with one or more characters |
| WP1, WP2 | Character(s) Out-of-order | Swap positions of WP1 and WP2 |

1) 'DEL', to label a character detected as the type of 'Redundancy'.
2) 'INS', to label a character before which a 'Character Missing' error is detected.
3) 'REP', to label a character detected as the type of 'Misuse'.
4) 'WP1' and 'WP2', to label two character sequences that need to swap positions.

'INS' (or 'REP') does not specify what characters should be inserted. In fact, the characters to be inserted may have multiple choices. For example, if an adverb of degree is missing before an adjective, '非常(very)' and '很(quite)' can be two possible options.

In addition to the above edit tags, there is another tag 'NOE' to represent 'No Error'. The tag set of RefTag is denoted as $TS = \{$'DEL', 'INS', 'REP', 'WP1', 'WP2', 'NOE'$\}$.

Given a pair of parallel sentences $(\boldsymbol{e}, \boldsymbol{c})$, in which $\boldsymbol{e}$ is a ungrammatical sentence and $\boldsymbol{c}$ is the corresponding correct sentence, edit tags can be generated based on an Edit Tag Annotation algorithm based on Levenshtein distance [8] as shown in Fig. 2.

### 3.2 Editing and Character Filling Based Error Correction (ECFCor)

ECFCor has two main steps: tag based editing (TbE) and character slot filling (CSF).

1 https://github.com/ymcui/LERT.

**Algorithm 1: Edit Tag Annotation**

```
Input: e = (e1, ..., em), c = (c1, ..., cn)
while e != c
    leventags ← Levenshtein-Distance (e, c)
    if chars1 and chars2 swap positions
        chars1 ← WP1, chars2 ← WP2
    if chars1 moved before chars2
        chars1 ← WP1
        chars between chars1 & chars2 ← WP2
        if chars2 deleted
            chars2 ← DEL
    for each insert(ei, cj) in leventags:
        ei ← INS
    for each replace(ei, cj) in leventags:
        ei ← REP
    for each delete(ei, cj) in leventags:
        ei ← DEL
    for each other token ei in e
        ei ← NOE

    t ← set of generated edit tags
    e ← revise e based on t

Output: t = (t1, ..., tm)
```

**Fig. 2.** Algorithm of Edit Tag Annotation.

**Algorithm 2: Character Slot Filling**

```
Input: a sentence with n char-slots
Fill each char-slot with a [MASK] tag
for j = 1 to n              # char-slot
    for k = 1 to λc         # number of [MASK]
        Fill jth char-slot with k [MASK] tags
        Dk ← Beam search top x predictions
    Dj = union(D1, ... Dλc)
    Fill jth char-slot with best prediction in Dj

Output: D = {D1, ..., Dn}
```

**Algorithm 3: Beam Search Top x Predictions**

```
Input: a sequence with m continuous [MASK]
E ← top x predictions for the first [MASK]
for i = 2 to m
    Ei ← ∅
    for j = 1 to x
        Fill 1~(i-1) [MASK] with ej in E
        for e in top x predictions for ith [MASK]
            Append join(ej, e) to Ei
    E ← top x predictions in Ei

Output: E = (e1, ..., ex)
```

**Fig. 3.** Algorithm of Character Slot Filling.

**Tag Based Editing (TbE).** Based on the edit tags obtained by RefTag, TbE performs the following operations.

- Characters with tag 'DEL' are deleted;
- Characters with tag 'WP1' and 'WP2' in a clause will swith their positions.
- Each character with tag 'REP' is replaced with one or more '[MASK]' tags;
- For each character with tag 'INS', one or more '[MASK]' tags are added.

In a word, the errors of character redundancy and out-of-order are corrected in TbE, and for the errors of character missing and misuse, a preliminary step are conducted for their final correction, which will be performed in CSF.

**Character Slot Filling (CSF).** In TECF, the correction of 'INS' and 'REP' is transformed to a task of character slot filling, i.e., filling 'INS' and 'REP' with appropriate characters. Because the number of characters to be filled is unknown in advance, 1$\sim$ $\lambda_c$ characters are respectively filled to each 'INS' and 'REP' to select the best filling results. $\lambda_c$ is a hyperparameter.

To balance the computation, the accuracy of correction and the variety of correction results, an iterative beam search based CSF algorithm is designed as shown in Fig. 3.

In the CSF algorithm, a char-slot corresponds to an 'INS' or a 'REP' tag. Filling characters to a char-slot is realized by firstly adding [MASK] tags and then predict the 'masked' characters. The character prediction of a [MASK] tag is implemented based on character embeddings of LERT.

$$p_k = softmax(\hat{U}_k W^\top + b) \tag{1}$$

$$\hat{U}_k = LayerNorm(FFN(U_k)) \tag{2}$$

$p_k \in \mathbb{R}^{|V|}$ is probability distribution over the vocabulary $V$ for the $k$th token, which is a [MASK] tag, in the input sentence. $U_k \in \mathbb{R}^d$ is the last hidden layer representation of LERT for the $k$th token, and $d$ is the hidden size. $\hat{U}_k$ is the result of passing $U_k$ through a Feedforward layer and a layer normalization.

For a specific char-slot, the score of filling $m$ characters is calculated as:

$$s = \prod_{i=1}^{m} p_i(c_i) \tag{3}$$

$p_i$ is the probability distribution calculated by Formula (1), and $c_i$ is character used to replace the $i$th [MASK].

Though the character prediction of all [MASK] tags can be output at once, the CSF algorithm makes prediction of [MASK] tags one by one in order to improve the prediction accuracy of a [MASK] given other [MASK] tags has been predicted.

Multiple iterations of the workflow of RefTag and ECFCor can be performed to get the final corrections for those errors that cannot be corrected by one iteration. The maximum number of iterations is a hyperparameter denoted as $\lambda_t$.

## 4 Experimental Setup

### 4.1 Datasets

The experimental data was collected from CGED evaluations from 2015 to 2021[2]. After data denoising and deduplication, there are 52,343 parallel ungrammatical and error-free sentences. Besides, there are 17,193 grammatical correct sentences without parallel incorrect sentences. In sum, there are 69,536 instances of real data.

Meanwhile, 1,742,970 synthetic parallel sentences were produced based on the probability distribution of word-parts of speech (pos) combinations of grammartical errors in real-data sentences. The source corpus for synthetic data generation was collected from THUNews[3] and The People's Daily[4]. Besides, a synonym set[5] was used to produce the instances of character misuse by selecting a synonym to replace the original character. The threshold $\lambda_1$ and $\lambda_2$ were set to 10 and 0.075 respectively.

[2] https://github.com/blcuicall/cged_datasets.
[3] http://thuctc.thunlp.org/.
[4] http://data.people.com.cn/rmrb.
[5] http://www.ltp-cloud.com/download.

### 4.2 Evaluation Metrics

The performance of CGED models is evaluated from the following aspects. The precision, recall and F1 score of each level are evaluated respectively.

- Detection-level: Binary classification (correct or incorrect) of a given sentence.
- Identification-level: Multi-class categorization of error types in a sentence.
- Position-level: Judgement of the occurrence range of the grammatical error.
- Correction-level: Correction of error types of Character Misuse and Missing.

### 4.3 Implementation Details

**RefTag:** Training of RefTag sequence tagging was performed in two stages: 1) Pre-trained a RefTag model based on the synthetic dataset, which was randomly divided to 95% training data and 5% dev data; 2) Based on the pre-trained RefTag model, a final RefTag model was trained using the real dataset. The loss function was cross entropy, the max_seq_length was set to 200, the batch_size was 32, and the number of epochs was 5 and 20 in pre-training and final training respectively.

**ECFCor:** First of all, a Masked Language Model (MLM) Finetuning was performed on LERT based on the training data and the synthetic dataset. The masking mechanism in MLM Finetuning is presented below.

(1) If there are 'INS' or 'REP' tags in the input sentence, *m* [MASK] tags are added to the position of each 'INS' or 'REP' tag, in which *m* is the number of characters to be inserted or used for replacement.
(2) Otherwise, the default masking mechanism of LERT is used.

The default parameters of MLM in LERT was kept in the MLM finetuning. The max_seq_length was 512, the batch_size was 16, the number of epoch was 10 and the learning rate was 1e−5.

**Iterations:** The maximum number of iterations of error detection and correction, $\lambda_t$, was set to 3.

### 4.4 Comparison Methods

A GECToR-style CGED model, named CGEDToR, was used as the baseline model. The edit tag set of CGEDToR includes 'DEL', 'WP1', 'WP2', and 260 character-level 'INS' and 634 character-level 'REP'. The subtypes of 'INS' (or 'REP') with 10 or more occurrences are set to unique 'INS' (or 'REP') subtype tags (e.g., 'INS(的)'), while all the subtypes with less than 10 occurrences are set to a 'INS(others)' (or 'REP(others)') tag. CGEDToR cannot give correction for 'INS(others)' or 'REP(others)', which accounts for about 23% and 15% of the type 'INS' and 'REP' respectively. CGEDToR was also pre-trained on the synthetic data and finally trained on the real data.

Meanwhile, three representative models with good performance in the CGED2020 evaluation [4] were selected as the comparison models, including YD_NLP, XHJZ and Flying. YD_NLP has the best performance in precision, XHJZ has the highest recall and Flying has the highest F1 score. To make a fair comparison, the performance of TECF was also evaluated on the test data of the CGED2020 evaluation.

# 5 Experimental Results

## 5.1 Overall Performance on CGED2020 Dataset

The overall performance of each model on the CGED2020 test data is shown in Table 3.

**Table 3.** Overall performance of different models on CGED2020 test data.

| Model | Detection Level | | | Identification Level | | | Position Level | | | Correction Level (Top1 & Top3)[a] | | |
|---|---|---|---|---|---|---|---|---|---|---|---|---|
| | Pre. | Rec. | F1 | Pre. | Rec. | F1 | Pre. | Rec. | F1 | Pre. | Rec. | F1 |
| YD_NLP | **0.9319** | 0.8565 | 0.8926 | **0.7623** | 0.5678 | 0.6508 | **0.5145** | 0.2965 | 0.3762 | **0.3386**<br>**0.3217** | 0.1259<br>0.1333 | 0.1836<br>0.1885(2.7%↑) |
| XHJZ | 0.8062 | **0.9730** | 0.8818 | 0.5669 | **0.6714** | 0.6147 | 0.2993 | 0.2655 | 0.2814 | 0.1764<br>0.1764 | 0.1646<br>0.1646 | 0.1703 0.1703 |
| Flying | 0.9101 | 0.8800 | 0.8948 | 0.7356 | 0.6213 | **0.6736** | 0.4715 | 0.3536 | 0.4041 | 0.2290<br>0.2290 | 0.1575<br>0.1575 | 0.1867 0.1867 |
| CGEDToR | 0.9028 | 0.8424 | 0.8716 | 0.6243 | 0.5002 | 0.5554 | 0.4458 | 0.3073 | 0.3638 | 0.2193<br>0.2193 | 0.1708<br>0.1708 | 0.1803 0.1803 |
| TECF | 0.9109 | 0.8902 | **0.9004** | 0.6855 | 0.6594 | 0.6721 | .4430 | **0.3847** | **0.4118** | 0.2467<br>0.3103 | **0.2220**<br>**0.2792** | **0.2337**<br>**0.2939(25.8%↑)** |

[a] In the cells below, the first row is the result of Top1 correction, and the second row is the result of Top3 correction.

The most significant difference between TECF and the comparison models is the performance in correction. In addition to a higher F1 score, TECF got a 25.8% increase in F1 of the Top3 correction result compared with that of the TOP1 result, whereas YD_NLP only has a 2.7% increase (the results of Top1 and Top3 of other models are the same). Although the design details of the CGED2020 models are unknown, we can reason out that, based on the results of CGEDToR, the GECToR-style pre-defined tag set is the major limiting factor of performance improvement in the Top3 corrections, as the second drawback of GECToR mentioned in Sect. 2. A separate evaluation of the CSF module shows that the F1 value of the Top1, Top3 and Top5 predictions for 'INS' and 'REP' tags are 0.575, 0.723 and 0.782 respectively.

In terms of error detection, Fig. 4 shows a comparison of the performance in the position level between TECF and CGEDToR. On the types of 'INS' and 'REP' with scattered subtypes, the performance of GECToR has a significant decline compared with that of TECF. (Detailed analysis can be found in Sect. 5.3).

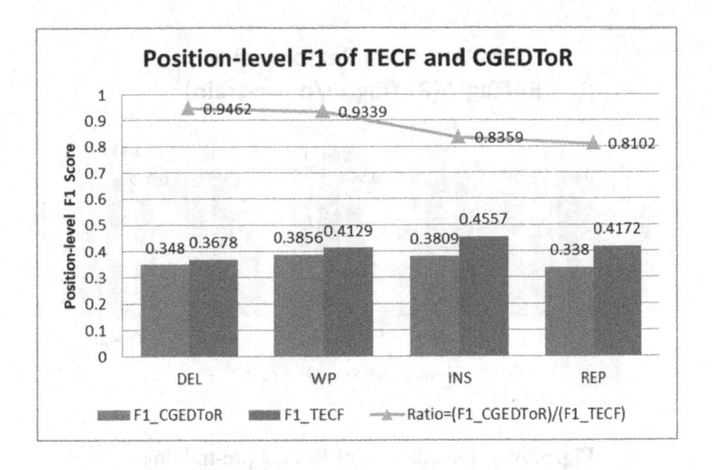


**Fig. 4.** Position-level performance comparison.

The following conclusions can be obtained based on the above analysis.

1) Refinement of edit tags is helpful for improving the performance of error detection.
2) CSF is helpful for giving more accurate and diverse correction suggestions.

### 5.2 Effects of Model Pre-training

To measure the effects of model pre-training and finetuning, the following ablation experiments were conducted.

**RefTag w/o pre-train** is a variation of RefTag for which the step of pre-training based on the synthetic data is removed.

**ECFCor w/o MLM finetune** is a variation of ECFCor for which the step of MLM finetuning based on the synthetic data is removed.

Figure 5 shows the performance comparison of position-level error detection of RefTag with or without pre-training. The values in the figure are the ratios of the performance of RefTag to that of RefTag w/o pre-train. RefTag pre-training is helpful for improving the recall for all error types, while it also leads to lower precision because more false positive predictions were made.

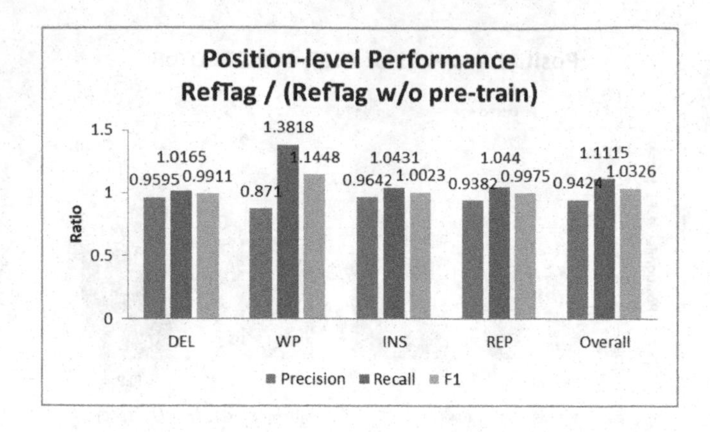


**Fig. 5.** Ablation analysis of RefTag pre-training.

Table 4 shows the F1 value of error correction (for error types 'INS' and 'REP') of ECFCor with or without MLM finetuning. MLM finetuning is conducive to advancing the ranking of effective corrections. The performance of Top1 corrections has a higher increase than that of Top3 and Top5.

**Table 4.** F1 of ECFCor with or w/o MLM finetune.

| Model | Top1 | Top3 | Top5 |
|---|---|---|---|
| ECFCor w/o MLM finetune | 0.536 | 0.694 | 0.761 |
| ECFCor | 0.575(7.28%↑) | 0.723(4.18%↑) | 0.782(2.76%↑) |

## 5.3 Analysis of Model Generalization

To analyze model generalization, an experiment of evaluating the model performance on different classes of data was conducted. The subtypes of character missing and character misuse are categorized into 9 classes based on their occurrences in the dataset. Class '>500' contains the subtypes with at least 500 occurrences, and class '300–500' contains the subtypes with at least 300 and up to 500 occurrences, and so on. The recall on the position level and correction level of each class was evaluated separately. The recall on the correction level was evaluated based on the tokens that were correctly labeled on the position level, i.e., it is in fact the recall of the CSF model. The experimental results are shown in Fig. 6.

The performance of models on the position level, CGEDToR_Pos and TECF_Pos, does not have much difference in the trends on different classes of data. However, on the correction level, the recall of CGEDToR (CGEDToR_Cor in the figure) has a signigicant decline in the classes of subtypes with few occurrences, while the performace of TECF (TECF_Cor(Top3) in the figure) is more stable. Especially for class '<10',

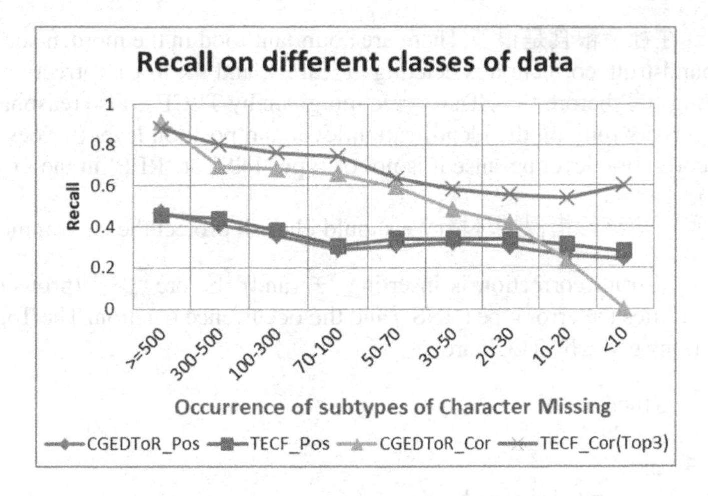


**Fig. 6.** Model generalization comparison.

which contains subtypes with less than 10 occurrences, CGEDToR cannot give corrections because the corresponding edit tag, 'INS(others)' or 'REP(others)', does not contain correction suggestions.

Another serious problem for CGEDToR is that it cannot give correction suggestions for the subtypes of character missing and character misuse that never appear in the training dataset, thus it does not perform well in the open domain, while TECF can still have a consistent performance. A study of correction-level performance in the open domain was conducted. A set of sentences, in which each case of character missing or character misuse has 0 occurrence in the dataset of CGED evaluations, was constructed based on the synthetic dataset. Such set of sentences contains 36,359 cases of character missing and 80,376 missing. TECF achieved a recall of the Top1, Top3 and Top5 predictions for 'INS' and 'REP' tags are 0.2015, 0.2513 and 0.2755 respectively.

Based on the above analysis, we come to the following conclusion: TECF has a more stable performance both in the long-tail scenarios and in the open domain, thus it has better model generalization, compared with GECToR-style CGED models.

### 5.4 Case Study of Correction Diversity

Compared with GECToR which achieved a F1 value of 62.5% for correcting English grammatical errors [3], TECF has a much lower performance for CGED. An important factor of the low F1 values of CGED models is that the correction of Chinese grammatical errors has multiple choices in many cases, i.e., the ground-truth answers are often not the unique solution. A study of the perplexity (PPL) value comparison was conducted, which shows that ratios of the PPL value of the ground-truth answer lower than the corrections of TECF, if they are not identical, are: 1.37% (for Top1 correction), 6.13% (for Top3 correction) and 15.83% (for Top5 correction).

Two examples of diverse corrections for CGED are shown below.

1) Input: '现在社会粮食是很多(There are abundant food in the morden society.)'
   The ground-truth correction is deleting '是(are)', and the Top1 correction of TECF is inserting '的' before '。'. The correction given by TECF is also reasonable, but it is considered wrong on the identification level and position level (it does not count for the correction level because it is not of type 'INS' or 'REP' in the ground-truth tagging).
2) Input: '我们应该珍惜保护环境(We should cherish protect the environment.)'

The ground-truth correction is inserting '并(and)' before '保护(protect)'. TECF correctly identifies the error type ('INS') and the occurrence position. The Top5 correction suggestions given by TECF are:

- No. 1: '并且(and also)'
- No. 2: '和(and)'
- No. 3: '并(and)'
- No. 4: '并共同(and also together)'
- No. 5: '它并且(it and also)'

The No. 1 and No. 2 suggested words to insert are synonyms of '并(and)', and the No. 4 and No. 5 suggestions provide effective correction and slightly expand the semantics of the original text.

The diverse correction options and the relatively limited ground-truth answers are the bottleneck for improving the performance of CGED models. In practical scenarios, people often have different choices for correcting an ungrammatical sentence, based on their understanding of the sentence and pragmatic habits. The flexibility of TECF for giving multiple correction suggestions makes it more suitable in industrial applications.

## 6 Conclusions

In this paper, we put forward a strategy of Chinese grammatical errors diagnosis by task decomposition and transformation, namely turning the problem into an error tag based sentence editing and character filling task. Based on our evaluation on CGED datasets, the proposed approach has achieved better performance, higher generalization and flexibility in error detection and error correction tasks. However, there are still much room for improvements in terms of accuracy of identification of error types and determination of error locations.

In future work, a fundamental and primary task is to construct a higher-quality and larger-scale evaluation dataset, as well as a more effective evaluation methodology of the correction results. In addition, the study of more types of grammatical errors, such as improper collocation of sentence components, is also a valuable attempt.

## References

1. Rozovskaya, A., Chang, K.-W., Sammons, M., Roth, D., Habash N.: The illinois-columbia system in the CoNLL-2014 shared task. In: Proceedings of the 18th Conference on Computational Natural Language Learning, pp. 34–42. ACL, Baltimore (2014)

2. Awasthi, A., Sarawagi, S., Goyal, R., Ghosh, S., Piratla, V.: Parallel iterative edit models for local sequence transduction. In: Proceedings of the 2019 Conference on Empirical Methods in Natural Language Processing (EMNLP), pp. 4260–4270. ACL, Hong Kong (2019)
3. Omelianchuk, K., Atrasevych, V., Chernodub, A., Skurzhanskyi, O.: GECToR-grammatical error correction: Tag, not rewrite. In: Proceedings of the 15th Workshop on Innovative Use of NLP for Building Educational Applications, pp. 163–170. ACL, Seattle (2020)
4. Rao, G., Yang, E., Zhang, B.: Overview of NLPTEA-2020 shared task for Chinese grammatical error diagnosis. In: Proceedings of the 6th Workshop on Natural Language Processing Techniques for Educational Applications, pp. 25–35. ACL, Suzhou (2020)
5. Liang, D., et al.: BERT enhanced neural machine translation and sequence tagging model for Chinese grammatical error diagnosis. In: Proceedings of the 6th Workshop on NLPTEA, pp. 57–66. ACL, Suzhou (2020)
6. Chen, M., Ge, T., Zhang, X., Wei, F., Zhou, M.: Improving the efficiency of grammatical error correction with erroneous span detection and correction. In: Proceedings of the 2020 Conference on Empirical Methods in NLP, pp. 7162–7169. ACL (2020)
7. Chollampatt, S., Wang, W., Ng., H.-T.: Cross-sentence grammatical error correction. In: Proceedings of the 57th Annual Meeting of the Association for Computational Linguistics, pp. 435–445. ACL, Florence (2019)
8. Levenshtein, V.-I.: Binary codes capable of correcting deletions, insertions, and reversals. Soviet Phys. Doklady **10**, 707–710 (1966)
9. Lichtarge, J., Alberti, C., Kumar, S., Shazeer, N., Parmar, N., Tong, S.: Corpora generation for grammatical error correction. In: Proceedings of the 2019 Conference of the North American Chapter of the Association for Computational Linguistics: Human Language Technologies, vol. 1, pp. 3291–3301. ACL, Minneapolis (2019)
10. Malmi, E., Krause, S., Rothe, S., Mirylenka, D., Severyn, A.: Encode, tag, realize: High-precision text editing. In: Proceedings of the 2019 Conference on Empirical Methods in Natural Language Processing and the 9th International Joint Conference on Natural Language Processing (EMNLP IJCNLP), pp. 5054–5065. ACL, Hong Kong (2019)
11. Sutton, C., McCallum, A.: Composition of conditional random fields for transfer learning. In: Proceedings of Human Language Technology Conference and Conference on Empirical Methods in Natural Language Processing, pp. 748–754. ACL, Vancouver (2005)
12. Yue, T., Liu, S., Cai, H., Yang, T., Song, S., Yu, T.: Improving Chinese grammatical error detection via data augmentation by conditional error generation. In: Findings of the Association for Computational Linguistics: ACL 2022, pp. 2966–2975. ACL, Dublin (2022)

# Conversational Search Based on Utterance-Mask-Passage Post-training

Shiyulong He, Sai Zhang, Xiaowang Zhang(✉), and Zhiyong Feng

College of Intelligence and Computing, Tianjin University, Tianjin 300350, China
{hsyl1104,zhang_sai,xiaowangzhang,zyfeng}@tju.edu.cn

**Abstract.** Conversational Search (CS) aims to retrieve relevant passages from multiple documents based on the questions given by the user in conversations. Since users ask questions and get answers step by step through multiple rounds of conversations, conversation history is usually utilized to enhance retrieval accuracy. Existing methods rely too heavily on conversation history to retrieve while ignoring the semantics of the current question. However, the conversation history is not fully relevant to the current question, which includes some document information irrelevant to the current question. It is challenging to extract utterances information relevant to the current question from the conversation history to facilitate retrieval without breaking semantic coherence. We propose the reranker based on the Utterance-Mask-Passage (UtMP) post-training method to address this challenge, which includes three training tasks: passage relevance classification, utterance correlation classification, and context mask. Our method decomposes complex conversation history into short contexts, learns fine-grained semantic associations between utterances and document passages through three training tasks based on multi-task learning, and further learns correlations between conversation history and document passages based on contrastive learning. On the MultiDoc2Dial dataset, our results are 1.1% and 1.2% higher than the SOTA on the Recall@1 and MRR@10 metrics, respectively, which verifies the improvement of our method on retrieval performance. Extensive experiments show that our method helps deal with conversation histories with multiple documents information.

**Keywords:** Conversational Search · multiple documents · post-training

## 1 Introduction

In recent years, with the development of natural language processing technology, researchers have shown widespread interest in developing conversational systems applicable to various fields. Conversational Search (CS) is considered to be an important task in this domain. CS aims to retrieve the most relevant passage from multiple documents based on the conversation history of multiple rounds of natural language interactions between the user and the agent, so as to better


H. Wang et al. (Eds.): CCKS 2023, CCIS 1923, pp. 148–159, 2023.
https://doi.org/10.1007/978-981-99-7224-1_12

meet the information seeking needs of users [5,8]. Compared with single round question answering tasks or machine reading comprehension tasks given only a single document (passage), CS assumes that users ask questions and get answers step by step in a conversation. Therefore, the conversation history is crucial to understand the user's current question and performing relevant passage retrieval [11].

As the questions asked by the user change, different documents may need to be retrieved during the conversation to answer the user's question. Therefore, the conversation history will include information related to multiple documents. However, it will also contain information about documents that are not relevant to understanding the user's current question (interfering documents), and this information does not actually provide useful help or even has a negative impact [4]. Most of the current approaches are inspired by the field of information retrieval [18], Bansal et al. [1] use sparse retrievers to retrieve based on the entire conversation history in a lexical-based manner. In [9, 12, 16], authors flatten and concatenate the conversation history based on a dense retriever to fully utilize the information in the conversation history. In addition, Tran et al. [14] use discourse segmentation to delete utterances irrelevant to the current question from the conversation history and perform retrieval to avoid the influence of interfering documents.

However, these approaches do not address the problem of semantic coherence. During the retrieval process, on the one hand, the current model may tend to rely on partial historical information to retrieve passages, ignoring the current question, resulting in poor retrieval performance when multiple documents are involved in the conversation history. On the other hand, the discourse segmentation method may destroy the conversation's structure, resulting in incoherent semantic information, making it difficult to improve retrieval performance. Therefore, how to extract utterances related to the current question from the conversation history without breaking semantic coherence is challenging.

To address these issues, we propose the reranker based on the Utterance-Mask-Passage (UtMP) post-training method. Considering that complex conversation histories involve information from multiple documents, we propose a series of training tasks to enhance the model's ability to distinguish passages in interfering documents. First, we split the entire conversation history into multiple short context segments and introduce the passage relevance classification task. At the same time, we introduce the utterance correlation classification task and the context mask task to enable the model to learn the semantic information inside the utterance to enhance the modeling of complex conversation history. In addition, we use a multi-task learning method to train multiple tasks and integrate the feature information learned in each task. Finally, we fine-tune the model using contrastive learning to better learn the correlation between conversation history and passages.

Our contributions can be summarized as follows:

- To enhance the model's ability to model complex conversation histories and identify passages in interfering documents, we propose the Utterance-Mask-

Passage (UtMP) post-training method based on passage relevance classification, utterance correlation classification task, and context mask task;
- To integrate the feature information learned in different tasks, we comprehensively learn multiple training tasks based on the multi-task learning method and dynamically adjust weights to balance the learning speed of different tasks;
- To better learn the correlation between conversation histories and passages, we employ contrastive learning to fine-tune the model.

## 2 Related Works

### 2.1 Conversational Search

Inspired by recent developments in retriever architectures [18], some methods use bi-encoder DPR [10] as the retriever in single-stage retrieval. Feng et al. [5] further use BM25 to construct negative samples, and Jang et al. [7] explore strategies such as data augmentation using synonym enhancers. To better achieve recall and rerank performance, CPII-NLP [12] expands the retrieval into two phases, using DPR as the retriever, and the reranker is an ensemble of three cross-encoder models, including BERT, RoBERTa, and ELECTRA. UGent-T2K [9] divides the retrieval into two stages: document retrieval and passage retrieval. The model uses the LambdaMART [2] algorithm combined with TF-IDF and term-matching techniques to synthesize passage scoring and ranking. R3 [1] replaces the dense retriever with a sparse retriever based on DistilSplade, adds a RoBERTa-based cross-encoder passage reranker, and uses MS-MARCO and pseudo-labels from the cross-encoder model as additional training data. Besides, Tran et al. [14] assume that retrieval performance will be improved when only utterances related to the current question are included in the conversation history, and proposes PC (Passage Checking model), which deletes utterances irrelevant to the current question in the conversation history during retrieval and reuses the model probability score as a re-ranking indicator.

Although our work also employs a two-stage retrieval framework, we focus on improving the reranker's learning ability for complex conversation histories, thereby enhancing the model's ability to handle conversation histories that contain multiple documents information.

### 2.2 Response Selection

The retrieval-based response selection task aims to select the most appropriate response from a set of response candidates based on the conversation history to construct an open-domain conversational system. DAM (Deep Attention Matching network) [17] captures complex dependency information in utterances, using self-attention mechanisms and cross-attention strategies to learn representations of context and response candidates. PHMN (Personalized Hybrid Matching Network) [11] considers that the existing model is stuck in the dilemma of learning matching signals from context and response, introducing the wording behavior of a specific user in the conversation history and personalized attention weights for candidate responses as additional information. CFC (Contextual Fine-to-Coarse) [3] argues that although various complex models have been proposed to calculate the fine-grained similarity [6,15] between queries and candidates, their performance will still be limited by the quality of the candidate list and proposes to use coarse-grained retrieval.

But these methods aim to improve the performance on response selection tasks, where conversation histories and responses have similar representations. However, we focus on the retrieval of passages, where the conversation history and the expression of passages have greater differences containing different amounts of information, and bringing greater challenges.

# 3 Method

As shown in Fig. 1, our method mainly includes two parts: (1) Utterance-Mask-Passage post-training. We enhance the model's ability to learn complex conversational histories and to identify passages in interfering documents through multiple tasks, and integrate the feature information learned in different tasks based on multi-task learning. (2) Contrastive fine-tuning. Our model further learns the correlation between conversation history and passages to improve the model's retrieval performance on downstream tasks.

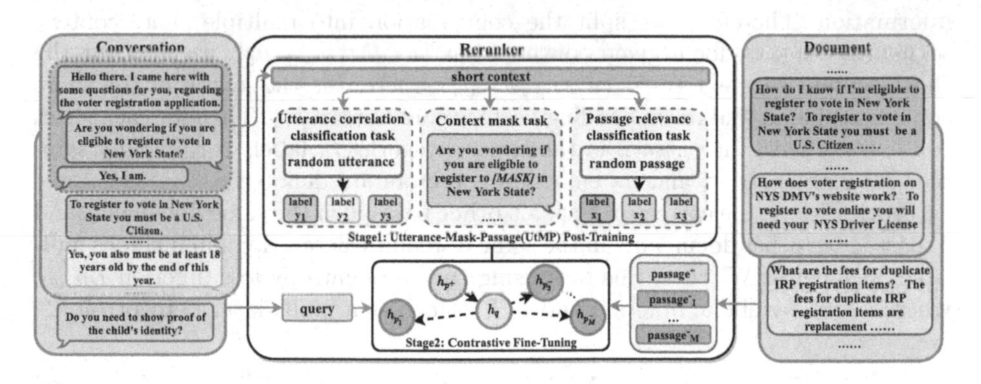


**Fig. 1.** The overview framework.

## 3.1 Preliminaries

Given the conversation history $\{u_1, \ldots, u_{t-1}\}$, the user's current question $u_t$ and the collection of multiple documents $D = \{D_0, \ldots, D_m\}$, where $m$ is the number of documents. Each document is divided into passages based on the structure of the document or a fixed-size sliding window. That is, for each document $Di \in D$, the segment set $P_i = \{p_{i0}, ..., p_{ij}\}$ is obtained, where $j$ is segmentation number of passages of document $i$ . The goal is to retrieve the most relevant passage to $u_t$ from the collection of passage sets $P = P_1 \cup \cdots \cup P_m$ obtained from all document segmentations based on $\{u_1, \ldots, u_{t-1}\}$ and $u_t$.

### 3.2 Utterance-Mask-Passage Post-training

To improve the retrieval performance of the model when multiple documents are involved in the conversation history, the Utterance-Mask-Passage (UtMP) post-training method is proposed, which enables the model to learn semantic associations between complex conversation history and passages in interfering documents. Our model learns features at a fine-grained level within conversations, between conversational contexts and passages through the passage relevance classification task, the utterance correlation classification task and the context mask task. At the same time, we integrate the feature information learned by different tasks based on multi-task learning.

**Passage Relevance Classification.** We innovatively propose this training task to enable the model to better learn the semantic associations between conversation history and passages containing information from multiple documents. Our approach is based on the following insight, in the conversation, the closer the utterances are to the current question, the more relevant they are. This characteristic is more evident when the conversation contains multiple documents information. Therefore, we split the conversation into multiple short context segments. That is, for a given conversation $\{u_1, \ldots, u_{t-1}, u_t\}$, we segment the short context segment $C_j = \{u_j, \ldots, u_{j+d}\}$, where $d$ is the length of the short context segment, and its embedding vector represents as $h_{c_j}$. Then for each short context segment, the passage in the three categories of labels is randomly sampled from the corpus, and its embedding vectors are denoted as $h_p$. The three categories are: 1. the correct passage, 2. other passage in the same document, and 3. passage in other documents in the same domain, where a domain includes multiple documents. We train this task using the cross-entropy loss function $L_{PRC}$, where $x_i$ is a symbolic function and $MLP$ denotes a multi-layer perceptron.

$$L_{PRC} = -\sum_{j=1}^{t-d}\sum_{i}^{3} x_i log(MLP(h_{c_j}, h_p)_i) \tag{1}$$

**Utterance Correlation Classification.** To learn more fully the semantic associations and transitions between internal utterances in a multi-round conversation history, we introduce the task of utterance correlation classification. Similar to the passage relevance classification task, first, we split the conversation into multiple short context segments, and randomly sample the next utterance corresponding to three categories: 1. the correct next utterance, 2. a random utterance in the same conversation, 3. utterance in other conversations in the same domain.

It enhances the modeling of complex conversation histories by learning irrelevant or coherent relations between two utterance segments. We train this target using the cross-entropy loss function $L_{UCC}$, where $h_u$ denotes the embedding vector of sampled utterance and $y_i$ is a symbolic function.

$$L_{UCC} = -\sum_{j=1}^{t-d}\sum_{i}^{3} y_i log(MLP(h_{c_j}, h_u)_i) \tag{2}$$

**Context Mask.** To more fully learn the contextual representation between conversation utterances, the Whole Word Masking method is introduced. By randomly masking the entire word, the model learns long-distance context dependence and obtains more semantic information. The corresponding loss function is denoted as $L_{CM}$. Where $C$ represents a short context segment, and $mask(C)$ represents a set of words using $[MASK]$ token random masked.

$$L_{CM} = -\sum_{\hat{C} \in mask(C)} logP(\hat{C}|C_{\setminus mask(C)}) \tag{3}$$

**Multi-task Post-training.** By utilizing the similarity and internal correlation between different tasks, we adapt a multi-task learning strategy to jointly optimize the above training tasks, and integrate the feature information. We use the following loss function to optimize the model.

$$L_{UtMP} = w_1 L_{UCC} + w_2 L_{CM} + w_3 L_{PRC} \tag{4}$$

where $w_i, i \in \{1, 2, 3\}$ are the weights of the corresponding tasks, we use Dynamic Weight Averaging (DWA) [13] to adaptively and dynamically adjust the weights, so as to balance the learning speed of different tasks. Specifically, using the following formula to calculate the change of the continuous loss of each task, which is used as the learning speed of the task, and the weight of each task is obtained after normalization.

$$r_i(s-1) = L(s-1)/L(s-2) \tag{5}$$

$$w_i(s) = Nexp(r_i(s-1)/T)/\sum_j exp(r_j(s-1)/T) \tag{6}$$

where $L \in \{L_{UtMP}, L_{UCC}, L_{PRC}\}$. $s$ is an iteration index. $N$ represents the number of tasks, and $T$ represents a temperature which controls the softness of task weighting. As $T$ increases, the weights between tasks will gradually become uniform. When $T$ is large enough, $w_i \approx 1$, at this time, each task is weighted equally.

### 3.3 Contrastive Fine-Tuning

For enabling the model to better learn the similarity between conversation histories and passages, we introduce the contrastive learning to optimize it. By pulling similar samples closer and dissimilar samples farther away, our model can learn semantic representations that are more suitable for similarity matching, which helps to improve the retrieval performance of the model on downstream tasks.

Specifically, for $\{u_1, \ldots, u_{t-1}\}$ and current question $u_t$, we concatenate them together in reverse order. $u_t$ and $\{u_1, \ldots, u_{t-1}\}$ are separated by $[SEP]$ special token, inside the conversation history is separated by "||", and information related to the current role is added before each utterance.

$$query = [CLS]\, u_t\, [SEP]\, role : u_{t-1} || \ldots || role : u_1\, [SEP] \tag{7}$$

During fine-tuning, for each query, we first obtain top-K results from the retriever, from which $M$ hard-negative passages are sampled. Then use the model trained after UtMP to encode the query, positive and negative passages, and the obtained vectors are respectively recorded as $h_q$, $h_{p^+}$ and $\left\{h_{p_1^-}, \ldots h_{p_M^-}\right\}$. Then, using the transformer self-attention mechanism to calculate the cross-attention between the query and each passage, obtain the aggregate representation to calculate the score, which is denoted as $score(h_q, h_p)$. The results are optimized using the following contrastive loss function.

$$L_{reranker} = -log \frac{exp(score(h_q, h_{p^+}))}{\sum_{i=1}^{M} exp(score(h_q, h_{p_i^-})) + exp(score(h_q, h_{p^+}))} \tag{8}$$

## 4 Experiments

### 4.1 Experimental Settings

**Dataset.** We consider that other relevant datasets lack fluent conversations, or are based only on single documents or web pages. In order to better verify the retrieval performance of our method in conversations containing multiple documents information, we choose the MultiDoc2Dial [5] as the evaluation dataset. The dataset contains a total of 488 documents and 4796 conversations from four domains (ssa, va, dmv, and studentaid), each conversation contains an average of 14 turns, and each document contains an average of about 800 words. We follow MultiDoc2Dial's method of dividing and constructing the dataset, and follow CPII-NLP [12] to preprocess the document data to ensure the fairness of the evaluation results. The specific data information is shown in Table 1.

**Evaluation Metrics.** In order to quantify the retrieval performance, we use Recall@k (k = 1/5/10), and MRR@k (k = 5/10) as the evaluation metrics. Recall@k is a measure to evaluate how many correct passages are recalled at top K results. MRR@k is a measure to evaluate the position of the most relevant passage in the top K ranking result.

**Table 1.** MultiDoc2Dial data statistics. #doc and #dial represent the number of documents and conversations in the corresponding domain respectively. single-doc, two-doc and >two-doc represent the number of conversations containing different documents.

| domain | #doc | #dial | single-doc | two-doc | >two-doc |
|---|---|---|---|---|---|
| ssa | 109 | 1191 | 302 | 701 | 188 |
| va | 138 | 1337 | 198 | 648 | 491 |
| dmv | 149 | 1328 | 290 | 781 | 257 |
| studentaid | 92 | 940 | 158 | 508 | 274 |
| total | 488 | 4796 | 948 | 2638 | 1210 |

**Parameter Settings.** We use the BERT-base model to implement our retriever and reranker. We adopt a representation-based approach to construct the retriever, using in-batch samples as negative samples. During the training process of the retriever, we use a batch size of 32 and a maximum sequence length of 256. And we use a learning rate of 2e-05 using Adam, linear scheduling with warmup and dropout rate of 0.1. In the post-training and fine-tuning stages of the reranker, we both use a learning rate of 1e-05 using Adam, linear scheduling with warmup and dropout rate of 0.1. The batch size is 8 and the maximum sequence length is 512. In the post-training stage, the short context segment length d = 3, temperature T = 2.0 for adjusting the weight of loss. And M = 7, K = 100 for the fine-tuning stage.

### 4.2 Performance Comparison

We compare the evaluation results with the following models on this dataset:

**RAG** [5] uses the bi-encoder model DPR pre-trained on the NQ dataset as the retriever, and **G4** [16] uses another dense retrieval bi-encoder model ANCE; **CPII-NLP** [12] follows the baseline setting, uses DPR as the retriever, and uses a collection of three cross-encoder models as the reranker, and achieves the SOTA results on the dataset; **PC** [14] builds a classification model to retain the utterances related to the current question in the conversation history for retrieval, and reuses the probability score of the classification model as a reranking index; **R3** [1] replaces the dense retriever with a sparse retriever based on DistilSplade, and adds a cross-encoder passage reranker based on RoBERTa; **UGent-T2K** [9] divides the retrieval part into document retrieval and passage retrieval, using LambdaMART algorithm combined with TF-IDF and other methods to score comprehensively.

### 4.3 Experimental Results

We experiment on the validation set of MultiDoc2Dial to verify the retrieval effect of our model, and the results are shown in Table 2.

**Table 2.** Results on MultiDoc2Dial validation set.

| Models | Recall@1 | Recall@5 | Recall@10 |
|---|---|---|---|
| G4 | 0.395 | 0.685 | 0.773 |
| RAG | 0.490 | 0.723 | 0.800 |
| PC | 0.525 | 0.754 | 0.823 |
| R3* | 0.558 | 0.767 | 0.847 |
| UGent-T2K | 0.570 | 0.821 | 0.883 |
| CPII-NLP* | 0.614 | 0.821 | 0.881 |
| **UtMP(Ours)** | **0.625** | **0.837** | **0.892** |

* indicates the results we reproduced based on the parameters in their paper.

The results show that our model achieves improvements on all metrics. This means that our model effectively enhances the ability to learn complex conversation histories and identify passages of inferring documents, allowing for more efficient retrieval of passages.

In order to verify the improvement of our post-training method on the effect of reranking, we further compared with the SOTA model CPII-NLP on the MRR@k metric, and the results are shown in Fig. 2.

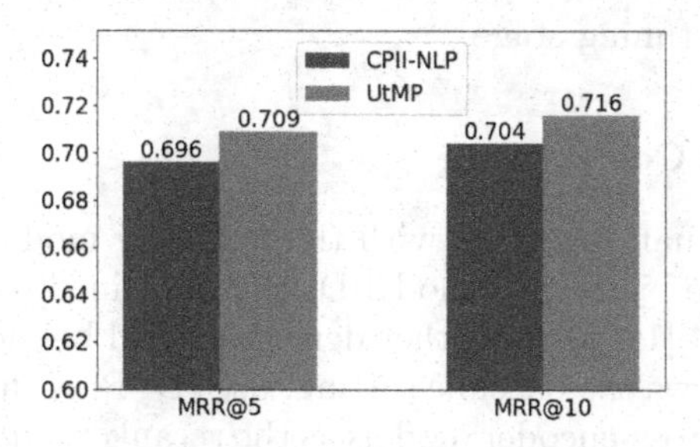


**Fig. 2.** Comparison with the SOTA model CPII-NLP at MRR@k (k=5/10).

The results show that our model outperforms CPII-NLP, which means our method can rank more relevant passages at higher positions.

### 4.4 Ablation Experiment

To further explore the effects of each training task in the UtMP post-training method, as shown in the Table 3, we use the model that has not been post-trained and directly fine-tuned as the baseline, and add our training tasks before fine-tuning one by one.

According to the Table 3, it can be seen that the utterance correlation classification task has the greatest impact on the results, significantly improving

**Table 3.** Multi-task ablation experiment. +UCC indicates that only the utterance correlation classification task is used; +CM and +PRC indicate that the context mask task and passage correlation classification task are further used, respectively.

| Models | MRR@10 | Recall@1 | Recall@5 | Recall@10 |
|---|---|---|---|---|
| BERT | 0.665 | 0.579 | 0.791 | 0.859 |
| +UCC | +2.6% | +2.6% | +2.8% | +1.6% |
| +CM | +0.6% | +0.8% | +0.6% | +0.6% |
| +PRC | +1.9% | +1.2% | +1.2% | +1.1% |

the four metrics, indicating that better learning of utterance interactions within the conversation history can make better use of implicit semantic information between utterances. The passage relevance classification task brings slightly less improvement, but also shows that the method is helpful. The improvement brought by the context mask task is smaller, probably because its main role is to assist in modeling the utterance interaction inside the correlation history.

### 4.5 Further Analysis

**Improvement Effect of Multiple Documents CS.** To verify the improvement of our method when the conversation history contains multiple documents information, we divide the Mutidoc2dial dataset into several subsets according to the number of documents involved in the conversation. The number of documents involved only indicates how many documents have been used in the conversation history, and the entire conversation process may repeatedly switch among these documents. We use the model that has not been post-trained and the model that has been post-trained to perform retrieval on the subset and count the retrieval results.

The results are shown in Fig. 3(left). Our method has better boosting effect when the conversation history involves multiple documents. This indicates that our post-training method enhances the model's ability to identify passages in inferring documents and reduces the impact of irrelevant conversation history on retrieval.

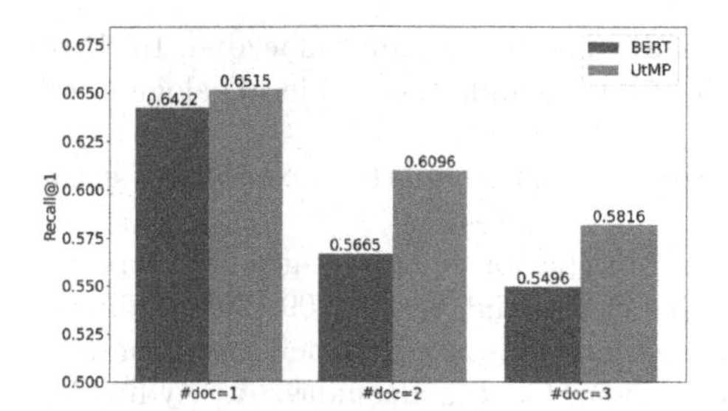


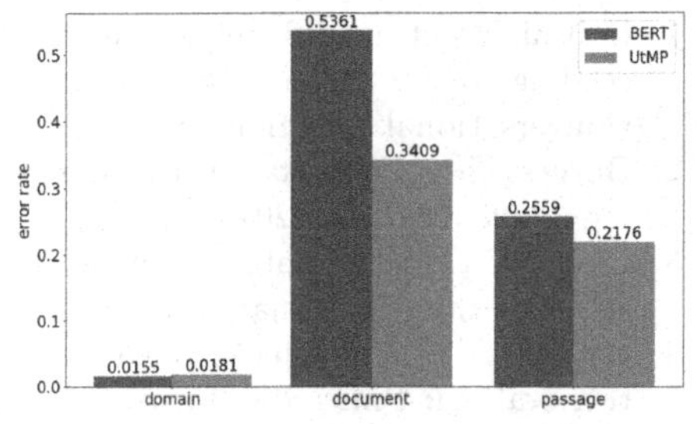


**Fig. 3.** Improvement effect of multiple documents CS (left) and proportion of error types changes after post-training (right).

**Proportion of Error Types Changes After Post-training.** We divide the retrieval errors into three parts: *domain retrieval errors* (i.e. incorrect retrieval results in passages from other domains), *document retrieval errors* (i.e. incorrect retrieval results in passages from other documents under the same domain), and *passage retrieval errors* (i.e. incorrect retrieval results in passages under the same document). We use the model without post-training as a comparison and make statistics on the three types of errors in the top-1 results, as shown in Fig. 3(right). The result demonstrates a significant decrease in the number of *document retrieval errors* and a certain decrease in the number of *passage retrieval errors* for the post-trained model. However, there is a slight increase in the number of *domain retrieval errors*, which may be due to the fact that we focus more on improving the ability to distinguish between passages under the same domain during the post-training process.

## 5 Conclusion

For enhancing the model's ability to learn complex conversation histories and to identify passages in interfering documents, we propose a reranker based on the Utterance-Mask-Passage (UtMP) post-training method. By completing the passage relevance classification, utterance correlation classification and context mask training tasks, our model is able to learn features between conversation history and passages at a fine-grained level, which in turn improves retrieval performance. In addition, multi-task learning and contrastive learning are introduced to dynamically adjust the task weights to balance the learning rate of different tasks and to better learn the relevance between conversation histories and passages, respectively. Experiments on the MultiDoc2Dial demonstrate the effectiveness of our approach and reflect that a fine-grained exploitation of conversation history is beneficial for improving retrieval accuracy in the CS of multiple documents situations. In future work, we will further investigate more effective post-training task, as well as explore more complex cases where there is more than one correct passage.

## References

1. Bansal, S., et al.: R3: refined retriever-reader pipeline for multidoc2dial. In: Proceedings of the Second DialDoc Workshop on Document-grounded Dialogue and Conversational Question Answering, pp. 148–154 (2022)
2. Burges, C.J.: From Ranknet to Lambdarank to Lambdamart: an overview. Learning **11**(23–581), 81 (2010)
3. Chen, W., et al.: Contextual fine-to-coarse distillation for coarse-grained response selection in open-domain conversations. arXiv preprint arXiv:2109.13087 (2021)
4. Du, M., et al.: Topic-grained text representation-based model for document retrieval. In: Pimenidis, E., Angelov, P., Jayne, C., Papaleonidas, A., Aydin, M. (eds.) Artificial Neural Networks and Machine Learning – ICANN 2022. ICANN 2022. LNCS, vol. 13531. Springer, Cham (2022). https://doi.org/10.1007/978-3-031-15934-3_64

5. Feng, S., Patel, S.S., Wan, H., Joshi, S.: MultiDoc2Dial: modeling dialogues grounded in multiple documents. arXiv preprint arXiv:2109.12595 (2021)
6. Han, J., Hong, T., Kim, B., Ko, Y., Seo, J.: Fine-grained post-training for improving retrieval-based dialogue systems. In: Proceedings of the 2021 Conference of the North American Chapter of the Association for Computational Linguistics: Human Language Technologies, pp. 1549–1558 (2021)
7. Jang, Y., et al.: Improving multiple documents grounded goal-oriented dialog systems via diverse knowledge enhanced pretrained language model. In: Proceedings of the Second DialDoc Workshop on Document-grounded Dialogue and Conversational Question Answering, pp. 136–141 (2022)
8. Jeffrey, D., Chenyan, X., Jamie, C.: CAsT 2019: the conversational assistance track overview. In: TREC (2019)
9. Jiang, Y., Hadifar, A., Deleu, J., Demeester, T., Develder, C.: UGent-T2K at the 2nd DialDoc shared task: a retrieval-focused dialog system grounded in multiple documents. In: Proceedings of the Second DialDoc Workshop on Document-grounded Dialogue and Conversational Question Answering, pp. 115–122 (2022)
10. Karpukhin, V., et al.: Dense passage retrieval for open-domain question answering. arXiv preprint arXiv:2004.04906 (2020)
11. Li, J., et al.: Dialogue history matters! personalized response selection in multi-turn retrieval-based chatbots. ACM Trans. Inf. Syst. (TOIS) **39**(4), 1–25 (2021)
12. Li, K., et al.: Grounded dialogue generation with cross-encoding re-ranker, grounding span prediction, and passage dropout. In: Proceedings of the Second DialDoc Workshop on Document-grounded Dialogue and Conversational Question Answering, pp. 123–129 (2022)
13. Liu, S., Johns, E., Davison, A.J.: End-to-end multi-task learning with attention. In: Proceedings of the IEEE/CVF Conference on Computer Vision and Pattern Recognition, pp. 1871–1880 (2019)
14. Tran, N., Litman, D.: Getting better dialogue context for knowledge identification by leveraging document-level topic shift. In: Proceedings of the 23rd Annual Meeting of the Special Interest Group on Discourse and Dialogue, pp. 368–375 (2022)
15. Whang, T., et al.: Do response selection models really know what's next? Utterance manipulation strategies for multi-turn response selection. In: Proceedings of the AAAI Conference on Artificial Intelligence, vol. 35, pp. 14041–14049 (2021)
16. Zhang, S., Du, Y., Liu, G., Yan, Z., Cao, Y.: G4: grounding-guided goal-oriented dialogues generation with multiple documents. In: Proceedings of the Second DialDoc Workshop on Document-grounded Dialogue and Conversational Question Answering, pp. 108–114 (2022)
17. Zhou, X., et al.: Multi-turn response selection for chatbots with deep attention matching network. In: Proceedings of the 56th Annual Meeting of the Association for Computational Linguistics (Volume 1: Long Papers), pp. 1118–1127 (2018)
18. Zhu, F., Lei, W., Wang, C., Zheng, J., Poria, S., Chua, T.S.: Retrieving and reading: a comprehensive survey on open-domain question answering. arXiv preprint arXiv:2101.00774 (2021)

# Knowledge Graph Applications

# Financial Fraud Detection Based on Deep Learning: Towards Large-Scale Pre-training Transformer Models

Haitao Wang[1,2](✉), Jiale Zheng[3], Ivan E. Carvajal-Roca[2], Linghui Chen[2], and Mengqiu Bai[2]

[1] School of Cyber Science and Engineering, Southeast University, Nanjing 210096, China
wanghaitao@seu.edu.cn
[2] Zhejiang Laboratory, Hangzhou 311121, China
[3] Huawei Noah's Ark Lab, Shenzhen 518129, China

**Abstract.** Fraud detection is a critical issue in the field of finance, as it can help to prevent fraud and minimize losses caused by fraud. Deep learning techniques learn the intrinsic knowledge of huge data, build explainable transaction knowledge graphs, and effectively predict potential fraudulent transactions, making it an essential technique in financial fraud detection. In this paper, we systematically review the existing financial fraud detection technologies, focusing on deep learning-based financial fraud detection methods. To the best of our knowledge, our work is the first to systematically introduce financial fraud detection methods based on transformer models, including the most recent pre-training transformer models, which can be thought of as parametric knowledge. Finally, we also analyze and summarize the challenges of financial fraud detection research, to promote its future development of research.

**Keywords:** Deep Learning · Financial fraud Detection · Pre-training transformer model · Knowledge

## 1 Introduction

Financial fraud is a problem that has a wide impact on social development and daily life. It will not only affect people's sense of trust in all walks of life but also damage their living security. Moreover, it will cause huge economic losses and even serious damage to the economy and society. West et al. [1] defined financial fraud as "the intentional use of illegal methods or practices for the purpose of obtaining financial gain". Achieving efficient and accurate fraud detection ahead of various fraud methods of criminals is a relevant research problem for the improvement of the financial sector and society [2,3]. To solve this problem, financial fraud detection techniques have become an important technical means of fraud prevention.

The evolution of financial fraud detection technologies can be classified into three stages: manual audit and expert guidelines-based methods, conventional


H. Wang et al. (Eds.): CCKS 2023, CCIS 1923, pp. 163–177, 2023.
https://doi.org/10.1007/978-981-99-7224-1_13

machine learning-based methods, and deep learning-based methods. Manual audit and expert standards-based approaches mainly depend on prior information and are unable to adapt to intricate and shifting fraud patterns. Conventional machine-learning techniques [4] like logistic regression, decision trees, support vector regression, and others are gradually being used in the area of financial fraud detection. In accordance with conventional machine learning techniques, a sizable amount of user and transaction data is mined, statistical features are extracted from various aspects, such as user profiles and historical behaviour, and the internal laws and representations of sample data are learned. This allows for establishing models such as classification prediction to achieve fraud identification. However, when data becomes more complex, diverse, and large-scale, fraud data is even more unbalanced, and statistical machine-learning methods struggle to provide reliable fraud detection.

Deep learning techniques, which use multi-layer neural networks to learn implicit representations of data, have been shown to perform much better when the data is large and more complex. With the explosive growth of financial transaction data and financial fraud becoming more hidden, deep learning techniques have consistently become a popular topic and emphasize financial fraud detection. Convolutional neural networks are one of the earliest neural networks used in financial fraud detection, using convolution to record local features of transactions, but they lack a long-term memory mechanism; recurrent neural networks, on the other hand, can memorize long-term sequences of transaction records and have achieved great success in fraud detection. Graph neural network uses graph structure to represent transaction network topology and entities, which can better capture the internal relationships in graph data, perform complex reasoning and prediction in the graph, and is widely used in financial fraud detection. The transformer was first employed as a sequence-to-sequence model in the field of natural language processing [5], modeling global context information with a self-attention mechanism. The pre-training model based on the transformer implements State-Of-the-Art performance in various fields [6]. Subsequently, the large-scale pre-training model generative pre-training transformer-3 (GPT-3) with a scale of 100 billion parameters has achieved amazing results, establishing the large-scale pre-training transformer model as an essential research direction in the field of artificial intelligence.

In this paper, we discuss the widely used deep learning techniques and their applications in financial fraud detection from the standpoint of deep learning. Fraud detection has been studied in reviews [7,8], which cover credit card fraud, insurance fraud, financial statement fraud, and cryptocurrency fraud. However, they mainly integrate and analyze research on machine learning models in artificial intelligence (AI). There are few current survey articles on fraud detection with neural network-related deep learning, specifically transformer-based models. Compared with the existing related works, our main significant contributions in this review paper are as follows:

– We provide a comprehensive overview of financial fraud detection based on deep learning, which focuses on the latest research on deep learning in fraud

detection and makes up for the lack of comprehensive discussion on financial fraud detection based on deep learning in previous reviews.
- To the best of our knowledge, our work is the first to systematically introduce financial fraud detection methods based on transformer models, we also present the latest pre-training transformer models for financial fraud detection.
- We discuss the difference between our work and the previous surveys, the challenges, and the future research directions regarding AI-oriented financial fraud detection.

The remainder of this paper is structured as follows: Sect. 2 focuses on the development of deep learning-based financial fraud detection technology, and sorts out six types of deep learning models. Section 3 introduces the applications of deep learning fraud detection methods in the financial field. Section 4 summarizes some publicly available data sets for financial fraud detection research. Section 5 expounds on the challenges and prospects. Finally, Sect. 6 summarizes this article.

## 2 Deep Learning Financial Fraud Detection Methods

With the diversification of transaction forms, the scale and structure of the analyzed transaction data are becoming larger and more complex, which brings great challenges to traditional fraud detection methods. Deep learning techniques can process large-scale and complicated data more efficiently than classic fraud detection methods. This opens new research directions and opportunities for developing financial fraud detection technologies. Deep learning-based financial fraud detection research has grown in importance as a subject of financial fraud detection research in recent years due to the fast growth of deep learning theory. The following describes deep learning models and how they are used to detect financial fraud.

### 2.1 Auto-encoder

Auto-encoder (AE) is the general unsupervised and semi-supervised learning model. This goal is to learn and obtain the low-dimension feature from the information in the input data. The autoencoder hopes that the reconstructed data is as similar as possible to the original data, so its objective function is to minimize the Euclidean distance of the sum.

AE can be utilized to extract implicit information from data and reduce dimensionality to reduce the computation required for subsequent tasks. AE can also be applied to unsupervised learning to reduce data labeling costs. However, AE has the disadvantages of requiring a large amount of computation and a lengthy training period, and the training process may lead to overfitting and gradient disappearance. Based on AE, researchers further proposed a variety of self-encoding methods, which are widely used in numerous industries, such as risk assessment and the detection of financial misconduct. Kolli et al. [9] used a deep stack autoencoder based on Harris Gray Wolf to manage data imbalance and identify fraud in banking transactions. Kumar et al. [10] used deep belief network-based autoencoders for fraud detection on bank credit and insurance data. Gradxs et al. [11] suggested the HGW-Deep stacked autoencoder, which shows promise in the areas of precision for detecting credit card transaction fraud. To increase the effectiveness of fraud detection for credit card fraud, Fanai et al. [12] suggested a framework incorporating supervised deep learning algorithms, deep autoencoders as dimensionality reduction approaches, and unsupervised neural networks.

### 2.2 Convolutional Neural Networks

Convolutional Neural Networks (CNN) are feedforward neural networks evolved from multi-layer perceptrons. Typically, the input layer, convolutional layer, pooling layer, and fully connected layer make up a deep CNN model. The convolutional layer's goal is to extract specific local features from the data. The input data map is subjected to a dot product operation with the convolution kernel in the convolutional layer, and the convolutional feature mapping is then acquired through the activation function's output. Typically, the pooling layer follows the convolutional layer and corresponds to each convolutional layer individually. It down-samples the feature map after convolution, reduces the data dimension and the amount of calculation, and avoids overfitting. The CNN convolution operation's parameter-sharing feature significantly lowers the number of parameter optimizations and boosts training effectiveness.

CNN can either be used explicitly for end-to-end model training or as a feature extractor to combine with other classification, prediction, and other models. Singh et al. [13] used a hybrid CNN-MRFO model, where the CNN model's hyperparameters are tuned by an MRFO learning method, to forecast financial statement fraud. Abakarim et al. Zioviris et al. [14] proposed a multi-stage hybrid model that combines CNN and an autoencoder to identify fraudulent credit card transactions. [15] proposed a novel text2IMG conversion technique for producing small pictures, and deep features are used to minimize the computational complexity of machine learning classifiers. Illanko et al. [16] used a CNN to find fraudulent credit card transactions. Synthetic minority oversampling and random resampling technology were employed to lower false negative and false positive rates in order to address the issue of data imbalance. Gambo et al. [17] suggested a CNN model for identifying credit card fraud and resolving data imbalance with an adaptive synthesis sampling technique. Murugan et al. [18] proposed a CNN-based technique for identifying potentially fraudulent patterns in credit card transactions. This technique involved turning each transaction's data into a feature matrix so that the CNN model could detect any inherent flaws.

### 2.3 Recurrent Neural Networks

Recurrent Neural Networks (RNN) are well-known deep learning model for modeling sequence data, which can effectively deal with long-distance dependencies in sequences. RNN uses a time-based backpropagation algorithm for parameter training. These long-term relationships in the sequence can be effectively captured by an RNN, which also has the power to characterize the sequential information in the sequence, using a way of cyclically updating the hidden state. As the analyzed data sequence length increases, RNN is likely to have gradient disappearance or explosion problems in practical applications. Based on this, scholars introduced the gating mechanism into RNN. They proposed a long short-term memory model (LSTM) and a Gated Recurrent Unit (GRU) to make up for the defects of RNN.

Early research on credit card transaction fraud only considered single transactions. Jurgovsky et al. [19] focused on the characteristics of credit card transaction information over time and used LSTM to analyze user transaction sequence data to realize fraudulent behaviour and normal behaviour. Esenogho et al. [20] proposed a deep learning ensemble classifier utilizing the AdaBoost method and the LSTM neural network as the foundation learner.

Considering that users' transactions have the characteristics of staggered occurrence and irregular intervals, Branco et al. [21] proposed an interleaved GRU model that incorporates transaction intervals as features together with transaction data to realize the classification of credit card fraud users and normal users. Xie et al. [22] designed two time-aware gates in the recurrent neural network unit to learn the user's LSTM trading behaviours, respectively, and to record the variations in behaviour brought on by various time intervals between the user's ongoing transactions. According to experiments, this strategy is effective at separating fraudulent from legitimate conduct, which enhances the efficacy of credit card fraud detection according to several evaluation criteria.

Due to the limitations of a single model, the current trend in RNN-based fraud detection research is to combine multiple models. Roseline et al. [23] combined an RNN with LSTM to detect credit card fraud; LSTM replicates the sequential dependencies between credit card transactions, and over time, the hidden state of LSTM shifts between neural network nodes. Geetha et al. [24] proposed a hybrid method using CNN and RNN along with an adaptive feature selection method that has a good reliability and recognition rate for credit card transaction fraud detection. Xia et al. [25] proposed a deep learning model that incorporates CNN, LSTM, and DNN for detecting car insurance fraud. This model can capture more abstract features and avoid potential problems with typical machine learning methods, which are overly dependent on the intricate feature extraction procedures of domain specialists.

### 2.4 Graph Neural Network

RNNs and CNNs have produced successful results on Euclidean spatial data. Non-Euclidean spatial data, like social networks and information networks, exist in the actual world. Researchers developed the abstract graph in graph theory to represent non-Euclidean structured data in answer to this issue and proposed the Graph Neural Network (GNN) to mine it. GNN has the benefits of simple inference and interpretability and is a popular topic of research in deep learning today [26]. Researchers have proposed a number of GNN variants to develop and expand GNN, including graph convolution networks (GCN), graph attention networks (GAN), graph autoencoders, graph generation networks (GGN), CARE-GNN [27], PC-GNN [28], and graph spatial-temporal networks, etc.

Graph structure data also exists widely in the financial field, and the transaction behaviours of users and various institutions in banks constitute a large amount of graph data [29,30]. Mao et al. [31] employed the GNN to generate forecasts after connecting listed corporations and their linked parties in a heterogeneous graph to provide a complete transaction picture. Zhang et al. [32] presented health insurance data as dynamic heterogeneous graphs and introduced hierarchical multimodal fusion GCN, which can embed longitudinal and multimodal entities in addition to learning topological information, to enhance the effectiveness of fraud detection. The work of Wang et al. [33] provided a GNN with feature augmentation for isolated portions, which included the procedures of policy label propagation, graph creation, and graph splitting. All nodes are categorized into a number of groups based on where they are and how they are connected. Long et al. [34] proposed a GNN fraud detection approach with a neighbour sampler and attributes extractor, using under-sampling to address the issue of class imbalance. Pan et al. [35] proposed a federated graph learning two-stage method called 2SFGL, this framework uses the FedAvg method to train the GNN model on virtual fused graphs.

Due to the outstanding performance of GNN-type models, GNN-based methods are widely used in cryptocurrency trading networks for fraud detection. Li et al. [36] proposed a graph learning algorithm for detecting fraudulent Ethereum transactions that can learn the topological properties and payment amount characteristics of the network of transactions. Li et al. [37] proposed a GNN-based detection approach to effectively discriminate between legitimate accounts and phishing accounts in the Ethereum transaction network. Mo et al. [38] proposed the Motif-signed Temporal Graph Convolutional Network for Bitcoin transaction network fraud detection, which is a discrete temporal GCN model that simultaneously considers the temporal, local structure and balance in the signature network information. Qiao et al. [39] introduced an autoencoder-based dynamic graph representation learning technique. In order to create low-dimensional feature vectors, the encoder uses a GCN to collect local features. Hall et al. [40] modelled transaction records and systems as graph data, used semi-supervised graph convolutional networks for fraud detection, and verified the effect of the model in Bitcoin and Ethereum transaction networks.

### 2.5 Transformer

Transformer is a deep learning model that is built on the self-attention mechanism and was originally utilized in natural language processing tasks with great results. It is made up of encoding and decoding components, as well as connections between them. The encoding component is made up of a series of encoders, and the decoding component is made up of a series of decoders, with an equal number of encoders and decoders. Through the self-attention layer, the encoder pays attention to context information. A feed-forward neural network follows the self-attention layer, and each unit utilizes the same feed-forward network independently. The decoder additionally includes a self-attention layer and a feed-forward neural network. Through the self-attention technique, Transformer can perform fast parallel model training, overcoming the problem of slow RNN training. Transformer makes use of the DNN model's multi-level properties to deepen the network and improve model accuracy.

The use of transformers in the identification of fraud is becoming increasingly popular. Yuan et al. [41] proposed a two-stage fraud feature selection approach that integrates machine learning and deep learning, as well as a new multi-stage strategy that retains key characteristics while filtering out unimportant data in the first stage. The second level employs a feature extractor based on the transformer structure, and the last layer employs a binary classifier to predict whether or not there is fraud. Through the analysis of transaction data, Zhang et al. [42] discovered that fraudulent transactions are highly correlated with continuous non-discrete transaction activities, and proposed a dynamic graph embedding method DynGraphTrans, which uses smooth attention layers and time interval-aware relative position encoding to process time interval information and improve multi-head attention ability. Rodríguez et al. [43] trained the Transformer model using transaction data analogy and language data to predict the next transaction based on the user's past transaction records, compared the predicted transaction with the actual transaction, and flagged the actual transaction as fraudulent if the difference is greater than a threshold. Zhang et al. [44] proposed a hybrid model for credit card fraud detection that combines TabNet and XGBoost, in which TabNet implements feature selection using the attention transformer layer and then uses the Feature transformer layer to calculate the features chosen by the attention transformer layer. This strategy combines the benefits of model interpretability with sparse feature selection.

### 2.6 Pre-training Transformer Model

The pre-training transformer model first executes self-supervised learning on a large-scale data set to generate a pre-trained model that has nothing to do with downstream tasks, and then the downstream tasks just need to be fine-tuned to finish the learning. The pre-training model utilizes a large amount of unlabeled data and has the advantages of strong generalization ability and fast convergence speed. The effect of the model is far better than that of direct training on small data sets without prior knowledge. Therefore, it has achieved excellence in various tasks and has become a significant milestone in the development of artificial intelligence technology.

As Bert and other transformer-based pre-training models achieved considerable success in numerous disciplines, researchers began to employ pre-training models in the field of anomaly detection. Abakarim et al. [45] suggested combining CNN and pre-trained CNN models, stacking the CNN architecture in parallel with the CNN model utilizing bagged ensemble learning, extracting the results of the CNN model alone using an SVM classifier, and combining the results using the majority round-robin technique. When combined, these factors maximize accuracy and increase the effectiveness of vehicle insurance claim fraud detection. Padhi et al. [46] proposed the TabBERT pre-training model for learning the time series representation of the table, training the TabBERT model on a synthetic credit card transaction data set, and using the learned code to validate the model's effectiveness in the downstream transaction fraud detection task. TabBERT encapsulated transactions hierarchically, capturing internal links between table column data and temporal dependencies between table rows. Hewapathirana et al. [47] evaluated the performance of TabBERT model-based embedding in credit card transaction fraud detection using three supervised learning methods and two unsupervised learning algorithms, TabBERT-based embedding increased the fraud detection performance of supervised machine learning algorithms and some unsupervised techniques. Hu et al. [48] proposed the first general-purpose pre-training model BERT4ETH for Ethereum fraud detection. The model generated seven features for each transaction, including address, account type, input-output type, amount, count, timestamp, and location, using three effective strategies, namely reducing duplication, mitigating skewness, and modeling heterogeneity, to achieve significant improvements in phishing account detection and deanonymization tasks. Gai et al. [49] developed a large-scale pre-training model BLOCKGPT for Ethereum transaction anomaly detection using 68 million Ethereum transactions as training data. This model employs custom data encoding to convert transactions into vectors and transformers to learn the entire intrinsic structure of the transaction network, and the experimental results demonstrate that the model is quite effective at detecting fraudulent transactions.

## 3 Applications of Deep Learning Fraud Detection

Financial fraud detection requires analyzing and mining data from commercial organizations. Due to the different forms of business and transactions, there are various and complex financial frauds. In this section, we will mainly introduce the applications of deep learning fraud detection.

**Credit Card Fraud.** Credit card fraud is the most common and widespread fraud in the financial industry. Since the vast majority of credit card transactions are normal transactions, the proportion of fraudulent transactions is low. So solving the problem of data imbalance to detect credit card fraud is a top priority that cannot be delayed. Teng et al. [50] proposed a generative adversarial network framework called BalanceGAN to detect online bank fraud on highly unbalanced data. Langevin et al. [51] proposed utilizing generative adversarial networks to provide artificial sample augmentation data, a method for detecting credit card fraud. El et al. [52] utilized a new oversampling technique based on generative adversarial neural networks to enhance the processing of credit card transaction imbalance data.

**Money Laundering.** Money laundering negatively impacts national and global social security and financial management order. Wu et al. [53] introduced a general-purpose message-passing neural network protocol that can produce both node and edge representations on directed multigraphs at the same time. The research [54] used the DEGSO and LSTM bionic algorithms to identify money laundering threats. Krvzmanc et al. [55] found that gradient boosting and multilayer perceptrons have good performance in money laundering detection. Yu et al. [56] proposed a graph convolutional network model named MP-GCN using message passing for phishing scam detection on Ethereum. Tang et al. [57] advocated using message passing and a graph convolutional network model called MP-GCN for detecting phishing scams on Ethereum.

**Insurance Fraud.** Due to the large number of insured objects involved and the high or frequent compensation amounts, auto and health insurance have always been the focus of fraud detection research. Gangdhar et al. [58] used variational encoder-based methods for fraud detection of health insurance fraud and auto insurance data. Xia et al. [25] mixed DNN, LSTM, and CNN to present a deep learning model for detecting auto insurance fraud. Abakarim et al. [45] suggested mixing pre-trained CNN models to maximize accuracy and increase the effectiveness of auto insurance claim fraud detection.

**Financial Statement Fraud.** Financial statement fraud is the act of deceiving the users of relevant financial reports in financial accounting by related parties. Liu et al. [59] suggested using a two-way LSTM model to detect and warn of financial statement fraud threats, which can assist investors, audit departments, and state regulations departments in accurately identifying the financial fraud activity of publicly traded corporations. Fukas et al. [60] used GANs to detect fraud events on listed company financial datasets.

## 4 Financial Fraud Detection Datasets

For financial fraud detection studies to move forward, they need real and reliable research data. For privacy and safety reasons, banks, stock brokerages, and insurance companies often don't share information about their customers and their transactions. Therefore, scholars often cannot obtain real and reliable research data. Some studies [50,51] utilize artificially synthesized data for model training and to evaluate the performance of the proposed method. However, this cannot accurately reflect the performance and robustness of the proposed fraud detection method when applied to real complex transaction data. Through the review of numerous articles, this paper summarizes some publicly available real financial fraud datasets in Table 1.

**Table 1.** Datasets for financial fraud detection research.

| Dataset Name | Description | Source |
|---|---|---|
| European Credit Card Fraud Dataset | 284,807 samples, 31 features, 492 fraud cases | www.kaggle.com/mlg-ulb/creditcardfraud |
| IEEE CIS Fraud Detection Dataset | 590,000 samples, 434 features, 3.5% fraud cases | www.kaggle.com/c/ieee-fraud-detection |
| The Abstract credit card fraud dataset | 3075 samples, 12 features, 448 fraud cases. | www.kaggle.com/shubhamjoshi2130of |
| Mobile Money Transaction Data | 574,255 samples, 10 features, 271 fraud cases | www.kaggle.com/ealaxi/paysim1 |
| Target2-Slovenija Dataset | 8 million transactions from 2007 to 2017 | www.bsi.si/placila-in-infrastruktura/placilni-sistemi/target2-in-target2-slovenija |
| Bitcoin OTC Dataset | Bitcoin trading users, positive cases comprise 89% | snap.stanford.edu/data/soc-sign-bitcoin-otc.html |
| Bitcoin Alpha Dataset | Bitcoin users, positive samples is 93% | snap.stanford.edu/data/soc-sign-bitcoin-alpha.html |
| Ethereum Transactional Dataset | 20,000 Ethereum addresses, 7000 fraudulent accounts | www.kaggle.com/datasets/hamishhall/labelled-ethereum-addresses |
| Ethereum Dataset | 2,973,382 nodes, 1157 fraud cases | https://etherscan.io/accounts/label/phish-hack |
| Center for Medicare Services Dataset | 8304 samples, 8 features, 895 fraudulent samples | data.cms.gov/ |
| Vehicle Insurance Fraud Dataset | 15,420 samples, 923 fraudulent samples | www.kaggle.com/datasets/khusheekapoor/vehicle-insurance-fraud-detection |
| Vehicle Insurance Claim Fraud Detection | American insurance business and 15,420 samples | www.kaggle.com/datasets/shivamb/vehicle-claim-fraud-detection |
| Medical insurance data | 100,000 samples, 27 features | www.datafountain.cn/datasets/5068 |
| Financial Accounting Dataset | 146,000 samples, less than 1% fraudulent | github.com/JarFraud/FraudDetection |
| Financial Statement DataSet | 739 fraud cases, 4994 legitimate cases | www.kaggle.com/datasets/securities-exchange-commission/financial-statement-extracts |

## 5 Challenges and Prospects

Existing financial fraud detection methods have achieved good results, but there are still a lot of problems worthy of further thought and research. We have listed some problems and challenges, summarized as follows.

**Data Integrity.** Research on the detection model of financial fraud under the influence of missing data. In actual situations, user data collected by banks, securities companies, and other financial institutions often have certain deficiencies. The lack of user data may be caused by a variety of reasons. For example, users may submit personal information with different degrees of completeness to financial institutions according to their own circumstances and needs. Early users and new users of financial institutions may have different degrees of information integrity. Financial institutions have different levels of mastery over customer information at different levels, and so on. How to use missing user data for fraud detection is an important issue. Existing models such as GANs and VAE have attracted the attention of relevant scholars because of their data generation capabilities, and are used to fill in missing data. Furthermore, how to consider filling in data reliability and data defect locations to provide user-related information is worthy of follow-up exploration and thinking.

**Data Annotation.** Research on unsupervised and deeply weakly supervised financial fraud detection models. The label of the data directly determines the structure and evaluation indicators of the detection model, which is of great significance to the research of anti-fraud issues. The need for many tagged samples and the emphasis on detecting fraud category examples that have appeared in training are the foundations for the high accuracy of several present fraud detection techniques. Typically, data labelling for financial institutions requires expert determination, consuming time and money. Large-scale data labels are often not available. Therefore, for the research on financial fraud detection problems, the more challenging situation is how to use a lot of unlabeled data or a bit of abnormally labelled data to carry out various fraudulent behaviour detections.

**Data Quality.** With the rise of technologies like big data, artificial intelligence, and blockchain, financial fraud detection technology will develop in a direction that is more intelligent, credible, and explicable. In various disciplines, technologies such as ChatGPT have demonstrated their usefulness, and the detection of financial misconduct will make greater use of these technologies in the future. The pre-training model algorithm based on Transformer has high requirements on the amount of data. How to effectively generate the large amount of high-quality data required by the model is a current problem, especially when the data in the financial field is relatively precious. The existing generative adversarial networks solve this problem to a certain extent, and more methods will appear in the future.

Existing single-modality data cannot completely utilize the detection technology's benefits because fraudulent transactions are highly concealed. The detection of fraud using multimodal data, such as images, texts, and videos, will be a significant area of research. Due to the paucity of fraud samples, the combination of unsupervised and weak supervision is a crucial direction of research in financial fraud detection. Cryptocurrency has the characteristics of anonymity and decentralization, it plays an essential role in the financial transaction system, and it is receiving more and more attention. Moreover, detecting fraud on transaction networks is becoming a relevant research topic.

## 6 Conclusion

Financial fraud is a significant problem for the financial industry, and the research on detection technology has important social and economic implications for the development of the financial industry. This paper begins with a discussion of the financial fraud detection problem and the development route of financial fraud detection technologies. Furthermore, it explores the financial fraud detection technology based on deep learning and introduces typical deep learning model frameworks and their applications in various types of financial fraud detection tasks. Finally, we present the challenges in the field of financial fraud detection and discuss the development trend of fraud detection technology.

**Acknowledgments.** This work was supported by the National Key R&D Program of China (2021YFB2700500, 2021YFB2700501), the self-established Key Research Fund (119001-BB2201) of Intelligent Computing Theory and Method from Zhejiang Laboratory. The authors wish to acknowledge Dr. Fei Yu, Guilin Qi, and Yuandi Li for their contribution to the discussion and review. In addition, the authors wish to acknowledge the editor and anonymous reviewers for their insightful comments, which have improved the quality of this publication.

## References

1. West, J., Bhattacharya, M.: Intelligent financial fraud detection: a comprehensive review. Comput. Secur. **57**, 47–66 (2016)
2. Mangala, D., Soni, L.: A systematic literature review on frauds in banking sector. J. Financ. Crime **30**(1), 285–301 (2023)
3. Dyck, A., Morse, A., Zingales, L.: How pervasive is corporate fraud? Rev. Account. Stud., 1–34 (2023)
4. Chauhan, N.K., Singh, K.: 2018 A review on conventional machine learning vs deep learning. In: International Conference on Computing, Power and Communication Technologies (GUCON). IEEE, pp. 347–352 (2018)
5. Vaswani, A., et al.: Attention is all you need. In: Advances in Neural Information Processing Systems, vol. 30 (2017)
6. Hu, N., et al.: An empirical study of pre-trained language models in simple knowledge graph question answering. In: World Wide Web, pp. 1–32 (2023)
7. Alarfaj, F.K., Malik, I., Khan, H.U., Almusallam, N., Ramzan, M., Ahmed, M.: Credit card fraud detection using state-of-the-art machine learning and deep learning algorithms. IEEE Access **10**, 39 700–39 715 (2022)
8. Hemdan, E.E.-D., Manjaiah, D.: Anomaly credit card fraud detection using deep learning. In: Deep Learning in Data Analytics: Recent Techniques, Practices and Applications, pp. 207–217 (2022)
9. Kolli, C.S., Tatavarthi, U.D.: Money transaction fraud detection using Harris grey wolf-based deep stacked auto encoder. Int. J. Amb. Comput. Intell. (IJACI) **13**(1), 1–21 (2022)
10. Kumar, S., Ravi, V.: Explainable deep belief network based auto encoder using novel extended Garson algorithm, arXiv preprint arXiv:2207.08501 (2022)
11. Gradxs, G.P.B., Rao, N.: Behaviour based credit card fraud detection design and analysis by using deep stacked autoencoder based Harris grey wolf (HGW) method. Scand. J. Inf. Syst. **35**(1), 1–8 (2023)
12. Fanai, H., Abbasimehr, H.: A novel combined approach based on deep autoencoder and deep classifiers for credit card fraud detection. Expert Syst. Appl. **217**, 119562 (2023)
13. Singh Yadav, A.K., Sora, M.: Unsupervised learning for financial statement fraud detection using manta ray foraging based convolutional neural network. Concurr. Comput. Pract. Exp. **34**(27), e7340 (2022)

14. Zioviris, G., Kolomvatsos, K., Stamoulis, G.: Credit card fraud detection using a deep learning multistage model. J. Supercomput. **78**(12), 14 571–14 596 (2022)
15. Alharbi, A., et al.: A novel text2IMG mechanism of credit card fraud detection: a deep learning approach. Electronics **11**(5), 756 (2022)
16. Illanko, K., Soleymanzadeh, R., Fernando, X.: A big data deep learning approach for credit card fraud detection. In: Pandian, A.P., Fernando, X., Haoxiang, W. (eds.) Computer Networks, Big Data and IoT. LNDECT, vol. 117, pp. 633–641. Springer, Singapore (2022). https://doi.org/10.1007/978-981-19-0898-9_50
17. Gambo, M.L., Zainal, A., Kassim, M.N.: A convolutional neural network model for credit card fraud detection. In: 2022 International Conference on Data Science and Its Applications (ICoDSA), pp. 198–202. IEEE (2022)
18. Murugan, Y., Vijayalakshmi, M., Selvaraj, L., Balaraman, S.: Credit card fraud detection using CNN. In: Misra, R., Kesswani, N., Rajarajan, M., Veeravalli, B., Patel, A. (eds.) ICIoTCT 2021. LNNS, vol. 340, pp. 194–204. Springer, Cham (2022). https://doi.org/10.1007/978-3-030-94507-7_19
19. Jurgovsky, J., et al.: Sequence classification for credit-card fraud detection. Expert Syst. Appl. **100**, 234–245 (2018)
20. Esenogho, E., Mienye, I.D., Swart, T.G., Aruleba, K., Obaido, G.: A neural network ensemble with feature engineering for improved credit card fraud detection. IEEE Access **10**, 16 400–16 407 (2022)
21. Branco, B., Abreu, P., Gomes, A.S., Almeida, M.S., Ascensão, J.T., Bizarro, P.: Interleaved sequence RNNs for fraud detection. In: Proceedings of the 26th ACM SIGKDD International Conference on Knowledge Discovery & Data Mining, pp. 3101–3109 (2020)
22. Xie, Y., Liu, G., Yan, C., Jiang, C., Zhou, M.: Time-aware attention-based gated network for credit card fraud detection by extracting transactional behaviors. IEEE Trans. Comput. Soc. Syst (2022)
23. Roseline, J.F., Naidu, G., Pandi, V.S., Alias Rajasree, S.A., Mageswari, N.: Autonomous credit card fraud detection using machine learning approach. Comput. Electr. Eng. **102**, 108132 (2022)
24. Geetha, N., Dheepa, G.: A hybrid deep learning and modified butterfly optimization based feature selection for transaction credit card fraud detection. J. Posit. Sch. Psychol **6**(7), 5328–5345 (2022)
25. Xia, H., Zhou, Y., Zhang, Z.: Auto insurance fraud identification based on a CNN-LSTM fusion deep learning model. Int. J. Ad Hoc Ubiquitous Comput. **39**(1–2), 37–45 (2022)
26. Zhou, Y., Zheng, H., Huang, X., Hao, S., Li, D., Zhao, J.: Graph neural networks: taxonomy, advances, and trends. ACM Trans. Intell. Syst. Technol. (TIST) **13**(1), 1–54 (2022)
27. Dou, Y., Liu, Z., Sun, L., Deng, Y., Peng, H., Yu, P.S.: Enhancing graph neural network-based fraud detectors against camouflaged fraudsters. In: Proceedings of the 29th ACM International Conference On Information & Knowledge Management, pp. 315–324 (2020)
28. Liu, Y., et al.: Pick and choose: a GNN-based imbalanced learning approach for fraud detection. In: Proceedings of the Web Conference 2021, pp. 3168–3177 (2021)
29. Ren, J., Xia, F., Lee, I., Hoshyar, A.N., Aggarwal, C.C.: Graph learning for anomaly analytics: algorithms, applications, and challenges. ACM Trans. Intell. Syst. Technol. **14**, 1–29 (2022)
30. Rajput, N., Singh, K.: Temporal graph learning for financial world: algorithms, scalability, explainability & fairness. In: Proceedings of the 28th ACM SIGKDD Conference on Knowledge Discovery and Data Mining, pp. 4818–4819 (2022)

31. Mao, X., Liu, M., Wang, Y.: Using GNN to detect financial fraud based on the related party transactions network. Procedia Comput. Sci. **214**, 351–358 (2022)
32. Zhang, J., Yang, F., Lin, K., Lai, Y.: Hierarchical multi-modal fusion on dynamic heterogeneous graph for health insurance fraud detection. In: 2022 IEEE International Conference on Multimedia and Expo (ICME), pp. 1–6. IEEE (2022)
33. Wang, J., Guo, Y., Wang, Z., Wen, X., Ni, J.: Graph neural network with feature enhancement of isolated marginal groups. Appl. Intell. **52**, 1–13 (2022)
34. Long, J., Fang, F., Luo, H.: A novel GNN model for fraud detection in online trading activities. In: Lai, Y., Wang, T., Jiang, M., Xu, G., Liang, W., Castiglione, A. (eds.) ICA3PP 2021, Part II. LNCS, vol. 13156, pp. 603–614. Springer, Cham (2021). https://doi.org/10.1007/978-3-030-95388-1_40
35. Pan, Z., Wang, G., Li, Z., Chen, L., Bian, Y., Lai, Z.: 2SFGL: a simple and robust protocol for graph-based fraud detection. In: 2022 IEEE International Conference on Cloud Computing Technology and Science (CloudCom), pp. 194–201. IEEE (2022)
36. Li, R., Liu, Z., Ma, Y., Yang, D., Sun, S.: Internet financial fraud detection based on graph learning. IEEE Trans. Comput. Soc. Syst. (2022)
37. Li, P., Xie, Y., Xu, X., Zhou, J., Xuan, Q.: Phishing fraud detection on ethereum using graph neural network. In: Svetinovic, D., Zhang, Y., Luo, X., Huang, X., Chen, X. (eds.) BlockSys 2022. CCIS, vol. 1679, pp. 362–375. Springer, Singapore (2022). https://doi.org/10.1007/978-981-19-8043-5_26
38. Mo, C., Li, S., Tso, G.K., Zhou, J., Qi, Y., Zhu, M.: Motif-aware temporal GCN for fraud detection in signed cryptocurrency trust networks, arXiv preprint arXiv:2211.13123 (2022)
39. Qiao, C., Tong, Y., Xiong, A., Huang, J., Wang, W.: Block-chain abnormal transaction detection method based on dynamic graph representation. In: Fang, F., Shu, F. (eds.) GameNets 2022. LNICS, SITE, vol. 457, pp. 3–15. Springer, Cham (2022). https://doi.org/10.1007/978-3-031-23141-4_1
40. Hall, H., Baiz, P., Nadler, P.: Efficient analysis of transactional data using graph convolutional networks. In: Kamp, M., et al. (eds.) ECML PKDD 2021, Part II. CCIS, vol. 1525, pp. 210–225. Springer, Cham (2021). https://doi.org/10.1007/978-3-030-93733-1_15
41. Yuan, M.: A transformer-based model integrated with feature selection for credit card fraud detection. In: 2022 7th International Conference on Machine Learning Technologies (ICMLT), pp. 185–190 (2022)
42. Zhang, S., Suzumura, T., Zhang, L.: DynGraphTrans: dynamic graph embedding via modified universal transformer networks for financial transaction data. In: 2021 IEEE International Conference on Smart Data Services (SMDS), pp. 184–191. IEEE (2021)
43. Rodríguez, J.F., Papale, M., Carminati, M., Zanero, S., et al.: A natural language processing approach for financial fraud detection. In: Proceedings of the Italian Conference on Cybersecurity ITASEC 2022, Rome, Italy, 20–23 June 2022, vol. 3260, pp. 135–149. CEUR-WS.org (2022)
44. Cai, Q., He, J.: Credit payment fraud detection model based on TabNet and xgboot. In: 2022 2nd International Conference on Consumer Electronics and Computer Engineering (ICCECE), pp. 823–826. IEEE (2022)
45. Abakarim, Y., Lahby, M., Attioui, A.: A bagged ensemble convolutional neural networks approach to recognize insurance claim frauds. Appl. Syst. Innov. **6**(1), 20 (2023)

46. Padhi, I., et al.: Tabular transformers for modeling multivariate time series. In: ICASSP 2021–2021 IEEE International Conference on Acoustics, Speech and Signal Processing (ICASSP), pp. 3565–3569. IEEE (2021)
47. Hewapathirana, I., Kekayan, N., Diyasena, D.: A systematic investigation on the effectiveness of the Tabbert model for credit card fraud detection. In: 2022 International Research Conference on Smart Computing and Systems Engineering (SCSE), vol. 5, pp. 96–101. IEEE (2022)
48. Hu, S., Zhang, Z., Luo, B., Lu, S., He, B., Liu, L.: BERT4ETH: a pre-trained transformer for ethereum fraud detection. In: Proceedings of the ACM Web Conference 2023, pp. 2189–2197 (2023)
49. Gai, Y., Zhou, L., Qin, K., Song, D., Gervais, A.: Blockchain large language models, arXiv preprint arXiv:2304.12749 (2023)
50. Teng, H., Wang, C., Yang, Q., Chen, X., Li, R.: Leveraging adversarial augmentation on imbalance data for online trading fraud detection. IEEE Trans. Comput. Soc. Syst. (2023)
51. Langevin, A., Cody, T., Adams, S., Beling, P.: Generative adversarial networks for data augmentation and transfer in credit card fraud detection. J. Oper. Res. Soc. **73**(1), 153–180 (2022)
52. El Kafhali, S., Tayebi, M.: Generative adversarial neural networks based oversampling technique for imbalanced credit card dataset. In: 2022 6th SLAAI International Conference on Artificial Intelligence (SLAAI-ICAI), pp. 1–5. IEEE (2022)
53. Wu, R., Ma, B., Jin, H., Zhao, W., Wang, W., Zhang, T.: Grande: a neural model over directed multigraphs with application to anti-money laundering. In: 2022 IEEE International Conference on Data Mining (ICDM), pp. 558–567. IEEE (2022)
54. Xia, P., Ni, Z., Zhu, X., He, Q., Chen, Q.: A novel prediction model based on long short-term memory optimised by dynamic evolutionary glowworm swarm optimisation for money laundering risk. Int. J. Bio-Inspired Comput. **19**(2), 77–86 (2022)
55. Kržmanc, G., Koprivec, F., Škrjanc, M.: Using machine learning for anti money laundering. Evaluation **12**, 13 (2022)
56. Yu, T., Chen, X., Xu, Z., Xu, J.: MP-GCN: a phishing nodes detection approach via graph convolution network for ethereum. Appl. Sci. **12**(14), 7294 (2022)
57. Tang, J., Zhao, G., Zou, B.: Semi-supervised graph convolutional network for ethereum phishing scam recognition. In: Third International Conference on Electronics and Communication; Network and Computer Technology (ECNCT 2021), vol. 12167, pp. 369–375. SPIE (2022)
58. Gangadhar, K., Kumar, B.A., Vivek, Y., Ravi, V.: Chaotic variational auto encoder based one class classifier for insurance fraud detection, arXiv preprint arXiv:2212.07802 (2022)
59. Liu, X., Fan, M.: Identification and early warning of financial fraud risk based on bidirectional long-short term memory model. Math. Probl. Eng. **2022** (2022)
60. Fukas, P., Menzel, L., Thomas, O.: Augmenting data with generative adversarial networks to improve machine learning-based fraud detection (2022)

# GERNS: A Graph Embedding with Repeat-Free Neighborhood Structure for Subgraph Matching Optimization

Yubiao Chang[1], Tian Wang[2], Chao Tian[1], Ding Zhan[1], Cui Chen[2], Xingyu Wu[1], Chunxia Liu[1](✉), Endong Tong[2], and Wenjia Niu[2]

[1] School of Computer Science and Technology, Taiyuan University of Science and Technology, Taiyuan 030024, China
{changyubiao,1521236687,zd334488}@stu.tyust.edu.cn, lcx456@163.com

[2] Beijing Key Laboratory of Security and Privacy in Intelligent Transportation, Beijing Jiaotong University, Beijing 100044, China

**Abstract.** Subgraph matching is used to determine whether a query graph exists within a target graph, and appears in a lot applications of domains including social sciences, chemistry, biology and database systems. Existing subgraph matching approaches can be broadly categorized into two types: exact matching and approximate matching. Due to allowing for slight variations between the target and query graphs, approximate matching has become a more practical solution with the introduce of graph neural networks (GNN). However, when dealing with large-scale target and query graphs, existing GNN-based approximate matching approaches still have to face the challenge that how to further improve the accuracy and promote the query efficiency. Therefore, we propose the GERNS, a graph embedding with repeat-free neighborhood structure for subgraph matching optimization. Through extracting subgraphs from the target graph based on a specified hop count limit, we incorporate and embed a repeat-free neighborhood structure using a two-layer GNN. Then we generate the relation constrains of subgraphs based on vector order embedding to form the embedding space. Finally, approximate subgraph matching can be realized based on graph embedding. Extensive experiments on both public graph datasets and real-world datasets show the effectiveness of our approach.

**Keywords:** Subgraph Matching · Approximate Match · Graph Neural Network · Graph Embedding

## 1 Introduction

Given a query graph, subgraph matching determines whether the query graph is isomorphic to a subgraph part of a larger target graph. If the graph data structure contains node and edge features, topology and features should match. Subgraph matching is a critical problem in many biological, social network


H. Wang et al. (Eds.): CCKS 2023, CCIS 1923, pp. 178–189, 2023.
https://doi.org/10.1007/978-981-99-7224-1_14

and knowledge graph applications [1–4]. For example, social and biomedicine researchers compute network structures of given graph data to study their critical subgraph structures [5]. In the knowledge graph, query the essential subgraph structure in the large-scale target graph [1,6] to extract subgraphs for research and analysis.

Existing methods mainly utilize efficient search algorithms [7–9], matching optimization strategies, efficient pruning operations, and specified pattern-matching strategies. But due to the NP-complete nature of the problem, the scalability is poor in large-scale cases, the cost of graph embedding is high, and the query time is expensive. When expanding to large-scale subgraph matching capabilities, the embedding space will have some repeated neighborhood structures of nodes, resulting in a suboptimal spatial structure for graph embedding, resulting in a longer query time. This section comprehensively analyzes the embedding process of the target graph $G_t$. In Fig. 1, it can be observed that the target graph $G_t$ is extracted into subgraphs within 2-hop and embedded with two layers of GNN. The constraint relationship between the subgraphs is guaranteed, but vectors with the same coordinates appear in the embedding space. Then, in the face of large-scale target graphs being embedded in this way, the accuracy rate may be reduced, and the query efficiency is affected.

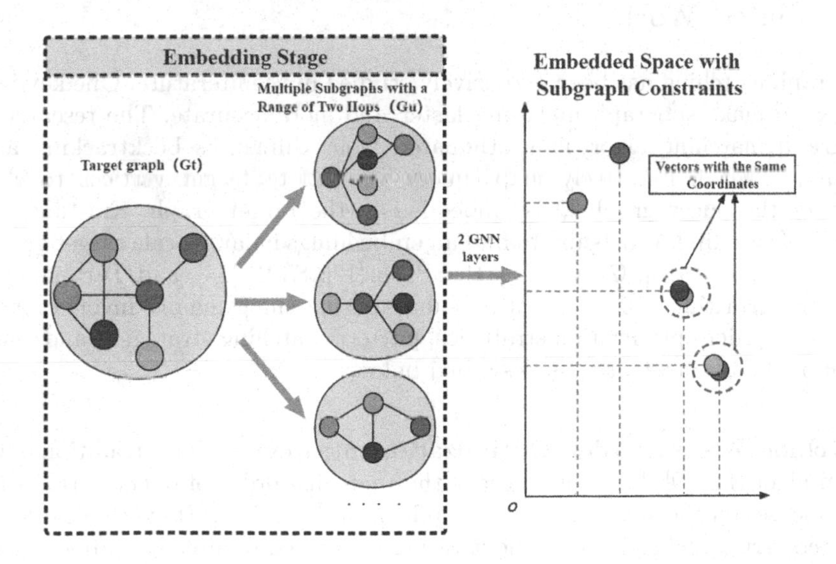


**Fig. 1.** Target graph embedding problem with repeated neighborhood structure.

In this paper, we propose to embed the space of repeat-free neighborhood structures for efficient subgraph matching query, combined with a graph neural network to optimize the embedding mechanism and conduct model training with subgraph relationship constraints on public datasets. Its core is to extract the target graph $G_t$ and combine GNN and our embedding optimization algorithm to quickly calculate whether the neighborhood structure between nodes needs to be embedded. Our method is mainly in the embedding stage, extracting the subgraph $G_u$ within the k-hop in the target graph $G_t$, using a 2-layer GNN and using a repetitive-free neighborhood structure algorithm to calculate $u_i(i = 1, 2, ......, n)$ and the neighborhood structure between $u_j(j = 1, 2, ..., k)$ to determine whether embedding is needed, to ensure the efficiency of subgraph matching.

Our contributions are as follows: 1) An efficient and accurate subgraph matching solution GERNS is proposed, which solves the problems existing in existing methods. 2) We propose a space-optimized algorithm for embedding repetitive-free neighborhood structures to speed up the time for subgraph matching. 3) Many experimental results show that GERNS and the current similar methods have remarkable performance.

The rest of the paper is organized as follows. Section 2 reviews common approaches to speed up subgraph matching and related work. Section 3 addresses methods for embedding repeating-free neighborhood structure spaces, and we implement algorithms for embedding repeating-free neighborhood structure spaces. Our proposed technique is evaluated in Sect. 4. Finally, we end with conclusions in Sect. 5.

## 2 Related Work

Subgraph matching has been extensively studied in the literature. One key issue is how to make subgraph matching faster and more accurate. The research on subgraph matching algorithms originated from Ullmann's backtracking algorithm [9], which iteratively maps query vertices to target vertices to check whether the query graph $Q$ is embedded in the target graph. Existing algorithms of this kind, which aim to find all embeddings in large-scale target graphs, include, QuickSI [10], GraphQL [11], SPath [12], STW [13], and TurboIso [14]. Existing algorithms for accelerating subgraph matching mainly involve matching access order optimization strategies, pattern matching strategies, and neural graph matching strategies, as described below:

**Matching Access Order Optimization Strategies.** The traditional Ullmann algorithm [9] does not consider the matching order of query vertices and only focuses on the structure of the graph itself. VF2 [7] starts with a randomly selected vertex and chooses the next vertex connected to already matched query vertices. QuickSi [10] utilizes a global statistical measure of vertex label frequencies and proposes a matching order that accesses query vertices with infrequent vertex labels as early as possible. In contrast to the global matching order selection in QuickSi, TurboIso [14] divides the candidate region into separate candidate regions and computes a local matching order for each candidate region. STW [13] and TurboIso both assign higher priority to query vertices with higher degrees and fewer labels.

**Pattern Matching Strategies.** Spath [12] proposed matching one graphical pattern at a time instead of the traditional vertex-based approach. The graph pattern used in Spath is a path. Turboiso [14] rewrites the query graph into an NEC tree, which matches query vertices with the same neighborhood structure simultaneously.

**Neural Graph Matching Strategies.** Early work [15] has demonstrated the capability of Graph Neural Networks (GNNs) in small-scale subgraph matching. Graph neural networks [16–18] have been proposed for subgraph matching and have achieved state-of-the-art results. However, since there is no one-to-one mapping between query and target graph nodes, these methods cannot be directly applied to subgraph matching.

Our approach is to improve the performance of subgraph matching, the core of which is to embed ordered and repetitive-free neighborhood structures to improve the performance of subgraph matching. Furthermore, similar works, including NeuroMatch, use GNN models to constrain subgraph relations. However, it does not consider the neighborhood structure issue in the embedding space. In contrast, our method uses GNN for model training to constrain the subgraph relationship in the embedding space and optimizes the algorithm of the graph embedding space so that it can be embedded in a space with a non-repeating neighborhood structure, thus speeding up the performance of subgraph matching.

# 3 Proposed Method

In this section, we propose the optimized embedding vector technique with repeated neighborhood structure to accelerate subgraph matching, improve the accuracy of similarity and reduce the cost of computation and storage, thus adding subgraph matching. The proposed optimization technique consists of three steps:

## 3.1 Embedding of Target Graph

The purpose of this method is to embed a given target graph $G_t$ into the space $Z_u$ of repeat-free neighborhood structure and accelerate the speed of subgraph matching. The embedding type adopts subgraph embedding in 2-hop, starting from any node in the graph, traversing and extracting and executing the embedding method until all nodes in the graph are traversed. Figure 2 shows an example of a method for extracting a target graph $G_t$ from the beginning to embedding a repetitive-free neighborhood space $Z_u$.

Let the given target graph be $G_t$, perform 2-*hop* subgraph extraction on the target graph, generate $G_1, G_2, ..., G_u$ subgraphs to be embedded, and continuously optimize our model by updating GNN parameters to improve our generalization performance and prediction ability. Use the high-dimensional sparse matrix $m1$ to reduce the dimension to the low-dimensional dense adjacency

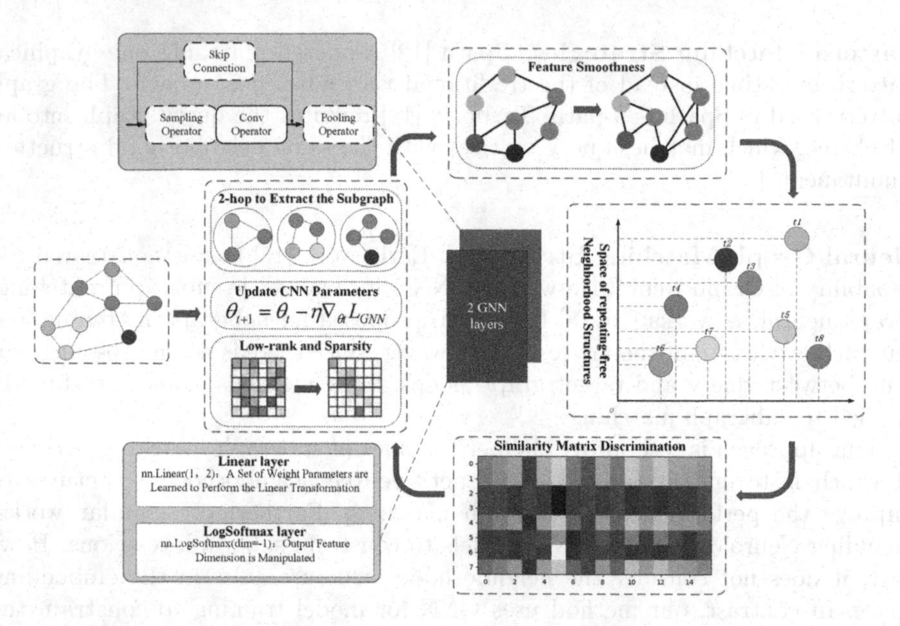


**Fig. 2.** An example of the process of extracting subclasses from the target graph to the embedding space of the repeat-free neighborhood graph.

matrix $m2$. Some useful inferences and predictions are made in graph data by using the feature smoothing properties. In addition, the feature smoothness property also provides a key optimization objective for GNNs to better learn feature propagation and information aggregation among nodes. Then, it is embedded in the space $Z_u$ of the repetitive-free neighborhood structure and recorded as a $t1$ vector. When the $t2$ vector subgraph is embedded, according to our proposed method, that is, similarity matrix discrimination, it is judged whether the next embedded subgraph neighborhood structure is embedded. During the entire embedding process, we also used two layers of GNN for embedding, one is the linear layer $Linear(1, 2)$, and the other is the $LogSoftmax$ layer ($dim$=-1). Through this process, the $G_u(u = 1, 2, ..., 7)$ subgraph is extracted from $G_t$, and each subgraph is constrained by the subgraph relationship in the specified order in the repeat-free neighborhood structure space $t1, t2, ..., t7$ vectors enter into the embedding space in turn.

## 3.2 Optimize Embedding Vectors

In the process of subgraph matching, the embedding stage lays the foundation for the response of subgraph matching and optimizes the embedding vector to improve the accuracy and time of subgraph matching. In the embedding process, the nodes in the target graph $G_t$ are first extracted by 2-*hop* subgraphs to form several small subgraphs $G_u$, which is not only a subgraph of $G_t$, but also a repeated neighborhood structure of u node in $G_t$, and the graph neural network is used to embed $G_1, ..., G_u$, GNN embeds the neighborhood structure $G_u$ of node u into Zu space. The details that need to be paid attention to in the design of embedding are: when embedding each node u, the actual embedded $G_u$ is also the neighborhood structure of $u$. In the comparison process, we only need to compare whether $u_1$ and $u_2$ are subgraphs, in other words, the matching relationship between the neighborhood structure of $u_1$ and $G_u$ and the neighborhood structure of $u_2$.

Commonly used algorithms for graph embedding, such as the Deepwalk algorithm, calculates the similarity between nodes based on whether there are edges connected. The LINE algorithm uses another calculation method, introducing first-order similarity and second-order similarity.

**First-order Similarity Optimization:** For each edge$(i, j)$ of the undirected graph $G(V, E)$, we define its first-order similarity for the two connected vertices $v_i$ and $v_j$:

$$p_i(v_i, v_j) = \frac{1}{1 + exp(-u_i^T \cdot u_j)} \tag{1}$$

For all vertices $V$, the KL divergence is introduced to measure the similarity between the empirical distribution and the first-order similarity:

$$d(\hat{p}_1, p_1) = -\sum_{(i,j)\in E} [\hat{p}_1(v_i, v_j) \log_{p_1} (v_i, v_j)] + \sum_{(i,j)\in E} [\hat{p}_1(v_i, v_j) \log_{\hat{p}_1} (v_i, v_j)] \tag{2}$$

**Second-order Similarity Optimization:** Applicable to both undirected and directed graphs. Its purpose is: an Embedding that a vertex itself wants to learn, and a vertex is an Embedding that a node is a neighbor of other nodes. Therefore, put this. Two vertices are represented by two vectors. So for any edge$(i, j)$, define the conditional probability of the content $v_j$ under $v_i$:

$$p_2(v_j \mid v_i) = \frac{exp(u_j'^T \cdot u_i)}{\sum_{k=1}^{|V|} exp(u_k'^T \cdot u_i)} \tag{3}$$

To make the graph model consider the first-order and second-order similarity at the same time, the vector representations of the two similarities are first trained separately and then combined (directly spliced) for use together. For the second-order similarity, direct optimization is relatively difficult. The negative sampling method is adopted, and the optimization objective function of each edge$(i, j)$ is transformed into:

$$loss = log\sigma(u_j'^T \cdot u_i) + \sum_{n=1}^{K} E_{v_n \sim P_n(v)} \left[log\sigma(-u_n'^T \cdot u_i)\right] \tag{4}$$

where $\sigma$ is the sigmoid function, K is the number of negative sampling edges, which are actually K target vertices, which are composed of query vertices $V_i$.

**Algorithm 1:** Optimizing Embedding Vectors

**input :** Target graph $G_t$, graph embeddings $Z_u$ of subgraph $G_u(u = 1, ..., s) \in G_t$
**output:** Optimized embedded vectors $Z_j$

```
1  for v in G.nodes() do
2      S ← Set()
3      neighbors1 ← G.neighbors(v)
4      S.add(frozenset([v]))
5      for u in neighbors1 do
6          S.add(frozenset([u, v]))
7          neighbors2 ← G.neighbors(u)
8          for w in neighbors2 do
9              if w ≠ v and w ≠ u and ({w, u}) ∉ S then
10                 S.add(frozenset([v, u, w]))
```

$P_n(v) \propto d_v^{3/4}$, $d_v$ is the out-degree of vertex $v$. Second-order similarity optimization is used as an undirected graph, which captures more global information. In addition, undirected graphs can use two methods to obtain two types of Embedding, which are used in Concat. The denominator needs to be calculated for the number of full nodes, and negative sampling is generally used for optimization. If you want the two distributions to be consistent, use KL dispersion optimization, and the optimization goal is:

$$min\left[-\sum_{(i,j)\in E} w_{ij}\log p_2\left(v_j \mid v_i\right)\right] \tag{5}$$

### 3.3 Vector Generation in Embedded Space

Converting graph data into embedding vectors is a common operation in graph neural networks. And this process is called the graph embedding process, which maps the nodes or subgraphs in the graph into a low-dimensional continuous vector space for subsequent analysis and prediction. The data nodes or subgraphs as graph structures are mapped into a low-dimensional continuous vector space, which is convenient for subsequent analysis and prediction. Subgraphs or nodes are transformed into embedding vectors. First, a graph neural network model suitable for graph embedding is designed. For example, models such as GraphSAGE, GCN, GAT, etc., can be used for graph embedding tasks. These models are able to aggregate and convey information through neighbor relationships between nodes. Model training using large-scale graph data.

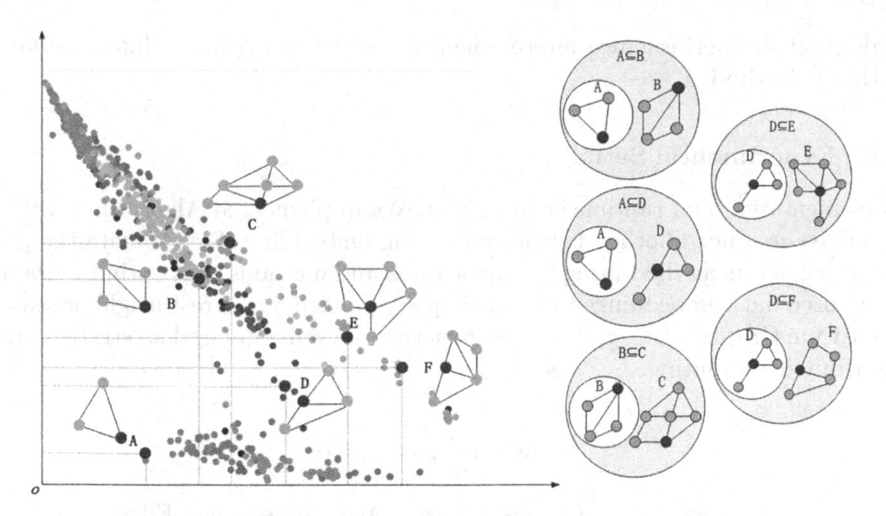


**Fig. 3.** Mapping between optimized embedding vector and subgraph relationship.

During the training process, the model will learn the constraint relationship of nodes or subgraphs, which makes similar nodes or subgraphs closer in the embedding space. After the model is trained, the nodes or subgraphs in the graph can be converted into embedding vectors using the already trained graph neural network model. For node embedding, you can directly use the output of the intermediate layers of the model as the representation of the nodes. For subgraph embedding, the relationship representation between subgraphs can be obtained by aggregating the embedding vectors of all nodes in the subgraph. The generated embedding vectors can be applied to different tasks such as similarity computation, clustering, classification, etc. We adopt a neural network model with the ability to constrain subgraph relationships and a neural network model for embedding and constraining relationships between subgraphs. The graph embedding model is used to convert the graph into an embedding vector, and the order relationship between the subgraphs is judged by calculating the difference between the embedding vectors.

In the embedding space, Fig. 3 shows the correct relationship between the subgraphs. Yellow points represent embeddings with larger graph sizes, and orange points represent embeddings with smaller graph sizes. Blue points represent our example, where we observe that generating vectors in the embedding space is able to carry out constraints on subgraph relations. This allows us to avoid extra overhead during the query phase.

## 4 Experiments

This section introduces our experiments. The purpose of experiment 2 is to evaluate (1) the accuracy and query efficiency of two subgraph matching methods, the SOTA method NeuroMatch [19] and our GERNS, in public datasets at different scales, (2) the performance improvement of backtracking algorithms combined with our method.

## 4.1 Experimental Setup

Implementation and runtime environment. We implemented Algorithm 1 with a repetitive-free neighborhood structure during embedding. We evaluate the performance of our method using five algorithms and methods that optimize graph-structured data embeddings to improve query efficiency: nearest neighbor search algorithm, GraphSAGE, graph index structure, incremental update strategy, and distributed computing.

**Table 1.** Graph datasets

| Datasets | Graphs | Classes | Avg. Nodes | Avg. Edges |
|---|---|---|---|---|
| Malonaldehyde | 893228 | R(1) | 9.00 | 36.00 |
| Benzene | 527984 | R(1) | 12.00 | 64.94 |
| Reddit-Threads | 203088 | 2 | 23.93 | 24.99 |
| Triangles | 45000 | 10 | 20.85 | 32.74 |

GraphSAGE is a common algorithm that we can use, and the overall performance is good, while distributed computing is a popular technique to deal with large-scale graph structure data and calculate it on multiple computers at the same time. These algorithms and methods have improved query efficiency. All the algorithms are implemented in Python language, and all the experiments are run on 128G memory, 2TB SSD, Intel Xeon Gold 6240 2.6GHz CPU and NVIDIA GeForce RTX 3080 Ti.

**Datasets.** We used four publicly available datasets in our experiments: Malonaldehyde, Benzene, Reddit-Threads, and Triangles. These datasets are sourced from a collection of benchmark datasets for graph classification and regression. We utilized these datasets for preliminary validation and evaluation. Table 1 provides an overview of the datasets.

**Query Sets.** We generated query graphs by randomly extracting subgraphs of different sizes from the target graph. We ensured that our query graphs were subgraphs of the target graph. The size of the query graphs can be set randomly or custom-defined, and we chose the custom-defined approach to specify their sizes. Each query set consists of ten query files, and each query file contains query graphs of different sizes.

### 4.2 Accuracy and Query Time Evaluation

We perform experiments by varying between 900,000 and 8 million vertices, that is, four data sets. The largest graph has 8 million nodes and 32.15 million edges. The evaluation metric is generated by utilizing our custom graph size generator. We use a generator to partition in scale for each graph dataset. Only one query set is used for each dataset to evaluate the precision of our subgraph matching.

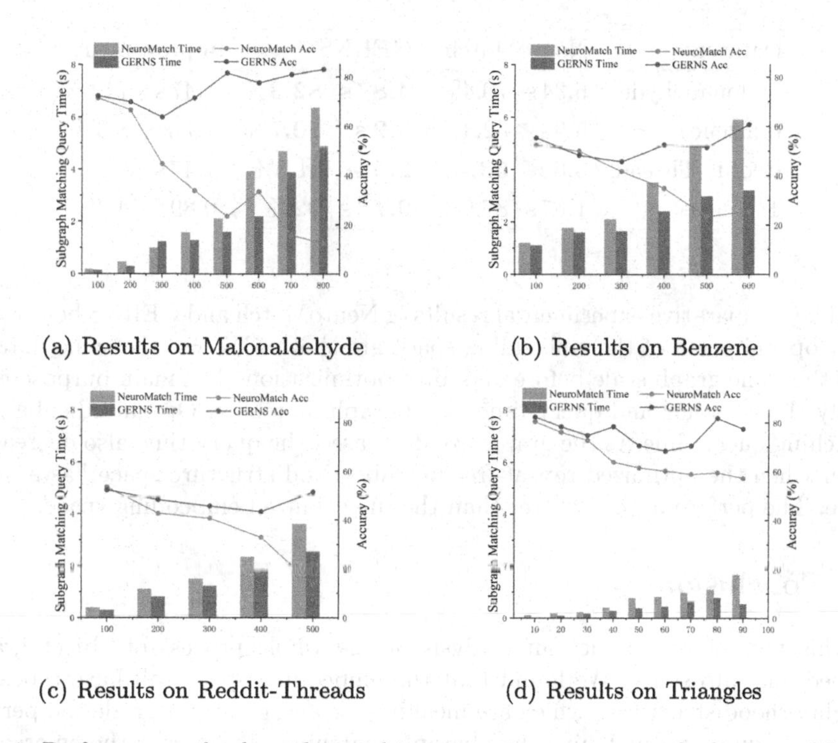


(a) Results on Malonaldehyde (b) Results on Benzene

(c) Results on Reddit-Threads (d) Results on Triangles

**Fig. 4.** Performance of subgraph matching execution on two methods, NeuroMatch and GERNS, on graph datasets. The x-axis represents different data set sizes.

Figure 4 shows the comparison of accuracy before and after optimization of neural subgraph matching under four datasets. Because we embed a space of repeat-free neighborhood structure, with the continuous increase of node scale, the accuracy of each dataset shows a linear growth trend. In the case of small scale, the accuracy of subgraph query optimization can almost keep very good. This shows that our method can maintain good accuracy regardless of the scale change in the space of embedding repetitive-free neighborhood structures.

### 4.3 Evaluation Results

We analyzed the comparison results of NeuroMatch and GERNS, two methods of subgraph matching on four data sets, mainly comparing the results of the two methods in terms of accuracy and query time. To comprehensively evaluate our experiments, each dataset is split by scale to verify the effectiveness of subgraph matching on GERNS under the two methods. The results of the overall performance after graph embedding optimization with repeat-free neighborhood structure are compared in Table 2.

**Table 2.** Comparison of Evaluation Results

| Datasets | NeuroMatch | | GERNS | | Improvement | |
|---|---|---|---|---|---|---|
| Malonadehyde | 6.34 s | 70.4% | **4.87 s** | **82.3%** | 1.47 s | 11.9% |
| Benzene | 5.9 s | 52.4% | **3.2 s** | **60.7%** | 2.7 s | 8.3% |
| Reddit-Threads | 3.6 s | **53.4%** | **2.54 s** | 51.5% | 1.47 s | — |
| Triangles | 1.67 s | 87.6% | **0.78 s** | **92%** | 0.89 s | 4.4% |

The comparative experimental results of NeuroMatch and GERNS before and after optimization of the embedding space are shown. We use the same dataset and the same graph scale before and after optimization. The main purpose is to verify the accuracy and query time of subgraph matching. Optimized subgraph matching query time As the graph size decreases, the query time also decreases. Even when the optimized repeat-free neighborhood structure space has a large scale, The performance is better than the unoptimized embedding space.

## 5 Conclusion

In this paper, we conduct an analysis of the whole process of object graph embedding into space. We found that the embedding space will have repeated neighborhood structures, which are meaningless and greatly degrade the performance of subgraph matching in subgraph matching. To address the emergence of this repeated neighborhood structure, we propose an algorithm for the optimization of graph embeddings with repeated-free neighborhood structures. Our formulated strategy extracts subgraphs from the target graph according to the specified hop limit and uses 2-layer GNNs to embed them in the repetitive-free neighborhood structure space. Using model training in this process to constrain the relationship between subgraphs in the embedding space, we refer to this optimized graph embedding method as graph embedding with repeat-free neighborhood structure for subgraph matching optimization.

Through extensive experiments, we show that optimization based on graph embeddings can speed up subgraph matching. Extensive experiments on public graph datasets and real-world datasets demonstrate the effectiveness of our approach.

**Acknowledgements.** The work was supported by the program under Grant No. 315087705.

## References


1. Gentner, D.: Structure-mapping: a theoretical framework for analogy. Cogn. Sci. **7**, 155–170 (1983)
2. Raymond, J.W., Gardiner, E.J., Willett, P.: Heuristics for similarity searching of chemical graphs using a maximum common edge subgraph algorithm. J. Chem. Inf. Comput. Sci. **42**, 305–316 (2002)
3. Yang, Q., Sze, S.-H.: Path matching and graph matching in biological networks. J. Comput. Biol. **14**, 56–67 (2007)
4. Dai, H., Li, C., Coley, C., Dai, B., Song, L.: Retrosynthesis prediction with conditional graph logic network. In NeurIPS (2019)
5. Alon, N., Dao, P., Hajirasouliha, I., Hormozdiari, F., Sahinalp, S.C.: Biomolecular network motif counting and discovery by color coding. Bioinformatics **24**, 241–249 (2008)
6. Plotnick, E.: Concept mapping: a graphical system for understanding the relationship between concepts. ERIC Clearinghouse on Information and Technology Syracuse, NY (1997)
7. Cordella, L.P., Foggia, P., Sansone, C., Vento, M.: A (sub) graph isomorphism algorithm for matching large graphs. PAMI **26**, 1367–1372 (2004)
8. Gallagher, B.: Matching structure and semantics: a survey on graph-based pattern matching. In: AAAI Fall Symposium (2006)
9. Ullmann, J.R.: An algorithm for subgraph isomorphism. J. ACM **23**, 31–42 (1976)
10. Shang, H., Zhang, Y., Lin, X., Yu, J.X.: Taming verification hardness: an efficient algorithm for testing subgraph isomorphism. PVLDB **1**(1), 364–375 (2008)
11. He, H., Singh, A.K.: Query language and access methods for graph databases. In: Aggarwal, C., Wang, H. (eds.) Managing and Mining Graph Data. Advances in Database Systems, vol. 40. Springer, Boston (2010). https://doi.org/10.1007/978-1-4419-6045-0_4
12. Zhao, P., Han, J.: On graph query optimization in large networks. PVLDB **3**, 340–351 (2010)
13. Sun, Z., Wang, H., Wang, H., Shao, B., Li, J.: Efficient subgraph matching on billion node graphs. PVLDB **5**, 788–799 (2012)
14. Han, W.-S., Lee, J., Lee, J.-H.: Turbo iso: towards ultrafast and robust subgraph isomorphism search in large graph databases. In: SIGMOD, pp. 337–348 (2013)
15. Scarselli, F., Gori, M., Tsoi, A.C., Hagenbuchner, M., Monfardini, G.: The graph neural network model. IEEE Trans. Neural Netw. 20, 61–80 (2008)
16. Kipf, T.N., Welling, M.: Semi-supervised classification with graph convolutional networks. In: ICLR (2017)
17. Hamilton, W., Ying, Z., Leskovec, J.: Inductive representation learning on large graphs. In: NeurIPS (2017)
18. Xu, K., Hu, W., Leskovec, J., Jegelka, S.: How powerful are graph neural networks? In: ICLR (2018)
19. Ying, Z., Wang, A., You, J., et al.: Neural subgraph matching (2020)

# Feature Enhanced Structured Reasoning for Question Answering

Lishuang Li(✉), Huxiong Chen, and Xueyang Qin

School of Computer Science and Technology, Dalian University of Technology, Dalian 116023, Liaoning, China
lils@dlut.edu.cn

**Abstract.** Answering complex natural language questions requires comprehensive reasoning about the question context and related knowledge. There are two main problems with the existing LM (language model) +KG (knowledge graph) methods. Firstly, they ignore the impact of negative words on Q&A inference. Secondly, they do not consider the effects of contextual entities on relation weight. Taking into account the above issues, we propose a method for Feature Enhanced Structured Reasoning (FESR) that exploits a two-branch graph neural network to improve the structured reasoning ability of question answering. Specifically, FESR first sets feature constraints and changes the attention scores between nodes, thereby strengthening the processing of negative-type question answering, and then optimizes the relation weights to enhance the effect of relations on question-answering inference by introducing contextual entities. We evaluate our model on three datasets in the fields of commonsense reasoning and medical question answering, and the experimental results indicate the effectiveness of our method.

**Keywords:** Question answering · Structured reasoning · Feature enhanced

## 1 Introduction

Question-answering reasoning is a challenging task that aims to reason about answers from various knowledge based on natural language questions. Currently, there are two approaches for question-answering tasks: using pre-trained language models to implicitly encode unstructured textual knowledge and combining pre-trained language models with knowledge graphs for comprehensive reasoning. For the former, for example, Petroni et al. [15] verify that pre-training language models on large text corpora can improve the accuracy of question-answering tasks. But the predictions obtained by such methods are not interpretable, and they perform poorly when handling structured reasoning.

Combining pre-training language models and knowledge graphs is the mainstream method for question-answering reasoning. The language model is used to obtain the initial representation of the question and answer, combined with


H. Wang et al. (Eds.): CCKS 2023, CCIS 1923, pp. 190–203, 2023.
https://doi.org/10.1007/978-981-99-7224-1_15

the knowledge graph for joint reasoning, and finally get a reasonable answer. Previous research have demonstrated the effectiveness of this approach. Bao et al. [1] propose constraint rules based on the knowledge graph, using the rules to constrain complex problem entities, splitting the problem into simple problems to get the answer. Sun et al. [19] combine knowledge base and entity-linked text, enabling question answering to extract answers from question-specific subgraphs containing text and knowledge base entities and relations. Feng et al. [4] constructs a subgraph from the knowledge graph and combines GNN to perform multi-hop reasoning on the subgraph. Similarly, Yasunaga et al. [23] propose a correlation score for subgraph redundant nodes, and introduce a message-passing mechanism in GNN. Zhang et al. [24] fuse encoded feature representations from pre-trained LMs and graph neural networks on multi-layer modality interaction operations to achieve reliable reason on structured knowledge. These methods effectively combine LM and KG, but face two challenges: on the one hand, attention scores between nodes do not consider specific contextual states (positive sentence or negative sentence). On the other hand, the effect of context on edge weight calculation is not fully considered.

Based on the above problems, we introduce a Feature Augmented Structured Reasoning Network, or FESR for short, that leverages a two-branch network for structured reasoning on question answering. Specifically, in one branch, we set two states of 0 and 1 for entity concept description, and introduce negation word features into node attention computation. In another branch, we weigh the context with the head and tail entities of an edge, so that the weight of an edge is related to the specific question answer. Finally, the features of the two branches are fused for final reasoning. Experiments on the three data sets of CommonsenseQA, OpenbookQA, and MedQA-USMLE show that the performance of FESR is better than other models with the same amount of parameters. Our contributions are as follows:

- We propose a method for Feature Enhanced Structured Reasoning (FESR) to perform question-answering reasoning, and experimental results on three datasets (CommonsenseQA, OpenbookQA, MedQA-USMLE) to verify the advancement of our approach.
- We propose a method to calculate the attention scores between nodes combined with the specific context type (positive or negative) and then update the node feature representation, improving the model's sensitivity to negative features.
- We weigh the context nodes and the related relations to obtain relation weights, and use a two-branch strategy to encode two parts of node features, enriching node feature information sufficiently.

## 2 Description

We use a pre-trained language model and a structured knowledge graph to answer natural language questions and perform interpretable reasoning on their

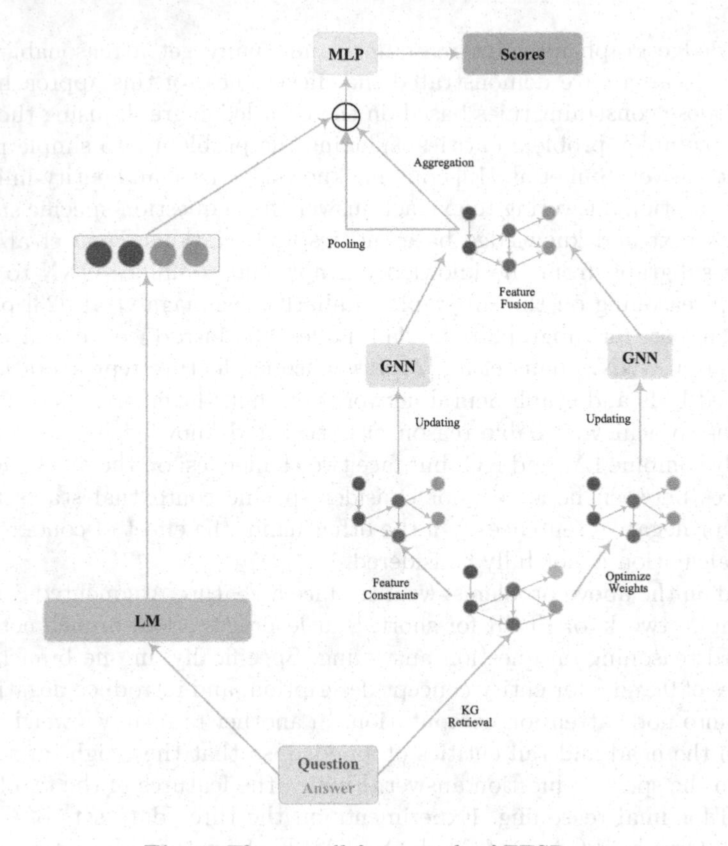


**Fig. 1.** The overall framework of FESR.

paths. For a question and answers in the context of question answering, we first obtain the conceptual description of the entity and then use the function $f_{concat}(x)$ to convert $x$ (the concatenated description of the entity and the question) into a vector form. Finally, we apply the function $f_{score}(x)$ to calculate a correlation score from this vector. This score can be used to delete the unrelated nodes of the knowledge graph. The knowledge graph $(KG)$ is defined as $G = (V, E)$, where $V$ is all entities in the knowledge graph. $E \subseteq V \times R \times V$ means a collection of edges, where R is a collection of relation types.

Correspond the entities $V_q$ and $V_a$ in question $q$ and answer $a$ to the knowledge graph to obtain the initial subgraph $(V_q, V_a \subseteq V)$. We use contextual information to compute node-to-question relevance scores, simplifying subgraphs. At the same time, feature constraints and recalculated edge weights are used to enhance the features of nodes. Finally, the pre-trained context, the context features of the GNN network, and the pooled reasoning path are comprehensively adopted to reason the question and answer, making the question and answer reasoning more reasonable and robust.

# 3 Method

Figure 1 is the main architectural diagram of our model. For a given natural language question $q$ and its options $a$, we take the question and options as context ($Con$). The specific process of FESR is as follows: We first retrieve the knowledge graph subgraph according to the context, calculate the relevance score of the node and the question, and delete irrelevant nodes in the subgraph based on this score to obtain the question answering subgraph ($G_{sub}$). To strengthen the structured reasoning of question answering, we design a two-branch GNN network to update node features. Finally, we use the pre-trained QA context features, the pooled inference path, and the QA context features of the GNN network to make the final prediction through the Multi-Layer Perceptron ($MLP$).

## 3.1 Subgraph Construction

To simplify the path of reasoning and improve the efficiency of model reasoning. We retrieve the QA subgraph according to the context, calculate the correlation score between the nodes and the context in the subgraph and obtain the highly relevant QA subgraph $G_{sub}$. Firstly, question and answer entities in the question answer context are linked to the knowledge graph KG, and all entities on the question-to-answer path are retrieved, up to four entities on one path. In this way, we get a QA subgraph with many entities, while some entities (paths) have little correlation with the QA. Then, we get the conceptual description $v_i^c$ of each node $v_i$ according to the external dictionary library Wiktionary. We use the pre-trained language model to calculate the correlation score between the concept description of the node and the context. The whole process is formalized as follows:

$$S_{v_i} = f_{score}(f_{concat}([Con; v_i^c])), \tag{1}$$

where the size of $S_{v_i}$ indicates how relevant this entity is to the question.

We only retain the top n (n = 200) scores nodes in the subgraph. We retrieve all the nodes ranked in the top n and then connect all the edges between any two nodes to obtain a knowledge graph subgraph ($G_{sub}$) that is highly relevant to the question and answer, where $G_{sub} = (V_r, E_r)$, and $V_r = \{V_q, V_a, V_{other}\}$ and $E_r$ means the edge after deletion. $V_{other}$ does not include head and tail nodes.

## 3.2 Two-Branch GNN Network

To perform better reasoning on $G_{sub}$, we propose a two-branch GNN Network, which is divided into feature constraints and weight optimization to update the feature representation of nodes. The former uses additional states of the context to constrain nodes, thereby affecting the attention score between nodes. The latter uses context to optimize the weight representation of edges, thereby affecting the features of edge-based nodes. The final node features obtained in this way not only fully consider local neighbor features, but also consider context and additional state features from a global perspective.

**Feature Constraints.** In this subsection, we additionally compute the feature influence of neighboring entities on the central entity by constraining words. Specifically, we first add an additional state: 0 or 1 to the conceptual descriptions of all entities in the context through context constraints (such as "not", "no" in negative sentences). For example, for the context: "If it is not used for hair, a round brush is an example of what ?", an extra state 0 is added after the concept description of "hair", and an extra state 1 is added after the concept description of "round brush". Non-contextual entity description in $G_{sub}$ is added an extra state in this way. Then, we calculate the constraint score $S_{ij}$ between two nodes according to the additional state, and obtain the node feature $N_{f1}$ based on the neighbor nodes. The whole calculation process is as follows:

$$g = f_{sore}(f_{concat}([v_i^c; v_j^c])), \tag{2}$$

$$\overrightarrow{v_i} = f_{concat}(v_i), \tag{3}$$

$$S_{ij} = Att(\overrightarrow{v_i}, \overrightarrow{v_j}) \times e^{\pm g}, \tag{4}$$

$$N_{f1} = \Delta_{k=1}^{K} \sum_{j \in N_i} S_{ij} W_k^l h_j^l, \tag{5}$$

where $\overrightarrow{v_i}$ is the vector of $v_i$ passed through the language model. $Att$ represents the dot product attention mechanism: first splicing $(\overrightarrow{v_i}, \overrightarrow{v_j})$ two vectors for feature dimension increase, and then with the defined feature matrix-vector of the same dimension $\overrightarrow{t}$ to do dot product. $\Delta_{k=1}^{K} x_i$ represents the concatenation of features from $x_1$ to $x_k$. $W_k^l$ represents the weight matrix in the attention mechanism, and $h_j^l$ is the feature of the corresponding neighbor node. $N_i$ represents the neighbor node of $v_i$. $K$ represents the number of layers of attention. $g$ means the score between the concept descriptions of two nodes, if the state is the same, it is positive, otherwise it is negative.

**Optimization Weights.** $G_{sub}$ consists of different entities and relations. For question answering, different relationships should have different weights when updating node features [5]. Relationships between entities are often related to head and tail entities. In this subsection, we first get the edge weight score $S_{ht}$ by weighting the head and tail entities of the edge and the context entity. Specifically, the calculation method is as follows:

$$G_h = \sum_{j=1}^{n} Att(\overrightarrow{v_h}, \overrightarrow{v_i}), \tag{6}$$

$$G_t = \sum_{j=1}^{n} Att(\overrightarrow{v_t}, \overrightarrow{v_i}), \tag{7}$$

$$S_{ht} = \frac{G_h + G_t}{2 * num}, \tag{8}$$

where $G_h$ is the normalized attention score of the head node and context nodes. $G_t$ is the normalized attention score of tail node and context nodes. $num$ represents the number of context entities. $v_h$ and $v_t$ represent the head node and tail

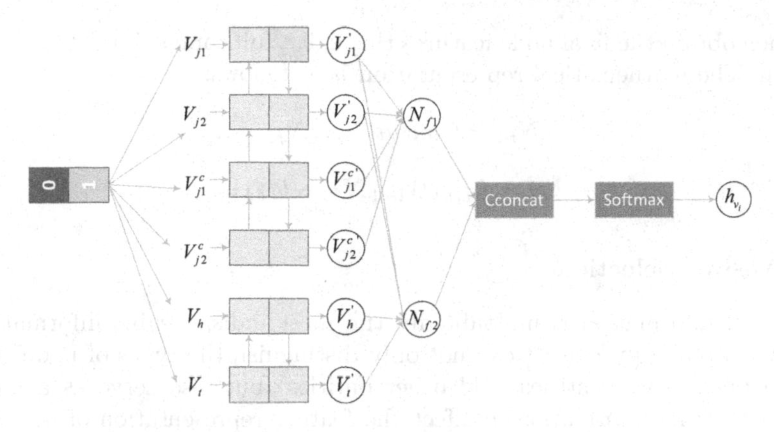


**Fig. 2.** The final feature representation diagram of nodes. $V_{j1}$ and $V_{j2}$ means the neighbor nodes of node $v_i$. Above all, our approach obtains the feature $N_{f1}$ between nodes based on the impact of the additional state on inter-node attention scores. Then, utilizing the influence of context on edge weights to obtain edge-based node features $N_{f2}$. Finally, the two parts are fused to get the final node feature $h_{v_i}$.

node of the edge, respectively. By computing the correlation between an edge (determined by its dependent head and tail nodes) and the context node, the weight score of the context node to the relation is obtained.

Then we get the attention scores of the relation and its dependent nodes and weight the two parts to obtain the final relation weight $G_R$. We finally obtain the node features $N_{f2}$ according to the relation weights. The formula is as follows:

$$S_{rel} = \sigma(f_{relu}(R_{ij}W_{m1} + b_{m1})W_{m2} + b_{m2}), \tag{9}$$

$$G_R = \gamma_1 \frac{\exp(S_{rel})}{\sum_{j=1}^{N_i} \exp(S_{rel})} + \gamma_2 S_{ht}, \tag{10}$$

$$N_{f2} = \Delta_{m=1}^{M} \sum_{j \in N_i} G_R W_m^l h_j^l, \tag{11}$$

where $\sigma$ represents sigmoid. $f_{relu}$ is the function representation of $relu$. $W_m^l$ is the transformation matrix parameter corresponding to the node. $m$ represents the number of attention layers. $R_{ij}$ represents the edge (relationship) between node $v_i$ and node $v_j$. $\gamma$ represents the feature parameter, assigning the proportion of the two-part score in the edge weight update, $\gamma \in (0, 1)$, $\gamma_1 + \gamma_2 = 1$.

### 3.3 Feature Fusion

As shown in Fig. 2, after the entity in the knowledge graph passes through the two-branch GNN network and performs a feature update, the final node feature $h_{v_i}$ can be calculated. The two parts of features are fused into a unified node feature, reducing the amount of parameters, so that updating the node features reasonably. To acquire the enhanced node features, we concatenate $N_{f1}$ and

$N_{f2}$, then obtain the final node features through a full connection and activation function. The mathematical representation is as follows:

$$N_{f3} = Concat[N_{f1}, N_{f2}], \tag{12}$$

$$h_{v_i} = f_{relu}(W_{l+1}N_{f3} + b_{l+1}). \tag{13}$$

### 3.4 Answer Selection

Contextual information is undoubtedly the most indispensable information in the QA reasoning process. It can not only distinguish the types of nodes (question entities, answer entities, and other entities) but also serve as a piece of important media information to affect the feature representation of nodes, and ultimately affect the correctness of QA reasoning. Some methods [23] apply context to the question-and-answer entities to continuously update the contextual representation. However, since we introduce the relevance score, each node in the subgraph is closely related to the context so that the context information can fully interact with other nodes. Therefore, we apply the context node to each node, thereby enhancing the features of each node, and enriching the representation of the context, making the information required for the final reasoning more comprehensive.

For a given natural language question and answer, we can aggregate information from the following three aspects and infer the probability $P(a|Q)$ that an option $a$ is the correct answer. Three aspects of information: (1) the contextual knowledge combined by the question and answer of the language model ($C^{LM}$), (2) a certain path pooled in the subgraph ($\mathsf{V}_{path}$), (3) the context obtained by aggregating the context features of all nodes on the path ($C^{GNN}$), at this time, the context information is fully integrated by the nodes on the local path, making the context information more abundant. The specific formula is as follows:

$$C^{LM} = f_{concat}(Con), \tag{14}$$

$$C^{GNN} = \tfrac{1}{t}\textstyle\sum_{i=1}^{t} V_i^f, \tag{15}$$

where $t$ means the number of nodes on the path, $V_i^f$ represents the contextual feature in each node on the path. The final probabilities are calculated as $P(a|Q) \propto \exp(MLP(C^{LM}, C^{GNN}, \mathsf{V}_{path}))$.

## 4 Experiment

### 4.1 Datasets and Knowledge Graph

**Datasets.** We evaluate our model FESR on three diverse datasets across two domains: CommonsenseQA (CSQA) [20], OpenBookQA (OBQA) [13], MedQA-USMLE [6]. CommonsenseQA is a question-answering dataset of 12,102 questions that require background commonsense knowledge beyond surface language understanding. OpenBookQA is a 4-way question-answering dataset that tests elementary scientific knowledge. It contains 5,957 questions along with an open book of scientific facts. We split the official 2018 data [14]. CommonsenseQA creates questions from ConceptNet entities and relations. OpenBookQA probes elementary science knowledge from a book of 1,326 facts. MedQA-USMLE, which consists of 12,723 questions, is a 4-way multiple-choice medical QA task. We split the original dataset using the method of [23].

**Table 1.** Performance comparison on CommonsenseQA in-house split. Experiments are controlled using same seed LM.

| Model | IHdev (%) | IHtest (%) |
|---|---|---|
| RoBERTa-large | 73.1 | 68.7 |
| + GconAttn [21] | 72.6 | 68.6 |
| + RGCN [2] | 72.7 | 68.4 |
| + KagNet [10] | 73.5 | 69.0 |
| + MHGRN [4] | 74.5 | 71.1 |
| + RN [17] | 74.6 | 69.1 |
| + QA-GNN [23] | 76.5 | 73.4 |
| + GREASELM [24] | 78.5 | 74.2 |
| **+ FESR (our)** | **79.2** | **76.1** |

**Knowledge Graph.** For our public domain QA datasets CommonsenseQA and OpenbookQA, we use a general domain knowledge graph ConceptNet [18] as our structured knowledge source to build a subgraph. It has 799,273 nodes and 2,487,810 edges in total. For MedQA-USMLE, we use a knowledge graph built by Yasunaga [23] in 2021. The knowledge graph contains 9958 nodes and 44561 edges. Entity descriptions are derived from the "original form" option in Wiktionary.

### 4.2 Baseline Methods

**Fine-Tuned LMs.** To explore the impact of knowledge graphs on question answering, we compare FESR with vanilla fine-tuned LMs without knowledge graphs. For the CommonsenseQA and OpenBookQA datasets, we use the language models RoBERTa [12] and AristoRoBERTa [3] for processing, respectively. For MedQA-USMLE, we adopt the state-of-the-art biomedical language model SapBERT [11].

**LM+KG Models.** We also evaluate FESR's ability to exploit its state constraints and edge weight optimization by comparing with existing LM+KG methods: GconAttn [21], KagNet [10], MHGRN [4] QA-GNN [23], Relation Network [17] RGCN [2], GREASELM [24], etc. Under the LM+KG method, the performance of GREASELM is the best above. The biggest difference between our method and these baseline models is that we not only consider the impact of negative words in computing node attention scores but also fully consider the impact of context on edge weights.

**Table 2.** Test accuracy comparison to public OpenBookQA model implementations (B = billion, M = million).

| Model | Test (%) | #Params |
|---|---|---|
| ALBERT+KB [10] | 81.0 | ~235M |
| HGN [22] | 81.4 | ≥361M |
| QA-GNN [23] | 82.8 | ~360M |
| T5 [16] | 83.2 | ~3B |
| GREASELM [24] | 84.8 | ~359M |
| UnifiedQA [8] | **87.2** | **~11B** |
| **FESR (our)** | 85.4 | ~360M |

### 4.3 Experimental Results

As shown in Table 1, FESR conducts experiments on the training set and test set of CommonsenseQA. FESR outperforms the fine-tuned language model by 7.4% and outperforms the state-of-the-art model GREASELM by 1.9%. Experimental results show that our method has some advantages when we use the same pre-trained model.

Table 2 shows that FESR is 4.4% higher than ALBERT+KB and 0.6% higher than the same type of method (GREASELM). Although there is a slight gap with the state-of-the-art model (UnifiedQA), our number of parameters is 1/30 of it. FESR is currently the best model with the same parameters.

Table 3 represents the performance of FESR on the medical dataset (MedQA-USMLE). Our model outperforms BioBERT-LARGE by 7.3% and outperforms GREASELM by 5.6%. The possible reason is that the external knowledge descriptions of entities in medical datasets are usually related to the answers. Moreover, in the medical knowledge graph, the correlation between edges and QA is easily demonstrated under the guidance of context. The attention scores between medical entities are more significantly affected by additional status words because the direction (directionality) between medical entities is obvious. So the improvement is obvious.

Table 4 demonstrates that when we use the same LM(AristRoBERTa), FESR achieves good results on OpenbookQA compared to other models. The boost over other methods suggests that FESR makes better use of KGs by optimizing edge weight and regulating inter-node attention scores using state constraints to perform joint reasoning than existing LM+KG methods. To some extent, it also proves the effectiveness of the method.

**Table 3.** Test Accuracy comparison on MedQA-USMLE.

| Model | Test (%) |
|---|---|
| BioBERT-base [9] | 34.1 |
| BERT-base [7] | 34.3 |
| BioBERT-LARGE [9] | 36.7 |
| SapBERT-Base [11] | 37.2 |
| QA-GNN [23] | 38.0 |
| GREASELM [24] | 38.5 |
| **FESR (ours)** | **44.0** |

**Table 4.** The results of the test on OpenBookQA. (The same seed LM).

| Model | Test (%) |
|---|---|
| AristoRoBERTa [3] | 78.4 |
| +RN [17] | 75.6 |
| +RGCN [2] | 74.6 |
| +QA-GNN [23] | 82.8 |
| +MHGRN [4] | 80.6 |
| +GREASELM [24] | 84.8 |
| **FESR (ours)** | **85.4** |

**Table 5.** Performance of FESR on the CommonsenseQA IH-dev set on negative questions.

| Model | Negation (%) |
|---|---|
| RoBERTa-large (w/o KG) | 63.8 |
| QA-GNN [23] | 66.2 |
| GREASELM [24] | 69.9 |
| **FESR (our)** | **71.1** |

Table 5 shows our FESR's ability to handle negative issues on CommonsenseQA. Obviously, FESR outperforms LM+KG models with the same architecture (QA-GNN and GREASELM) in dealing with negative problems, demonstrating its excellent structured reasoning ability. One of the most important reasons is that state constraints regulate the attention scores between nodes.

### 4.4 Ablation Experiment and Analysis

Table 6 and Table 7 summarize the ablation study conducted on the two parts of whether to introduce the two-branch GNN network, and the optimization of edge weights in GNN to verify our work effectiveness further, using the dev set of the CSQA.

**Table 6.** Ablation experiments of edge weights.

| Ablation Part | Dev-acc (%) |
|---|---|
| w/o edge weights | 76.5 |
| w/o context entity | 77.1 |
| **FESR** | **79.2** |

**Table 7.** Ablation experiment of two-branch GNN Network.

| Ablation Part | Dev-acc (%) | Neg-acc (%) |
|---|---|---|
| Only Feature Constraints | 77.9 | 68.7 |
| Only Optimize Weights | 77.4 | 67.9 |
| **FESR** | **79.2** | **71.1** |

**Optimization of Edge Weights.** The experimental results in Table 6 mean that when GNN updates node features without considering edge weights, the experimental results drop by 2.7%. When considering the edge weight, the performance of the model will be improved. We can observe that the model performance improves by 0.6% when using edge weights compared to not using edge weights. On this basis, we introduce the influence of context information on edge weights, and the model performance is improved by 2.1%, indicating that context can enhance node features by optimizing edge weights, thereby improving the accuracy of inference.

**Two-Branch GNN Network.** As shown in Table 7, when we only use feature constraints, the model performance drops by 1.3%, and for negative sentences, the model performance drops by 2.4%. When only edge weights are used to enhance node features, the model performance drops by 1.8%, and for negative sentences, the model performance drops by 3.2%. We believe that when the extra state is adopted, the LM model can handle negative sentences well. Then, after the subsequent feature fusion, the reasoning path of negative sentences is strengthened, and the accuracy of the model is improved. The enhancement of node features by edge weights also strengthens the reasoning of the model.

### 4.5 Case Analysis of Structured Reasoning

As shown in the first example in Table 8, when neighbor nodes are employed to enhance node features, if there is no feature constraint, that is, no state is added, the model tends to choose "bored". When feature constraints are used and negative words are in the context, the corresponding "you" status will be set to 0. Since the status of "dry book" is 1, the status word limits the attention score between nodes and then enhances node features through edge weights, causing the model to tend to the correct answer "interested". The second example is similar to the first example, but since the "dry book" and "you" state words are both 0, it has no effect on the attention score between nodes, making the model still biased towards the correct answer "bored".

This case illustrates the process of setting the state words to constrain the node features, strengthening the structured reasoning of the question-answering path, which indirectly demonstrates the effectiveness of our method.

**Table 8.** Example in Commonsense for case study. FESR correctly handles negative question answering.

| **CSQA Question(Negation ver1):** |
| --- |
| If you have to read a book that is very dry you may not become what? |
| **Some Options:** |
| A:interested B:bored |
| **CSQA Question(double negation):** |
| If you have to read a book that is not very dry you may not become what? |
| **Some Options:** |
| A:interested B:bored |

## 5 Conclusions

In this paper, we propose a method for Feature Enhanced Structured Reasoning (FESR) that exploits a two-branch graph neural network to improve the structured reasoning ability of question answering. In the two-branch graph neural network, one branch uses state features for attention score calculation between nodes, which strengthens the model's ability to handle negative questions and answers, and the other branch factors contextual knowledge into edge weights to optimize the edge calculation of weights. Extensive experimental results show that our method performs well on two public datasets and one domain-specific (medical) dataset compared to previous LM+KG and LM methods. In the future, we look forward to applying the proposed augmented structured reasoning network to some question answering related tasks.

**Acknowledgements.** This work is supported by grant from the National Natural Science Foundation of China (No. 62076048), the Science and Technology Innovation Foundation of Dalian (2020JJ26GX035).

## References


1. Bao, J., Duan, N., Yan, Z., Zhou, M., Zhao, T.: Constraint-based question answering with knowledge graph. In: Proceedings of COLING 2016, the 26th International Conference on Computational Linguistics: Technical Papers, pp. 2503–2514 (2016)
2. Chen, J., Hou, H., Gao, J., Ji, Y., Bai, T.: RGCN: recurrent graph convolutional networks for target-dependent sentiment analysis. In: Douligeris, C., Karagiannis, D., Apostolou, D. (eds.) KSEM 2019. LNCS (LNAI), vol. 11775, pp. 667–675. Springer, Cham (2019). https://doi.org/10.1007/978-3-030-29551-6_59
3. Clark, P., et al.: From 'f' to 'a' on the NY regents science exams: an overview of the aristo project. AI Mag. **41**(4), 39–53 (2020)
4. Feng, Y., Chen, X., Lin, B.Y., Wang, P., Yan, J., Ren, X.: Scalable multi-hop relational reasoning for knowledge-aware question answering. In: Proceedings of the 2020 Conference on Empirical Methods in Natural Language Processing (EMNLP), pp. 1295–1309 (2020)

5. Ishiwatari, T., Yasuda, Y., Miyazaki, T., Goto, J.: Relation-aware graph attention networks with relational position encodings for emotion recognition in conversations, pp. 7360–7370 (2020)
6. Jin, D., Pan, E., Oufattole, N., Weng, W.H., Fang, H., Szolovits, P.: What disease does this patient have? A large-scale open domain question answering dataset from medical exams. Appl. Sci. **11**(14), 6421 (2021)
7. Kenton, J.D.M.W.C., Toutanova, L.K.: BERT: pre-training of deep bidirectional transformers for language understanding. In: Proceedings of NAACL-HLT, pp. 4171–4186 (2019)
8. Khashabi, D., et al.: UnifiedQA: crossing format boundaries with a single QA system. In: Findings of the Association for Computational Linguistics: EMNLP 2020, pp. 1896–1907 (2020)
9. Lee, J., et al.: BioBERT: a pre-trained biomedical language representation model for biomedical text mining. Bioinformatics **36**(4), 1234–1240 (2020)
10. Lin, B.Y., Chen, X., Chen, J., Ren, X.: KagNet: knowledge-aware graph networks for commonsense reasoning. In: Proceedings of the 2019 Conference on Empirical Methods in Natural Language Processing and the 9th International Joint Conference on Natural Language Processing (EMNLP-IJCNLP), pp. 2829–2839 (2019)
11. Liu, F., Shareghi, E., Meng, Z., Basaldella, M., Collier, N.: Self-alignment pretraining for biomedical entity representations. In: Proceedings of the 2021 Conference of the North American Chapter of the Association for Computational Linguistics: Human Language Technologies, pp. 4228–4238 (2021)
12. Liu, Y., et al.: RoBERTa: a robustly optimized BERT pretraining approach. arXiv preprint arXiv:1907.11692 (2019)
13. Mihaylov, T., Clark, P., Khot, T., Sabharwal, A.: Can a suit of armor conduct electricity? A new dataset for open book question answering. In: Proceedings of the 2018 Conference on Empirical Methods in Natural Language Processing, pp. 2381–2391 (2018)
14. Mihaylov, T., Frank, A.: Knowledgeable reader: enhancing cloze-style reading comprehension with external commonsense knowledge. In: Proceedings of the 56th Annual Meeting of the Association for Computational Linguistics (Volume 1: Long Papers), pp. 821–832 (2018)
15. Petroni, F., et al.: Language models as knowledge bases? arXiv preprint arXiv:1909.01066 (2019)
16. Raffel, C., et al.: Exploring the limits of transfer learning with a unified text-to-text transformer. J. Mach. Learn. Res. **21**(1), 5485–5551 (2020)
17. Santoro, A., et al.: A simple neural network module for relational reasoning. In: Advances in Neural Information Processing Systems, vol. 30 (2017)
18. Speer, R., Chin, J., Havasi, C.C.: 5.5: an open multilingual graph of general knowledge. In: Proceedings of the Thirty-First AAAI Conference on Artificial Intelligence, pp. 4444–4451, December 2016
19. Sun, H., Dhingra, B., Zaheer, M., Mazaitis, K., Salakhutdinov, R., Cohen, W.W.: Open domain question answering using early fusion of knowledge bases and text (2018)
20. Talmor, A., Herzig, J., Lourie, N., Berant, J.: CommonsenseQA: a question answering challenge targeting commonsense knowledge. In: Proceedings of NAACL-HLT, pp. 4149–4158 (2019)
21. Wang, X., et al.: Improving natural language inference using external knowledge in the science questions domain. In: Proceedings of the AAAI Conference on Artificial Intelligence, vol. 33, pp. 7208–7215 (2019)

22. Yan, J., et al.: Learning contextualized knowledge structures for commonsense reasoning. In: Findings of the Association for Computational Linguistics: ACL-IJCNLP 2021, pp. 4038–4051 (2021)
23. Yasunaga, M., Ren, H., Bosselut, A., Liang, P., Leskovec, J.: QA-GNN: reasoning with language models and knowledge graphs for question answering. In: Proceedings of the 2021 Conference of the North American Chapter of the Association for Computational Linguistics: Human Language Technologies, pp. 535–546 (2021)
24. Zhang, X., et al.: GreaseLM: graph reasoning enhanced language models. In: International Conference on Learning Representations (2022)

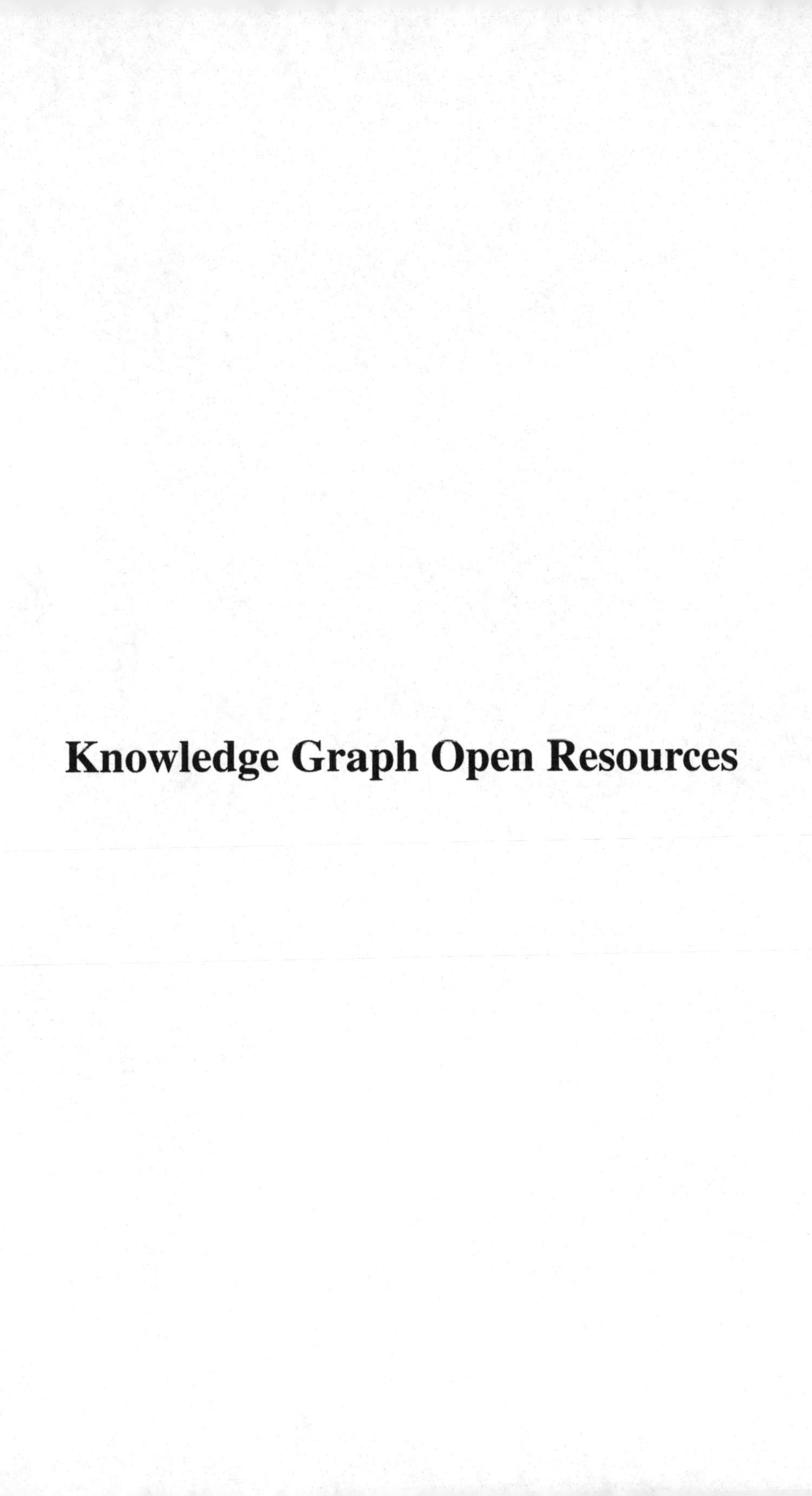

# Knowledge Graph Open Resources

# Conditional Knowledge Graph: Design, Dataset and a Preliminary Model

Yaojia Lv, Zihao Zheng, Ming Liu(✉), and Bing Qin

Research Center for Social Computing and Information Retrieval, Harbin Institute of Technology, Harbin, China
{yjlv,zhzheng,mliu,qinb}@ir.hit.edu.cn

**Abstract.** Facts are ***conditionally established*** in most cases. However, current Knowledge Graph (KG) techniques only focus on the modeling and representations of facts, neglecting the presence of conditions, which are necessary to establish the validity of facts. In this paper, we propose Conditional Knowledge Graph (***Conditional-KG***), which employs a three-layer hierarchical network to incorporate both facts and conditions. To facilitate research on the automatic construction of Conditional-KG, we manually annotate an innovative large-scale dataset named **HACISU**. Based on the Conditional-KG design and HACISU, we propose a simple construction model to benchmark HACISU. Experimental results show that our benchmark model outperforms several baselines but still has a considerable margin with human performance. We highlight the significance of HACISU, as it is the first carefully annotated dataset with conditional information. Our dataset is publicly available in http://101.200.120.155:5555/, hoping to serve as a challenging testbed and an ideal benchmark for Conditional-KG construction.

**Keywords:** Conditional Knowledge Graph · Knowledge representation · Open information extraction

## 1 Introduction

Conditions and facts are of equal gravity in providing knowledge. Without conditions, facts cannot be utilized concretely in the downstream tasks. However, current Knowledge Graphs (KGs) are simply constructed upon facts, resulting in what are known as **factual Knowledge Graphs** [1]. Mainstream Information Extraction (IE) techniques only extract facts from given text, while disregarding other attached information such as conditions [2]. Actually, facts are conditional rather than absolute. For instance, the fact "ginger syrup treats colds" only sticks under the condition "colds are wind-heat type". The condition here both acts as a fact and complements another fact. Inspired by this insight, we adopt a uniform structure, namely tuples, to depict conditions and facts separately.

Y. Lv and Z. Zheng — Equal contribution.


H. Wang et al. (Eds.): CCKS 2023, CCIS 1923, pp. 207–219, 2023.
https://doi.org/10.1007/978-981-99-7224-1_16

Note that KGs composed of facts without conditions provide limited service to knowledge-intensive tasks. In this scenario, KGs that incorporate conditions are desirable for comprehensive exploration. Many research [3,4] shows facts in the science domain are conditionally dependent, building upon which, we propose Conditional Knowledge Graph (***Conditional-KG***). However, the flat network structure of conventional KGs poses a challenge in involving both facts and conditions, necessitating a new representation. Meanwhile, previous research [3] have primarily focused on the scientific domain rather than the general domain. The relations between facts and their conditions have yet to be explicitly represented in the KGs, limiting practical applications. Therefore, we carefully design a hierarchical representation for the proposed Conditional-KG, with a specific framework in the same manner as an ordinary triple to involve conditions.

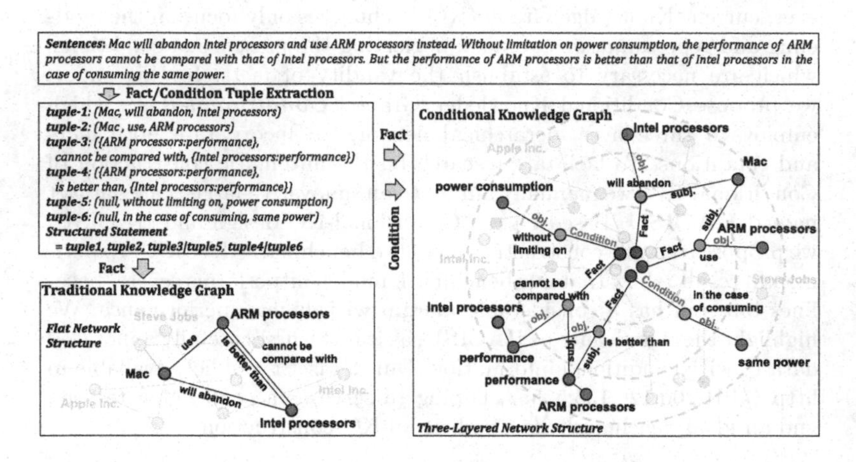


**Fig. 1.** An intuitive comparison between the traditional Knowledge Graph (left) and the Conditional Knowledge Graph (right). In the traditional KG, a severe conflict arises, as it is unclear whether the ARM processor is superior to the Intel processor. Yet the Conditional-KG can resolve the conflict by including the relevant condition.

We construct Conditional Knowledge Graph as three-layer networks, including ***outer layer***, ***middle layer*** and ***core layer*** (shown in Fig. 1). (1) Outer layer comprises ***entity nodes*** and ***attribute nodes***. Each entity node has multiple links to its attribute nodes. No links exist between any pair of entity nodes in this layer. Entity nodes are colored green and attribute nodes are colored red in Fig. 1. (2) Middle layer is composed of ***relation nodes***, where each relation serves as an individual node that connects the nodes in the outer layer as a subject or object (tagged 'subj.' and 'obj.'). The relation nodes are colored orange in Fig. 1. Using these links, we can obtain the information tuples (defined in Sect. 3) from the previous two layers. (3) Core layer is composed of Knowledge Units (defined in Sect. 3). It assigns a "fact" or "condition" tag to the information tuples by annotating the links to related nodes in the middle layer. Each

Knowledge Unit node, which is colored purple in Fig. 1, links one fact tuple and an arbitrary number of associated condition tuples.

We annotate a large-scale dataset to support the study of Conditional-KG, named HACISU, which stands for a Universal dataSet targetIng Conditional knowledge grApH. The nomination is a tribute to *Manchu*, because HACISU means "***condition***" in *Manchu*. HACISU comprises 5,141 sentences, 15,565 tuples and 8,451 conditional relation pairs. Each sentence is annotated with tuples and their conditional relations.

Based on the Conditional-KG framework and HACISU, we introduce the Conditional-KG construction task, which aims to extract tuples and their relevant conditional relations between tuples from a single sentence (as formally defined in Sect. 3). We propose a preliminary construction model to benchmark HACISU. Our model is inspired by [5] and comprises three modules: Predicate Extraction Module, Argument Extraction Module and Condition Discriminant Module. Experimental results demonstrate the effectiveness of our model.

## 2 Related Work

The limited verifiability and semantic integrity of factual knowledge hinder the extensive use and development of KGs in downstream tasks. Incorporating conditional information into Knowledge Graphs can enhance the verifiability of facts and improve semantic integrity. Prior research has recognized the significance of conditions. For instance, MinIE [6] supplements conditions to KGs by adding annotations to triples. Graphene [7] proposes adding a CONDITION field to triples to include conditions. NESTIE [8] employs a nested representation to model the relations between triples, including conditional relations. IMoJIE [9] utilizes a Ranking-Filtering framework to incorporate conditional annotations from multiple OPENIE systems. MacroIE [10] constructs a token span-based graph to extract factual and conditional triplets separately.

Prior works have proposed overly complicated structures to incorporate both facts and conditions, resulting in a significant disparity with the triple structure of current KGs. In contrast, our Conditional Knowledge Graph structure can seamlessly integrate with existing KGs. Additionally, previous methods rely on intricate rules to extract conditions, whereas we adopt an end-to-end model trained with our dataset, which is more efficient and generalizable.

## 3 Conditional Knowledge Graph

Informally, a ***Conditional Knowledge Graph*** (Conditional-KG) is a KG that incorporates both facts and their conditions in its network representation. To this end, we provide formal definitions of essential components of Conditional-KG.

**Definition 1 (Tuple).** Tuple is the primary unit of knowledge in Conditional-KG. An instanced tuple in a Conditional-KG is formally defined as follows:

$$t = (\{ent_1 : attr_1\}, p, \{ent_2 : attr_2\}) \quad (1)$$

where $t$ represents the instanced tuple, $\{ent_1 : attr_1\}$ and $\{ent_2 : attr_2\}$ denote the subject and object respectively, and $p$ is the predicate that indicates the relation between them. In this definition, $\{ent : attr\}$ represents an entity $ent$ and one of its attribute $attr$. We define $ent \in \{\text{null}\} \cup \mathcal{E}$ and $attr \in \{\text{null}\} \cup \mathcal{A}$, where $\mathcal{E}$, $\mathcal{A}$ are the sets of entities and attributes, respectively.

**Definition 2 (Conditional Relation).** We use the variable $r_{i,j}$ to denote a conditional relation. If tuple $t_i$ provides validation and supplementation to tuple $t_j$, there is a conditional relation from tuple $t_i$ to tuple $t_j$, and $r_{i,j}$ is assigned 1.

In traditional KG, a tuple can only refer to a fact. However, in Conditional-KG, we consider facts and conditions as **different roles** played by a tuple. Whether a tuple represents a fact or a condition depends on its role in the conditional relation. Given $r_{j,i} = 1$, $t_i$ is a ***fact*** and $t_j$ is a ***condition***.

**Definition 3 (Knowledge Unit).** A Knowledge Unit (KU) is an information carrier that contains a fact $t_i$ and all its conditions, providing complete knowledge about the fact. We formally define a KU as follows:

$$k_i = \{t_i, C_i\} \tag{2}$$

where $t_i$ represents the fact and $C_i = t_j | r_{ji} = 1$ represents the conditions for $t_i$.

**Definition 4 (KU-based Representation).** A sentence can be represented as a set of KUs, known as KU-based Representation (KU-Rep). We formally define KU-Rep as follows:

$$\mathit{KU\text{-}Rep} = \{k_1, k_2, k_3, ..., k_n\} \tag{3}$$

where $n$ represents the number of KUs in the sentence.

**Definition 5 (Conditional Knowledge Graph).** It organizes entities, attributes, tuples and Knowledge Units in a hierarchical manner. A Conditional-KG is formed as $G = \{L_1, L_2, L_3, E_{1,2}, E_{2,3}\}$, where $L_i$ denotes the $i$-th layer and $E_{i,j}$ denotes the connections between $L_i$ and $L_j$. Figure 1 is the visualization of the Conditional-KG.

- The outer layer is denoted by $L_1 = \{\mathcal{E}, \mathcal{A}, E_{\mathcal{E},\mathcal{A}}\}$. A link $(e, a) \in E_{\mathcal{E},\mathcal{A}}$ is created (tagged as "attr.") from the green entity node $e \in \mathcal{E}$ to the red attribute node $a \in \mathcal{A}$ if $a$ is an attribute of $e$.
- The middle layer is denoted by $L_2 = T$ where $T$ is the set of tuples. Each tuple is represented as an orange node tagged with its corresponding predicate. For a tuple $t_i$, $(subject, t_i) \in E_{1,2}$ is tagged as "subj." and $(object, t_i) \in E_{1,2}$ is tagged as "obj.", where *subject* and *object* are entity node or attribute node in $L_1$.
- The core layer is denoted by $L_3 = K$, where $K$ is the set of Knowledge Units. Each Knowledge Unit is represented as a purple node. For a Knowledge Unit $k_i$, $(t_i, k_i) \in E_{2,3}$ is tagged as "fact". $\{(t_j, k_i) \,|\, (t_j, k_i) \in E_{2,3}, r_{ji} = 1, i \neq j\}$ are tagged as "condition".

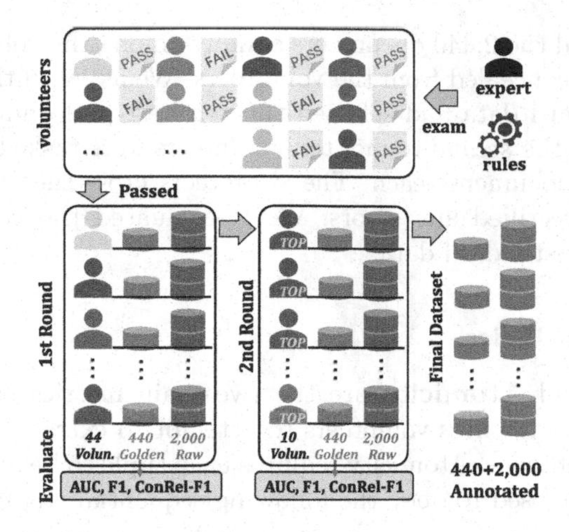


**Fig. 2.** The process of annotating Conditional Knowledge Graph datasets.

**Definition 6 (Conditional-KG Construction).** The construction of a large-scale Conditional-KG involves two essential steps: (1) extracting the KU-Rep of each sentence; and (2) merging the KU-Reps of all sentences to form a complete Conditional-KG. In this work, we consider the first step as the most critical step in Conditional-KG construction and treat it as equivalent to Conditional-KG construction. The second step is left for future works.

## 4 HACISU: a Universal Dataset

To facilitate the automatic construction of Conditional-KG, we annotate a large-scale dataset named HACISU. The annotated dataset aims to enable models to process general literature.

### 4.1 Data Annotation Process

Aiming at the annotation of conditional information in general literature, we selected ***AG News*** [11] due to its three advantages: (1) the data in AG News is presented in the form of titles and head sentences of news, containing an average of 2.1 sentences, 6.4 tuples, and 3.5 conditional relations. (2) The data in AG News has complete grammatical structures, with longer sentences containing compound structures and fewer informal language features. (3) AG News contains 127,600 pieces of training data and 7,600 pieces of test data, covering various domains.

We acquired 2,440 documents from AG News using stratified sampling, including 440 manually-annotated gold-standard documents. These documents served as expert ground truth to assess the annotating performance.

The annotation process is illustrated in Fig. 2, in which we employed a **two-round annotation strategy** to ensure the quality of the dataset. In the *first* round, we allocated the 2,440 documents equally among the volunteers, the location of which was concealed from the volunteers. We assessed their performance on the gold-standard data and selected the ten best-performing volunteers for the next round. In the *second* round, ten volunteers were tasked with annotating 44 gold-standard documents each. The volunteers cross-checked their previous annotations and rectified any errors. We re-evaluated the volunteers' performance on the gold-standard data.

### 4.2 Annotation Rules

**Completeness** and **Atomicity** are the two main metrics of the annotation rules. Completeness requires volunteers to attempt to extract all facts and conditions from the sentence. Atomicity requires each tuple to be an indivisible unit. Volunteers are supposed to obey the following sequential steps:

1. Identify all possible entities, attributes, and predicates in the given sentence to form fact or condition tuples (**Components Identification**).
2. Extract fact tuples for the given sentence using the obtained components, mainly from the main body of the sentence (**Facts Extraction**).
3. Extract condition tuples for the given sentence using the obtained components, which are mainly from the modified or conditional parts of the sentence (**Conditions Extraction**).
4. Recognize the conditional relations between fact tuples and condition tuples and annotate them for the given sentence (**Conditional relations Recognition**).

### 4.3 Analysis of Annotations

To assess the quality of the tuples, we employed the script proposed by [12], which utilizes two metrics: area under the curve (AUC) and F1-score (F1). AUC corresponds to the area under the receiver operating characteristic (ROC) curve, whereas F1-score is the harmonic mean of precision and recall values. We calculated the F1-score for the conditional relation (ConRel-F1) by comparing the volunteers' annotations with our gold-standard data. AUC and F1 provide a measure of the tuple annotation quality, while ConRel-F1 indicates the accuracy of conditional-relation labeling.

**Table 1.** Measure on the two annotation rounds.

| | AUC | F1 | ConRel-F1 |
|---|---|---|---|
| **Round 1.** | 73.4 | 87.6 | 89.0 |
| **Round 2.** | **77.9** | **94.3** | **96.7** |
| *Improvement* | 6.1%↑ | 7.6%↑ | 8.7%↑ |

**Table 2.** Comparisons of the three datasets used in the experiment.

| Dataset | #Sent. | #Tuples | #ConRel | Manual | Completeness |
|---|---|---|---|---|---|
| **OPENIE4** | 1,109,411 | 2,175,294 | ✗ | ✗ | ✗ |
| **CaRB** | 1,282 | 5,263 | ✗ | ✓ | ✗ |
| **HACISU** | 5,141 | 15,603 | 8,449 | ✓ | ✓ |

Table 1 presents the quality of the two annotation rounds. It reveals a significant improvement in accuracy in the second round, which reinforces the reliability of our dataset. These results demonstrate the effectiveness and necessity of adopting a two-round annotation strategy.

We conducted a comparative analysis of our new dataset with two popular OpenIE datasets in general domain. Table 2 provides an overview of the datasets. The OpenIE4 dataset is extracted by the OpenIE4 model [11]. The CaRB [12] is generated via human annotation based on the sentences in the OIE2016 [13] dataset. Notably, HACISU includes not only tuples but also conditional relations and aims for maximal information completeness.

## 5 Preliminary Construction Model

**Model Overview.** The input to our model is a sentence $W = (w_1, w_2, ..., w_n)$, where $w_i$ represents the $i$-th token, $n$ is the sequence length. Our model extracts a sentence's Conditional-KG in three steps. First, all predicates from the input sentence are identified using the Predicate Extraction (PE) module $f_{pred}$. Second, the Argument Extraction (AE) module $f_{arg}$ is used to extract entities and attributes associated with each identified predicate, which we refer to as **arguments** of the predicate [5]. Third, the Condition Discriminant (CD) module $f_{cond}$ is applied to extract the conditional relation between tuples. The PE and AE modules use the sequence labeling paradigm [14] to extract tuples by predicting a set of tags. The PE module predicts a predicate tagset $\hat{T}_{pred} = \{t_i | t_i \in \{P_B, P_I, P_O\}, i \in \{1, 2, 3, ...n\}\}$. The AE module extracts an argument tagset $Targ = \{t_i | t_i \in \{b_B, b_I, b_O\}, i \in \{1, 2, 3, ...n\}, b \in \{ent_1, attr_1, ent_2, attr_2\}\}$ based on $W$ and $\hat{T}_{pred}$. Finally, the CD module predicts the conditional relation $R$ between tuples based on $W$ and $\hat{T}_{pred}$. Our model maximizes the following log-likelihood formulation:

$$\sum_{i=1}^{n}(\log p(t_i^p | W; \theta_{pred}) + \sum_{i=1}^{n} \log p(t_i^a | W, \hat{T}_{pred}; \theta_{pred}, \theta_{arg})) + \sum_{i=1}^{n} \sum_{j=1, j \neq i}^{n} (\log p(r_{ij} | W, \hat{T}_{pred}; \theta_{pred}, \theta_{cond}))$$

where $\theta_{pred}$, $\theta_{arg}$ and $\theta_{cond}$ are trainable parameters from $f_{pred}$, $f_{arg}$ and $f_{cond}$ respectively. Note that $f_{pred}$ also contributes to the extraction of arguments and conditional relations, and the loss and gradients obtained from these two modules are propagated to $\theta_{pred}$.

### 5.1 Predicate Extraction Module

The input is a sentence tokenized by SentencePiece, denoted as $W = w_1, w_2...w_n$. The BERT model [15] encodes $W$ and the final hidden states $H \in R_{n \times d}$ are fed to a predicate classifier. The predicate classifier is composed of a position-wise feed-forward layer(two linear transformations surrounding a ReLU normalization) and a softmax layer that generates logits for each token being $\{P_B, P_I, P_O\}$. The predicted tagset $\hat{T}_{pred}$ is obtained by applying the argmax operation to the softmax outputs. Finally, the per-token cross-entropy loss $L_{pred}$ is calculated.

### 5.2 Argument Extraction Module

The Argument Extraction (AE) Module is designed to extract arguments for one predicate at a time, and the process is repeated for multiple predicates. To ensure stable training, we use the golden predicate tagset $T_{pred}$ instead of $\hat{T}_{pred}$.

The AE module consists of a Transformer Encoder [16] and an argument classifier with the same structure as the predicate classifier discussed in Sect. 5.1. The module takes $H$ and $E_P$ as inputs, where $H$ is the same as the last hidden states of BERT, as discussed in Sect. 5.1, and $E_P$ is a positional embedding of binary values that indicate whether each token is included in the predicate span [5]. The two features are concatenated to obtain $X \in \mathbb{R}^{n \times d_{mh}}$, where $d_{mh} = d + d_{pos}$, and $d_{pos}$ is the dimension of the position embedding $E_P$. Then, $X$ is divided into query and key-value pairs and fed to the Transformer Encoder, with $X$ serving as the query and subsets of $X$ derived from predicate positions serving as the key-value pairs. The output of the Transformer Encoder is then fed into the argument classifier. The process for obtaining a predicted argument tagset $\hat{T}_{arg}$ and the corresponding argument loss $L_{arg}$ is the same as that described in Sect. 5.1.

### 5.3 Condition Discriminant Module

The Conditional Discriminant (CD) module is designed to target one predicate pair $(p_i, p_j)$ at a time, and the process is repeated for multiple predicate pairs.

The CD module consists of a Transformer Encoder and a binary classifier. Similar to the AE module, we concatenate $H$ and $E_P$ to obtain $Y \in \mathbb{R}^{n \times d_{mh}}$, where $d_{mh} = d + d_{pos}$. $d_{pos}$ is the position embedding of ternary values that indicates whether each token is included in the first or the second predicate span. We then divide $Y$ into query and key-value pairs and feed them to the Transformer Encoder, with $Y$ serving as the query and subsets of $Y$ derived from predicate pair positions serving as the key-value pairs. The arithmetic mean vector of the Transformer Encoder's output is calculated and fed to the binary classifier. We obtain the predicted relation label $\hat{r_{ij}}$ using the argmax operation and calculate the corresponding cross-entropy loss, denoted as $L_{cond}$.

The final loss for parameter updating is the summation of $L_{pred}$, $L_{arg}$, and $L_{cond}$, and the training process follows a multi-task learning schema.

**Table 3.** The results of the 10-fold cross-validation of CaRB and HACISU. Statistics show that the consistency of HACISU is better than that of CaRB.

| Dataset | Measure | AUC | F1 | ConRel-F1 |
|---|---|---|---|---|
| **CaRB** | *mean* | 37.4 | 60.2 | - |
| | *variance* | 3.9 | 4.8 | - |
| **HACISU** | *mean* | 32.9 | 51.3 | 69.9 |
| | *variance* | **1.7** | **2.9** | **0.9** |

**Table 4.** The performance of our benchmark model and baselines on HACISU. The higher the score, the better the performance.

| Method | AUC | F1 | ConRel-F1 |
|---|---|---|---|
| Stanford | 11.0 | 18.8 | - |
| ClausIE | 15.2 | 27.1 | - |
| OpenIE4 | 17.5 | 31.5 | - |
| SpanOIE | 28.6 | 45.4 | 48.8 |
| BERT+BiLSTM | 29.7 | 46.2 | 51.3 |
| Multi$^2$OIE | 30.9 | 48.3 | 57.4 |
| **Our model** | **32.2** | **50.7** | **66.9** |
| **HUMAN performance** | **71.6** | **84.2** | **85.8** |

# 6 Experiments

## 6.1 Experimental Setup

**Dataset & Validation.** As discussed in Sect. 4, we use HACISU dataset for Conditional-KG construction. We split the dataset into training, validation, and test sets in a ratio of 6:1:3, based on the number of sentences. We repeat the experiment five times and report the average results on the test set.

**Evaluation Metrics.** We evaluate the performance of Conditional-KG construction models based on the extracted tuples and conditional relations. For tuple evaluation, we use AUC and F1 metrics mentioned in Sect. 4.3. For evaluating conditional relation extraction, we use ConRel-F1 mentioned in Sect. 4.3. The evaluation is conducted by comparing the model's extraction results with the golden data.

**Table 5.** The results of cross-domain experiments of our model under different categories. The higher the score, the better the performance.

| Source ⇒ Target | AUC | F1 | ConRel-F1 |
|---|---|---|---|
| THREE ⇒ World | 30.4 | 46.5 | 63.5 |
| THREE ⇒ Sports | 29.6 | 44.0 | 61.1 |
| THREE ⇒ Business | 29.9 | 46.7 | 62.9 |
| THREE ⇒ Sci/Tech | 30.3 | 45.3 | 62.3 |

**Implementation Details.** We select the best performing model on the validation set and evaluate it on the test set. The Transformer Encoder consists of six multi-head attention blocks with eight attention heads, and a 64-dimensional position-embedding layer. We train our model with the AdamW optimizer [17], using a base learning rate of 2e−5 and a batch size of 128. The dropout rate used in our experiments is 0.1.

### 6.2 Data Consistency

Inspired by cross-validation in model selection [18], we employ our preliminary model to conduct 10-fold cross-validation on HACISU and report the results in Table 3. We inspect the comparison with CaRB since it was manually annotated. We have three main observations:

***First***, the variance of F1 and AUC on HACISU is significantly smaller than that of CaRB, indicating that the consistency of HACISU is superior. ***Second***, the ConRel-F1 score on HACISU has a higher mean and a smaller variance, which demonstrates that our model can effectively identify the conditional relations between tuples. ***Third***, the F1 and AUC scores obtained on HACISU are slightly lower than those on CaRB. One possible explanation is that constructing a Conditional-KG is more challenging than constructing a general graph because it requires extracting more tuples, including both facts and conditions.

### 6.3 Comparison of Construction Models

**Baselines.** For unsupervised open IE methods, namely Stanford OpenIE [19], ClausIE [20], OPENIE4 [11], ConRel-F1 metrics are not applicable since they cannot generate conditional information between tuples. Conversely, supervised open IE models like SpanOIE [21], BERT+BiLSTM [15] and Multi$^2$OIE [5], can only extract facts and cannot detect conditional relations. Therefore, we resort to conditional annotations to train the Condition Discriminant Module, extract the facts, and then detect the conditional relations.

**Human Upper Bound.** For an estimate of human performance on HACISU, we recruit two volunteers who were not involved in the annotation process. They are asked to extract tuples and conditional relations from the test set. As shown in Table 4, the volunteers achieved results of 71.6% and 84.2% for AUC and F1, respectively, and 85.8% for ConRel-F1. These results serve as a reference for the upper bound of performance that can be achieved on our dataset.

**Results.** We report the performance of our model on the task of Conditional-KG construction in Table 4, with AUC, F1, and ConRel-F1 metrics. Our model outperforms the strongest baseline by 1.3%, 2.4% on AUC and F1, and achieves significant promotion on the ConRel-F1 measure by 9.5%. This can be attributed to two factors: (1) Our model effectively solves the entity overlap problem. (2) Our model employs a multi-task learning schema, which allows for mutual benefits between tuple extraction and conditional relation extraction. The performance gaps between our model and human performance are 39.4% and 33.5% in AUC and F1, and 18.9% in ConRel-F1, indicating significant room for improvement.

### 6.4 Generalization Analysis on HACISU

We are concerned about the generalization of HACISU, as it is declared as a universal dataset in general domain. To investigate whether a model trained on HACISU can perform well on unknown domain literature, we conduct experiments where the model is trained on three categories of literature and test on the remaining category. Results are reported in Table 5, with two findings observed.

***First***, the performance across different categories is fairly consistent, and comparable to the results obtained by our model in Table 4 where training and testing are done in the same domain. ***Second***, our model performs best on the World field and worst on the Sports field. One possible reason is that the data in the sports category differs from the other categories in the feature space. However, the performance differences between categories are not significant.

### 6.5 Case Study

We utilize our model to annotate sentences in OPENIE4 dataset, and obtain a Conditional-KG comprising of 1,023,602 fact tuples and 500,898 conditional tuples. Figure 3 illustrates one case constructed by the Conditional-KG. In this particular case, the fact “Intel is still winning” is constrained by the condition “in gaming and mainstream computing”. This example serves as evidence of the effectiveness of our proposed model in extracting tuples and conditional relations, which can be applied to other datasets and real-world scenarios.

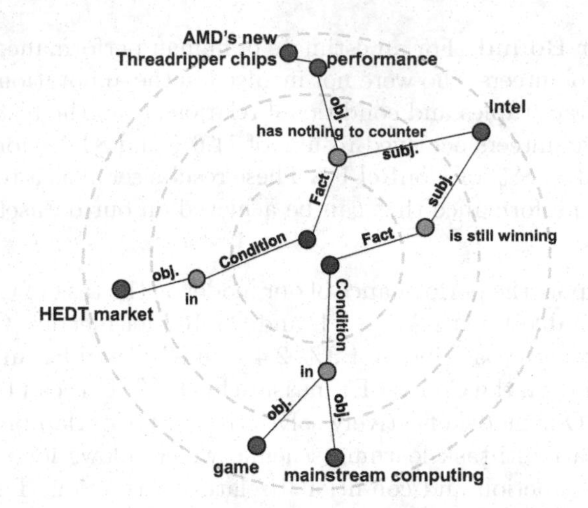


**Fig. 3.** A case constructed by the Conditional-KG.

# 7 Conclusion

The importance of conditions in knowledge representation cannot be overlooked. However, current KGs only focus on representing facts while ignoring their corresponding conditions. This paper proposes the Conditional-KG that takes into account both facts and conditions, which is a departure from existing KGs. To facilitate further research in this area, we manually annotate a large-scale dataset named HACISU, which is the first to annotate conditional information on general domain literature. We also propose a preliminary construction model trained on HACISU to demonstrate the feasibility of constructing a Conditional-KG. The results indicate that our model outperforms the strongest baselines, providing an improved approach to constructing KGs that captures both facts and conditions. Overall, the proposed Conditional-KG provides a promising direction for future research in knowledge representation and reasoning.

# References

1. Wu, F., Weld, D.S.: Open information extraction using wikipedia. In: Proceedings of ACL, pp. 118–127 (2010)
2. Ji, S., Pan, S., Cambria, E., Marttinen, P., Philip, S.Y.: A survey on knowledge graphs: Representation, acquisition, and applications. IEEE Trans. Neural Netw. Learn. Syst. **33**(2), 494–514 (2021)
3. Jiang, T., Zhao, T., Qin, B., Liu, T., Chawla, N., Jiang, M.: Multi-input multi-output sequence labeling for joint extraction of fact and condition tuples from scientific text. In: Proceedings of EMNLP-IJCNLP, pp. 302–312 (2019)
4. Jiang, T., Zhao, T., Qin, B., Liu, T., Chawla, N.V., Jiang, M.: The role of "condition" a novel scientific knowledge graph representation and construction model. In: Proceedings of ACM SIGKDD, pp. 1634–1642 (2019)

5. Ro, Y., Lee, Y.: Multi2OIE: multilingual open information extraction based on multi-head attention with BERT. In: Findings of EMNLP, pp. 1107–1117 (2020)
6. Gashteovski, K., Gemulla, R., del Corro, L.: MinIE: minimizing facts in open information extraction. In: Proceedings of EMNLP, pp. 2630–2640 (2017)
7. Cetto, M., Niklaus, C., Freitas, A., Handschuh, S.: Graphene: a context-preserving open information extraction system. In: Proceedings of COLING, pp. 94–98 (2018)
8. Bhutani, N., Jagadish, H.V., Radev, D.: Nested propositions in open information extraction. In: Proceedings of EMNLP, pp. 55–64 (2016)
9. Kolluru, K., Aggarwal, S., Rathore, V., Mausam, Chakrabarti, S.: IMoJIE: iterative memory-based joint open information extraction. In: Proceedings of ACL, pp. 5871–5886 (2020)
10. Yu, B., Wang, Y., Liu, T., Zhu, H., Sun, L., Wang, B.: Maximal clique based non-autoregressive open information extraction. In: Proceedings of EMNLP, pp. 9696–9706 (2021)
11. Mausam: Open information extraction systems and downstream applications. In: Proceedings of IJCAI, pp. 4074–4077 (2016)
12. Bhardwaj, S., Aggarwal, S., Mausam, M.: CaRB: a crowdsourced benchmark for open IE. In: Proceedings of EMNLP-IJCNLP, pp. 6262–6267 (2019)
13. Stanovsky, G., Dagan, I.: Creating a large benchmark for open information extraction. In: Proceedings of EMNLP, pp. 2300–2305 (2016)
14. Stanovsky, G., Michael, J., Zettlemoyer, L., Dagan, I.: Supervised open information extraction. In: Proceedings of NAACL, pp. 885–895 (2018)
15. Devlin, J., Chang, M.W., Lee, K., Toutanova, K.: BERT: pre-training of deep bidirectional transformers for language understanding. In: Proceedings of NAACL, pp. 4171–4186 (2019)
16. Vaswani, A., et al.: Attention is all you need. In: Proceedings of NeurIPS, pp. 5998–6008 (2017)
17. Loshchilov, I., Hutter, F.: Decoupled weight decay regularization. In: Proceedings of ICLR (2019)
18. Arlot, S., Celisse, A.: A survey of cross-validation procedures for model selection. Stat. surv. **4**, 40–79 (2010)
19. Angeli, G., Johnson Premkumar, M.J.: Leveraging linguistic structure for open domain information extraction. In: Proceedings of ACL, pp. 344–354 (2015)
20. Corro, L.D., Gemulla, R.: ClausIE: clause-based open information extraction. In: Proceedings of WWW '13, pp. 355–366 (2013)
21. Zhan, J., Zhao, H.: Span model for open information extraction on accurate corpus. In: Proceedings of AAAI, pp. 9523–9530 (2020)

# ODKG: An Official Document Knowledge Graph for the Effective Management

Bingjie Lu[1], Mingxin Lu[2], Yuyang Bian[3], Wenbo Zhou[4], Haowei Zhang[3], Gui Qiu[2], and Weizhuo Li[4](✉)

[1] Fintech Research Center, Zhejiang Lab, Hangzhou, China
lubj@zhejianglab.com

[2] Jiangsu Jinling Sci and Tech Group Co.,Ltd, Nanjing, China

[3] School of Computer Science, Nanjing University of Posts and Telecommunications, Nanjing, China

[4] School of Modern Posts, Nanjing University of Posts and Telecommunications, Nanjing, China
liweizhuo@amss.ac.cn

**Abstract.** Effective management of massive electronic documents is one of the hot topics for social services. There exist several knowledge bases of documents published for researchers to explore downstream applications such as PubLayNet and DocBank. Nevertheless, these datasets are mainly designed for document layout analysis and do not consider the linkages among documents. To improve this issue, in this paper, we present an official document knowledge graph, namely ODKG, which aims to collect the offical documents for effective management. We design a lightweight ontology of official documents. It can bring a well-defined schema of collected documents so that they could share more linkages with each other. We present the algorithms of element extraction, document archiving, and knowledge alignment during the process of ODKG construction, and further evaluate the corresponding algorithms based on our constructed datasets. Experimental results show that several algorithms can be competent to above tasks to some extent. Finally, we list three use cases of ODKG that are helpful for managers to improve the efficiency of their document management.

**Keywords:** Official Document · Knowledge Graph · Element Extraction · Document Archiving · Timeliness Verification · Association Discovery

**Resource type:** Dataset
**OpenKG Repository:** http://www.openkg.cn/dataset/odkg
**Permanent URL:** http://www.odkg.com.cn/odkg
**Github Repository:** https://github.com/chocoiii/ODKG

## 1 Introduction

Official documents are an important resource for social service. With the rapid growth of electronic documents, the traditional management of massive docu-


H. Wang et al. (Eds.): CCKS 2023, CCIS 1923, pp. 220–232, 2023.
https://doi.org/10.1007/978-981-99-7224-1_17

ments stored in the database is not enough to satisfy the various requirements such as document circulation, document approval, and auxiliary decision-making [1]. However, there exist few researchers that pay attention to intelligent processing and knowledge discovery tailored for official documents. Meanwhile, high-quality official documents with both labeled elements and textual information are still insufficient. Therefore, it is essential to construct an official document knowledge graph that freely provides for the research community so that more approaches will be further investigated and improved.

There exist several knowledge bases of documents that are constructed in recent years such as PubLayNet [2], DocBank [3]. On the other hand, several works focused on the relation extraction for governmental documents. Cui et al. [4] proposed a method based on distant supervision for element relation extraction, which combined ALBERT pre-training language model with a capsule network to extract the person names and position relationships in the official documents. Xu et al. [5] designed a model based on reorganizing and extracting the elements according to the structural logic of official documents. In these ways, the performances of element extraction have been improved to some extent.

Although these works are proposed for the intelligent processing of official documents, they still suffer from several limitations. For PubLayNet and DocBank, the main purposes of them are to provide a fair comparison for document layout analysis and other downstream applications, so they are not knowledge bases tailored for official documents. On the other hand, the data sets of the works that focus on document element extraction are not open source. Besides, the timeliness of these official documents is not considered.

To fill the above gaps, we dedicate a continuous effort to collect official documents from three domains and construct a official document knowledge graph, namely ODKG, for the research of effective documents management. Precisely, we design a lightweight ontology that brings a well-defined schema of collected documents, including 39 basic classes, 10 relations and 30 properties. It not only can make official documents from different sources share more linkages, but also can bring better services for effective management such as document circulation, document recommendation and so on. We design corresponding algorithms for element extraction and document archiving during the process of ODKG construction and generate lots of structured triples. Moreover, we utilizes knowledge alignment models to generate correspondences among values of documents from different documents and employ high-quality reasoner to verify the timeliness of offical documents so that more reliable linkages of documents can be shared. We evaluate the designed algorithms based on our constructed datasets and analyze their performances. Finally, three use cases are listed based on ODKG, which are helpful for users to improve the efficiency of document management.

## 2 Related Work

Zhong et al. [2] is one of the first groups that constructed the dataset called PubLayNet dataset for document layout analysis, which contains more than one million PDF articles that are publicly available on PubMed Central$^{TM}$. The authors demonstrated its value for downstream applications, including recognizing the layout of scientific articles by deep neural networks and their transferability on a different document domain by transfer learning.

Subsequently, Li et al. [3] developed a benchmark dataset that contains 500K document pages with fine-grained token-level annotations for document layout analysis, called DocBank. The authors constructed it in a simple way with weak supervision from the LATEX documents published on arXiv.com. Benefited from it, various proposed models from different modalities could be evaluated fairly, and more multi-modal approaches would be further investigated and optimized.

Zhang and Wu [6] collected 26,660 science and technology policies, and constructs the corresponding knowledge graph. The authors employed Bi-LSTM deep learning model to extract structured triple, and utilized the graph database to store knowledge and achieve graphical retrieval. This constructed method can enhance an alternative idea of KG for the science and technology policies domain.

In addition, several researchers focused on the relation extraction for governmental documents. Cui et al. [4] proposed a method based on distant supervision for entity relation extraction, which combined ALBERT pre-training language model with capsule network to extract the names of persons and position relationships in official documents. Experiments showed that it could effectively improve the performances of relation extraction with fewer labeled official documents. Xu et al. [5] designed a structure tree model to reorganize and extract the elements of official documents according to the structural logic of documents. The authors put forward a structured graph network to realize the extraction and management of these documents.

Nevertheless, the above research efforts still suffer from some limitations. PubLayNet and DocBank are not knowledge bases tailored for official documents. Their main purpose is to provide a fair comparison for document layout analysis and other downstream applications. Although other research works focus on the relation extraction of documents, their related data sets are not open source. In addition, the timeliness of these official documents is not considered. To the best of our knowledge, ODKG is the first open-source knowledge graph of official documents, in which the issue of timeliness among official documents is solved.

## 3 The Construction of Official Document Knowledge Graph

Figure 1 shows the framework of ODKG construction, which contains four layers, including the data layer, extraction layer, knowledge layer and application layer.

- **Data Layer**. The primary problem of knowledge graph construction is to collect massive suitable data. To achieve this goal, we combine real projects and focus on the domains of military, security and technology, whose documents are crawled from the Ministry of Industry and Information Technology,

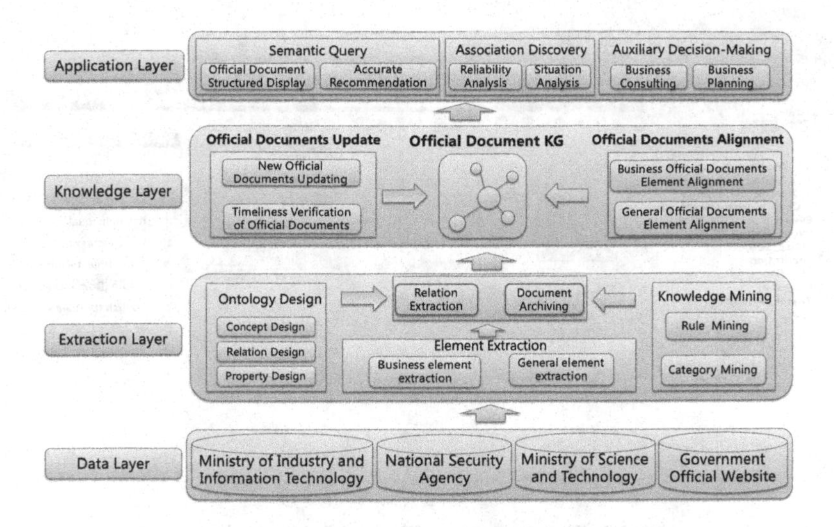


**Fig. 1.** The framework of ODKG construction.

the Nuclear Safety Administration, the Ministry of Science and Technology, and the Government Official Website.

- **Extraction Layer**. It mainly includes ontology design, knowledge mining and element extraction. The ontology is mainly based on the elements in view of different classifications for official documents. The task of knowledge mining is to generate fine-grained categories and rules. Element extraction is designed to identify the value of specified elements defined in official documents.
- **Knowledge Layer**. It is based on the methods of relation extraction and document archiving, which transforms the contents of document to structured triples. Further, the ODKG will continuously update and align the elements from new crawled official documents by the high-quality reasoner and alignments models.
- **Application Layer**. It mainly provides services through ODKG, including semantic query, association discovery, and auxiliary decision-making.

### 3.1 Ontology Design

To model a well-defined schema of official documents for effective document management, we focus on the domain of military, security, and technology because of the requirements of real projects. We discuss with related managers for document management and conclude two kinds of official documents, which are general official documents and business ones.

For general official documents, there exist fifteen sub-categories that cover twenty elements. Relatively, for business ones, they are divided into some special domains that focus on several elements in view of the event conceptual, including subject, object, location, finished time, purpose/task, and origin. Therefore, we choose to conceptualize these categories and elements, and further define a set

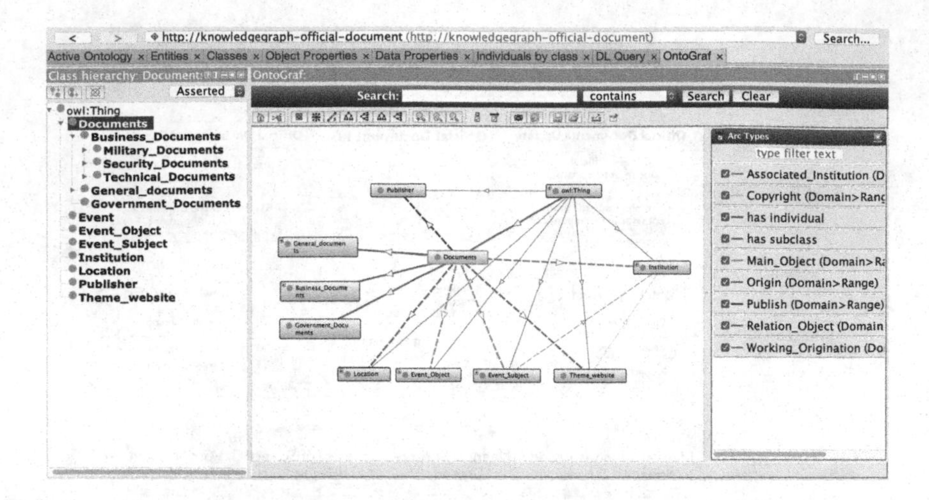


**Fig. 2.** The overview of lightweight ontology

of relations and properties by protégé[1] to build the relationships among them, which can enrich instantiate extracted triples for our knowledge graph.

Figure 2 shows an overview of our lightweight ontology, where blue ones represent *subclassof* that are basic relations. Overall, we define 39 basic concepts, 10 relations and 30 properties in the ontology. Benefited from a well-defined schema, It not only can make official documents from different sources share more linkages, but also bring better services for effective management such as document archiving, document recommendation, auxiliary decision-making.

### 3.2 Documents Crawling

With the help of scrapy framework, we mainly crawl the descriptive information of documents that have been published on official websites, such as the Ministry of Industry and Information Technology[2], Nuclear Safety Administration[3], Ministry of Science and Technology[4], and Government Official Website[5]

Finally, we crawled 1335 official documents from the above origins and divide them into several sub-categories. Table 1 lists the statistics of these official documents.

### 3.3 Element Extraction

As most official documents are unstructured texts, it is hard to improve the efficiency of document management and satisfy the various requirements. Therefore,

[1] https://protege.stanford.edu/.
[2] https://www.miit.gov.cn/.
[3] https://nnsa.mee.gov.cn/.
[4] https://www.most.gov.cn/.
[5] http://www.gov.cn/.

**Table 1.** The statistics of offical documents

| Categories | Sub-categories | Number |
|---|---|---|
| Military | Construction Approval Documents | 116 |
| | Battle Rewards Documents | 59 |
| | Personnel Appointment Documents | 9 |
| | Conscription Notice Documents | 7 |
| Security | Environmental Protection Documents | 397 |
| | Cyber Security Documents | 35 |
| | Quality Supervision Documents | 28 |
| Technology | Innovation Reform Documents | 198 |
| | New Energy Documents | 40 |
| | Technology Exploration Documents | 32 |
| Government Document | — | 414 |

we try to extract important elements and their value in official documents from the business perspective and general one.

For business elements, we select 6 elements (i.e., subject, object, location, finished time, purpose/task, origin) according to the designed ontology. Relatively, for general elements, we mainly consider 9 elements from 20 defined properties because more than half of the values of elements usually do not appear in our crawled documents, the final elements are title, issue-agency, document-number, index-number, finished-date, published-date, topic-words, website-type, general type.

We mainly consider the following strategies to parse official documents and obtain the values of elements in official documents as soon as possible.

- **Template-based extraction.** It is one of the effective methods to obtain huge values of general elements and some business ones, Moreover, we employ TextCNN [7] for sentence clusters among official documents, so that the number of manually defined templates could be reduced.
- **Named entity recognition.** For the defined elements (e.g., *Subject*, *Object*, *Purpose*) in our designed ontology, we evaluate several methods of named entity recognition [8] and select AutoNER [9,10] to capture the business elements of documents, which is a robust positive-only distant training method for phrase quality estimation to minimize the human efforts.
- **Relation extraction.** For several important elements (e.g., *issue-agency*, *published-date*) in our designed ontology, we employ promising relation extraction models from DeepKE [11] and select the best model according to our labeled corpus to achieve our goal.

Finally, we obtained a total of 3394 business elements and 4820 general ones from crawled official documents.

### 3.4 Document Archiving

Document archiving is the one of key functions of official document management. As shown in Table 1, we need to archive the crawled documents to one of the defined sub-categories. To reduce the heavy burden of managers, we design an automatic archiving method based on a pre-training model tailored for Chinese texts[6], which mainly includes two steps:

- **Feature Selection.** Except for the main body of official documents, we select some of the general elements usually asserted in documents and refine them to a set of features. In this way, the performances of methods could be improved to some extent.
- **Feature Learning.** We feed labeled corpus on the pre-training model BERT [12] to learn the feature presentation of documents and optimize it for the task of Chinese text classification.

Finally, we train a model to achieve document archiving, whose average accuracy is more than 90% (see Sect. 4.1).

### 3.5 Knowledge Alignment

As the topics and sources of official documents are different, so some extracted values with the same semantics are heterogeneous. Therefore, it is essential to find the correspondences among these elements, which can share more linkages and enhance the performances of some applications for document management. To achieve this goal, we try to employ the following approaches for knowledge alignment.

- **Rule miner method** [13]. It is a semi-supervised learning algorithm to iteratively refine matching rules and discover new matches of high confidence based on these rules.
- **Knowledge graph embedding-based method** [14]. It encodes entities and relations of the knowledge graph in a continuous embedding space and measures entity similarities based on the learned embeddings.

Based on the above models of knowledge alignment, we finally obtain the alignments among extracted values of elements (e.g., subject) and build a synonym dictionary to maintain these alignments.

### 3.6 Knowledge Storage and Update

After we utilized the extracted values to instantiate the properties based on our designed ontology, we transform them into structured triples $\{(h, r, t)\}$ with a specified URL by Jena[7]. For knowledge storage, we employ Neo4j[8] to store

[6] https://github.com/NLPScott/bert-Chinese-classification-task.
[7] http://jena.apache.org/.
[8] https://neo4j.com/.

the transformed triples, which is one of the efficient graph bases for storing the RDF triples and provides one convenience query language called Cypher. After we finished the normalization of the value of elements by our built synonym dictionary, we obtain 7332 entities and save 16133 triples in Neo4j.

To keep ODKG in sync with the websites, we try to periodically crawl the updated official documents and record the updated logs. Moreover, we employ the high-performance reasoning engine Drools[9] to verify the timeliness of official documents. Correspondingly, two rules are defined as follows.

$$\text{R1: } (x, \text{state}, \textit{invalid}) : -(x, \text{FinishDate}, t_1) \wedge (x, \text{CurrentDate}, t_2) \wedge (t_1, <, t_2)$$

$$\text{R2: } (x, \text{state}, \textit{invalid}) : -(x, \text{RelatedDocName}, n) \wedge (y, \text{RelatedDocName}, n)$$
$$\wedge ((y, \text{TimelessKeyWords}, \text{``}\textit{abolish}\text{''}) \vee (y, \text{TimelessKeyWords}, \text{``}\textit{cease}\text{''})$$

The first rule is verifying the timeliness of official documents by their finished date. If the current date is less than the finished date, then this document is invalid. The second rule is abolishing the old official document when a new one with the same name is published with some trigger words (e.g., "abolish", "cease").

# 4 Evaluation and Application

## 4.1 Evaluation

In this section, we mainly evaluate techniques of element extraction, document archiving, and document recommendation that are utilized to obtain abundant structured triples of documents and enrich ODKG. The evaluation is conducted on a desktop computer with an Intel(R) CoreTM i9-12900KF (3.4 GHz) which has 64GB memory and RTX 3080 GPU.

To verify the effectiveness of above techniques, we randomly divide the labeled official documents into the training set and test set, in which there are 1054 official documents in the training set (79.0%), and the number of test data set is 281 (21.0%).

**Evaluation of Element Extraction.** For the element extraction, we employ precision, recall, and F1-measure for evaluation in the following tables. Table 2 and Table 3 list the results of business element extraction and general element extraction. From these two tables, we can observe that:

- Our designed method can identify the value of business elements existing in the texts of documents. The average precision of this task is 89.3%, the recall is 65.6%, and the F1-measure is 73.3%. Nevertheless, the extracted results of "**Object**" and "**Purpose**" are not well because there are more than one values for them in the documents that need to be extracted.

[9] https://www.drools.org/.

- The performances of general elements are better than the ones of business elements. We notice that there exist some semi-structural lists asserted in documents, which can complement the lacked values of general elements to some extent. Nevertheless, the performance of "**Published Date**" for the general element extraction is not satisfactory, the main reason is that the boundary recognition between "**Published Date**" and "**Finished Date**" is not clear, so it is easy to be confused during the general element extraction.

**Table 2.** The results of business element extraction

| Element Type | ♯Labeled | ♯Extracted | ♯Correct | Precision | Recall | F1-measure |
|---|---|---|---|---|---|---|
| Subject | 1280 | 1056 | 998 | 0.935 | 0.771 | 0.845 |
| Object | 3631 | 2481 | 2076 | 0.836 | 0.571 | 0.679 |
| Purpose | 1475 | 921 | 858 | 0.931 | 0.581 | 0.716 |
| Finished Time | 914 | 755 | 725 | 0.960 | 0.793 | 0.868 |
| Location | 203 | 255 | 197 | 0.875 | 0.970 | 0.920 |
| Origin | 199 | 255 | 198 | 0.880 | 0.995 | 0.934 |
| Total | 7702 | 5663 | 5042 | 0.893 | 0.656 | 0.733 |

**Table 3.** The results of general element extraction

| Element Type | ♯Labeled | ♯Extracted | ♯Correct | Precision | Recall | F1-measure |
|---|---|---|---|---|---|---|
| Title | 401 | 401 | 381 | 0.950 | 0.950 | 0.950 |
| Issue Agency | 401 | 389 | 373 | 0.958 | 0.930 | 0.944 |
| Document Number | 401 | 385 | 384 | 0.997 | 0.957 | 0.977 |
| Index Number | 205 | 205 | 205 | 1.000 | 1.000 | 1.000 |
| Finished Date | 187 | 271 | 102 | 0.376 | 0.545 | 0.445 |
| Published Date | 401 | 102 | 102 | 1.000 | 0.254 | 0.406 |
| Topic Word | 61 | 61 | 61 | 1.000 | 1.000 | 1.000 |
| Website Type | 225 | 225 | 225 | 1.000 | 1.000 | 1.000 |
| General Type | 401 | 294 | 277 | 0.942 | 0.690 | 0.797 |
| Total | 2683 | 2333 | 2110 | 0.904 | 0.752 | 0.827 |

**Evaluation of Document Archiving.** The results of document archiving are listed in Table 4. Overall, the average accuracy of this task for 10 sub-categories is 90.0%. We observe that the results of several sub-categories of "**Battle Rewards**", "**Conscription Notice**", "**Personnel Appointment**", "**Technology Explo-ration**" are satisfied because of their clear boundary. On the contrary, the performances of "**Quality Supervision** ", "**CyberSecurity**" and "**New Energy** " are not well. We analyze that the boundary of these categories are vague, so it is easy to be confused for document archiving. It will be optimized in our future work.

**Table 4.** The results of document archiving about our method

| Categories | Sub-categories | Test Number | Correct Number | Accuracy |
|---|---|---|---|---|
| Military | Construction Approval | 49 | 49 | 0.857 |
| | Battle Rewards | 17 | 17 | 1.000 |
| | Personnel Appointment | 3 | 3 | 1.000 |
| | Conscription Notice | 2 | 2 | 1.000 |
| Security | Environmental Protection | 121 | 120 | 0.992 |
| | Cyber Security | 7 | 4 | 0.571 |
| | Quality Supervision | 9 | 5 | 0.556 |
| Technology | Innovation Reform | 58 | 50 | 0.862 |
| | New Energy | 14 | 9 | 0.643 |
| | Technology Exploration | 1 | 1 | 1.000 |
| Total Number | – | 281 | 253 | 0.900 |

**Evaluation of Document Recommendation.** To further evaluate the performances of document recommendation, we employ several classic unsupervised methods, including TF-IDF, DeepWalk [15], LINE [16], Node2Vec [17]. In this task, we introduce two metrics from the field of information retrieval to evaluate that are formally defined as follows.

$$\text{Top@n} = \frac{1}{n}\sum_{i=1}^{n} f(R_i) = R_i \in S?1 : 0 \qquad \text{Rank}_{min} = \frac{1}{n}\sum_{i=1}^{n} \arg\min_{l} Rank_{il}.$$

The first metric is Top@n which denotes the proportion of correct documents ranked in the top n. If the ith similar document belongs to the standard set $S$, then $f(R_i) = 1$. Otherwise, the value is 0. The second metric, written by $\text{Rank}_{min}$, is defined as the minimum rank of similar documents in descending order for each given document. The larger Top@n is, the closer of the similar search list is to the ideal one. Relatively, the smaller $\text{Rank}_{min}$ is, the earlier people can see similar documents.

Table 5 lists the recommended results by several classic unsupervised methods with two existing strategies [18]. The main differences between these two strategies are whether to integrate the value of extracted elements. For one document in the test set, we need to obtain the vector representations of text and elements, receptively. For the vector representation of text, we employ the arithmetic mean of the learned vectors of entities in the text, denoted by $V_{text}$. Relatively, for the vector representations of extracted elements, we concatenate the vectors together, denoted by $V_E$.

$$V_{text} = \frac{1}{l}\sum_{i=1}^{l} V_{ent}^{i}, \qquad V_E = [V_{ele}^{1}\oplus, ..., \oplus V_{ele}^{m}],$$

where $V_{ent}^{i}$ and $V_{ele}^{m}$ are the vector representations of entities and extracted values in documents, $\oplus$ is a concatenate operation $\mathbb{R}^{a \times d} \oplus \mathbb{R}^{b \times d} \rightarrow \mathbb{R}^{(a+b) \times d}$, $l$ is the number of entities in the new document and $m$ is the number of extracted elements. Notice that, some extracted values for elements are empty, we set these values to zero vectors with the same dimension. For the strategy with text and extracted elements, we still need to concatenate $V_{text}$ and $V_E$ together, denoted by $[V_{text} \oplus V_E]$.

Table 5 lists comparison results in terms of Top@n and $\text{Rank}_{min}$. From the table, we can observe that:

- For the training set with single texts, the results of TF-IDF outperform the ones of LINE and Node2Vec. We discover that the constructed network on the segmented tokens is not enough to obtain fine-grained correlations among documents.
- For the training set with texts and extracted elements, the performances of TF-IDF are better than the original ones. Benefited from extracted elements, more than half of the performances of NE-based models are improved because more values of extracted elements can be shared in the constructed networks.
- Overall, the results of NE-based models are better than ones of TF-IDF with help of extracted elements, but their performances are not stable. We analyze that the ratio of extracted values of elements and the designed concatenation strategy may restrict the performances of NE-based models. It makes sense to optimize these two aspects for document recommendation.

**Table 5.** The results of document recommendation with two strategies

| Categories | Training Set | Test Set | $\text{Rank}_{min}$ | | | | Top 10 | | | |
|---|---|---|---|---|---|---|---|---|---|---|
| | | | TF-IDF | DeepWalk | LINE | Node2Vec | TF-IDF | DeepWalk | LINE | Node2Vec |
| Government Doc | 414 Texts | 339 | 7.60 | 8.28 | 7.76 | 8.13 | 1.40 | 1.57 | 1.60 | 1.41 |
| Environmental Doc | 393 Texts | 313 | 9.88 | 9.92 | 9.78 | 9.96 | 1.00 | 1.01 | 1.01 | 1.00 |
| Innovation Reform Doc | 198 Texts | 155 | 6.05 | 5.35 | 5.44 | 5.77 | 1.62 | 2.47 | 2.72 | 2.19 |
| Construction Approval Doc | 116 Texts | 92 | 9.25 | 9.63 | 8.75 | 9.46 | 1.21 | 1.13 | 1.54 | 1.23 |
| Battle Rewards Doc | 59 Texts | 48 | 9.36 | 9.36 | 8.91 | 9.45 | 1.81 | 1.00 | 1.00 | 1.00 |
| New Energy Doc | 40 Texts | 30 | 5.20 | 2.90 | 3.80 | 3.40 | 1.90 | 4.80 | 3.40 | 2.90 |
| CyberSecurity Doc | 35 Texts | 29 | 1.67 | 1.83 | 1.33 | 1.67 | 7.16 | 6.00 | 7.33 | 6.00 |
| Technology Exploration Doc | 29 Texts | 24 | 5.20 | 4.40 | 3.00 | 4.00 | 1.00 | 2.20 | 1.60 | 1.20 |
| Quality Supervision Doc | 28 Texts | 22 | 0.67 | 1.17 | 0.17 | 0.83 | 50.50 | 17.60 | 33.00 | 26.17 |
| Government Doc. | 414 Texts +Extracted Elements | 339 | 7.72 | 7.67 | 7.80 | 7.73 | 1.43 | 1.78 | 1.76 | 1.93 |
| Environmental Doc. | 393 Texts + Extracted Elements | 313 | 9.87 | 9.94 | 10.00 | 9.93 | 1.00 | 1.00 | 1.00 | 1.00 |
| Innovation Reform Doc. | 198 Texts + Extracted Elements | 155 | 6.13 | 4.76 | 5.12 | 4.93 | 1.60 | 2.72 | 1.93 | 2.00 |
| Construction Approval Doc. | 116 Texts + Extracted Elements | 92 | 9.45 | 9.83 | 9.88 | 9.83 | 1.13 | 1.04 | 1.00 | 1.00 |
| Battle Rewards Doc. | 59 Texts + Extracted Elements | 48 | 9.45 | 9.72 | 8.36 | 9.27 | 1.09 | 1.00 | 1.00 | 1.00 |
| New Energy Doc. | 40 Texts + Extracted Elements | 30 | 5.50 | 2.80 | 3.60 | 3.70 | 1.70 | 6.50 | 5.60 | 6.30 |
| CyberSecurity Doc. | 35 Texts + Extracted Elements | 29 | 1.33 | 2.16 | 1.67 | 2.00 | 8.50 | 4.33 | 5.33 | 4.17 |
| Technology Exploration Doc. | 29 Texts + Extracted Elements | 24 | 5.20 | 5.80 | 6.00 | 7.20 | 1.20 | 1.00 | 1.00 | 1.00 |
| Quality Supervision Doc. | 28 Texts + Extracted Elements | 22 | 0.67 | 0.17 | 0.33 | 0.17 | 45.33 | 51.83 | 33.50 | 45.50 |

## 4.2 Application

We list three use cases that benefited from ODKG (shown in Fig. 1) for effective management.

- **Semantic Retrieval**. It can achieve semantic retrieval. For example, if users query one document, ODKG can display its comprehensive structured information in the form of knowledge cards. Benefited from this service, managers can quickly understand the key contents of documents.
- **Association Discovery**. It can recommend some similar documents for managers, which is helpful for them to further evaluate the reliability of other news and analyze current situation.
- **Auxiliary Decision-Making.** Its comprehensive information and above services can assist managers to make better decisions. Nevertheless, it still depends on the scale of ODKG and the ratio of extracted values of defined elements.

## 5 Conclusion

In this paper, we presented an official document knowledge graph, namely ODKG. Our work is to collect documents from main domains (e.g., military, security, technology). According to our designed lightweight ontology, lots of structured triples are obtained by our designed methods of element extraction and document archiving. We further employ knowledge alignment models to generate correspondences among values from different documents so as to they are able to share more linkages. The result is a high-quality official document dataset, which is helpful to enhance the efficiency of document management for users and provides an open data resource to researchers for further investigating and optimizing the related methods for its potential services.

In the future work, our plan is to broaden the official documents, so that ODKG becomes more comprehensive and covers more topics. Besides, we try to explore promising algorithms for the use-cases so as to provide better services for managers.

**Acknowledgements.** This work was supported by the Natural Science Foundation of China (62006125), the Foundation of Jiangsu Provincial Double-Innovation Doctor Program (JSSCBS20210532), the NUPTSF (NY220171) and Key Research Project of Zhejiang Lab (2022NF0AC01).

## References

1. Zhao, H., Wang, F., Wang, X., Zhang, W., Yang, J.: Research on construction and application of a knowledge discovery system based on intelligent processing of large-scale governmental documents. J. China Soc. Sci. Tech. Inf. **37**(8), 805–812 (2018)
2. Zhong, X., Tang, J., Yepes, A.J.: Publaynet: largest dataset ever for document layout analysis. In: Proceedings of the 2019 International Conference on Document Analysis and Recognition, pages 1015–1022. IEEE (2019)
3. Li, M.: DocBank: a benchmark dataset for document layout analysis. In Proceedings of the 28th International Conference on Computational Linguistics, pp 949–960. International Committee on Computational Linguistics (2020)

4. Cui, C., Shi, Y., Yuan, B., Li, Y., Li, Y., Zhou, C.: Research on relation extraction method for government documents. Comput. Technol. Dev. **31**(12), 26–32 (2021)
5. Ruilin, X., Geng, B., Liu, S.: Research on structural knowledge extraction and organization formulti-modal governmental documents. Syst. Eng. Electron. **44**(7), 2241–2250 (2022)
6. Zhang, Yu., Jun, W.: Research on the construction of science and technology policy knowledge graph. Digital Libr. Forum **8**, 31–38 (2021)
7. Xu, J., et al.: Short text clustering via convolutional neural networks. In: Proceedings of the 1st Workshop on Vector Space Modeling for Natural Language Processing, pp. 62–69 (2015)
8. Li, J., Sun, A., Han, J., Li, C.: A survey on deep learning for named entity recognition. IEEE Trans. Knowl. Data Eng. **34**(1), 50–70 (2022)
9. Shang, J., Liu, L., Ren, X., Gu, X., Ren,T., Han, J.: Learning named entity tagger using domain-specific dictionary. In: Proceedings of the 2018 Conference on Empirical Methods in Natural Language Processing, pp. 2054–2064. Association for Computational Linguistics (2018)
10. Shang, J., Liu, J., Jiang, M., Ren, X., Voss, C.R., Han, J.: Automated phrase mining from massive text corpora. IEEE Trans. Knowl. Data Eng. **30**(10), 1825–1837 (2018)
11. Zhang, N.: DeepKE: a deep learning based knowledge extraction toolkit for knowledge base population. In: Proceedings of the 2022 Conference on Empirical Methods in Natural Language Processing, pp. 98–108. Association for Computational Linguistics (2022)
12. Devlin, J., Chang, M.-W., Lee, K., Toutanova, K.: BERT: pre-training of Deep Bidirectional Transformers for Language Understanding. In Proceedings of the 2019 Conference of the North American Chapter of the Association for Computational Linguistics: Human Language Technologies, pp. 4171–4186. Association for Computational Linguistics (2019)
13. Niu, X., Rong, S., Wang, H., Yu, Y.: An effective rule miner for instance matching in a web of data. In: Proceedings of the 21st ACM international conference on Information and knowledge management, pp. 1085–1094. ACM (2012)
14. Sun, Z., et al.: A benchmarking study of embedding-based entity alignment for knowledge graphs. Proc. VLDB Endow. **13**(11), 2326–2340 (2020)
15. Perozzi, B., Al-Rfou, R., Skiena, S.: DeepWalk: online Learning of Social Representations. In: Proceedings of the 20th ACM SIGKDD International Conference on Knowledge Discovery and Data Mining, pp. 701–710. ACM (2014)
16. Tang, J., Qu, M., Wang, M., Zhang, M., Yan, J., Mei, Q.: LINE: large-scale information network embedding. In: Proceedings of the 24th International Conference on World Wide Web, pp. 1067–1077. ACM (2015)
17. Grover, A., Leskovec, J.: node2vec: scalable Feature Learning for Networks. In: Proceedings of the 22nd ACM SIGKDD International Conference on Knowledge Discovery and Data Mining, pp. 855–864. ACM (2016)
18. Li, W., Zhang, B., Xu, L., Wang, M., Luo, A., Niu, Y.: Combining knowledge graph embedding and network embedding for detecting similar mobile applications. In: Zhu, X., Zhang, M., Hong, Yu., He, R. (eds.) NLPCC 2020. LNCS (LNAI), vol. 12430, pp. 256–269. Springer, Cham (2020). https://doi.org/10.1007/978-3-030-60450-9_21

# CCD-ASQP: A Chinese Cross-Domain Aspect Sentiment Quadruple Prediction Dataset

Ye Wang[1], Yuan Zhong[1], Xu Zhang[1], Conghui Niu[1], Dong Yu[1], and Pengyuan Liu[1,2](✉)

[1] Beijing Language and Culture University, Beijing, China
liupengyuan@pku.edu.cn
[2] National Language Resources Monitoring and Research Center for Print Media, Beijing, China
https://github.com/blcunlp/CCD-ASQP

**Abstract.** This work present CCD-ASQP, a cross-domain Aspect Sentiment Quadruple prediction(ASQP) dataset in chinese. Based on e-commerce scenario, this dataset lables 15,878 sentiment quadruples out of 3,700 reviews across 6 life domain and 10 product entities. Multiple baselines have been test on CCD-ASQP in terms of ASQP task, and performances have been compared with ChatGPT. Deep learning models' dramatic decline of accuracy of when shifting to out-of-distribution data shows the lack of domain adaptiveness. ChatGPT achieves relatively consistent cross-domain performance in few-shot setup. Error analysis suggests effect of Chinese language forms on ASQP task. CCD-ASQP leaves great space for sentiment analysis tasks in Chinese language and perspectives from other disciplines are helpful.

**Keywords:** Fine-grained Aspect-Based Sentiment Analysis · Sentiment Quadruple · Dataset Construction · ChatGPT

## 1 Introduction

Traditional sentiment analysis don't distinguish the entity that the sentiment is ascribed to, which is necessary in practical use [8,10]. So, the task of aspect-based sentiment analysis (ABSA), a fine-grained sentiment analysis subtask has been proposed to mine sentimental information from opinion texts (social media tweets, product reviews...) together with its targeted entity and domain attribute (collectively referred to as aspect) [9,16]. The four key elements of ABSA task are (1) *aspect term* (2) *aspect category* (3) *opinion term* and (4) *sentiment polarity.*

As [6,15] pointed out that the correlation between the aspect term and the opinion term is helpful for better ABSA. Later studies in this domain include extracting multi-elements at one time with newly proposed PLMs [4,11,19,20]

The dataset can be obtained from https://github.com/blcunlp/CCD-ASQP.


H. Wang et al. (Eds.): CCKS 2023, CCIS 1923, pp. 233–245, 2023.
https://doi.org/10.1007/978-981-99-7224-1_18

and eventually develop into the Aspect Sentiment Quad Prediction (ASQP) task [21]. The ASQP task extracts 4 sentimental elements at one time and provide the most complete attributive sentimental information.

However, after reviewing previous ASQP tasks, we have found 2 major problems with data (details of ABSA and ASQP tasks and datasets provided in Part 2):

**Cross-Domain Issue:** most studies focus on single or few product domains, but they overlook that the same opinion term can express different sentiment in different domains. And there is no study or datasets discussing this phenomenon from socio-linguistic perspective. For example," big size" is basically positive when describing a car, but it is likely to be a negative emotional polarity when describing a laptop. **Lack of high-quality data in Chinese:** As shown in Table 1, the prevalent ASQP datasets ACOS and ABSA-QUAD are both based on English texts. There is no ASQP datasets in Chinese which extract 4 sentimental elements at one time.

To mitigate the effect of the mentioned problems, there is a need to test current approaches on a qualified Chinese multi-domain aspect dataset. This article presents CCD-ASQP, the first comprehensive cross-domain sentiment analysis dataset in Chinese. It comprises of 15,878 manually tagged sentiment quadruples based on both explicit and implicit sentiment texts. We also conduct ASQP experiments on our dataset with several common approaches of sentimental analysis and compare the result with the-state-of-the-art LLM, ChatGPT. Results have been analyzed in socio-linguistic perspectives. With the relatively unsatisfactory metrics, we finally conclude that CCD-ASQP stands a challenging and meaningful dataset which calls for further investigation.

The main contributions of this work are as follows: (1) We present an original and demanding ASQP dataset in Chinese. With 15,878 manually annotated sentiment quadruples, ASQP experiments were carried out on this dataset, the unsatisfactory performances of overall baselines show there is still a great space for improving current sentiment analysis approaches.

(2) Results demonstrate that model with fine tuning Paraphrase outperforms ChatGPT few-shot in training data, sweeping machine domain. But ChatGPT achieves consistent performance across all domains while fine-tuned models meet dramatic decline. It shows poor generalization ability of current approaches.

(3) we find the superior performance of CCD-ASQP. Additionally, we analyse this phenomenon from socio-linguistic views and find the effect of language structure and social phenomenon on sentimental analytic modeling.

## 2 Related Work

**Aspect-based sentiment analysis (ABSA)** [2] aims to predict sentiment polarities on all aspect categories mentioned in the text. Pontiki et al., 2015 defines aspect category as a combination of an entity type and an attribute type (e.g., Food/Style Options). Early studies focus on the prediction of a single element such as extracting the aspect term, for example, Bu et al. detect

**Table 1.** The prevalent sentiment analysis datasets

| Dataset | Language | Domain | Type |
|---|---|---|---|
| SemEval-2014 | English | Laptops, restaurants | ABSA |
| SemEval-2015 | English | Laptops, restaurants | ABSA |
| SemEval-2016 | Multilingual | Electronics, hotels, restaurants | ABSA |
| TOWE | English | Laptops, restaurants | ABSA |
| ASC-QA | Chinese | Bags, cosmetics, electronics | ABSA |
| MAMS | English | Restaurants | ABSA |
| ASAP | Chinese | Restaurants | ABSA |
| ACOS | English | Laptops, restaurants | ASQP |
| ABSA-QUAD | English | Restaurants | ASQP |

aspect categories in Chinese reviews [2]. More recently, the extraction of multiple emotional element tasks are proposed in ABSA. Peng et al. proposed the aspect sentiment triplet extraction task (ASE) to extract aspect items, opinion items and sentiment polarity in text [11], Wan et al. introduced the target aspect sentiment detection (TASD) task, aiming to predict the aspect category, aspect term, and sentiment polarity simultaneously [17].

**Aspect Sentiment Quad Prediction (ASQP)** [21] is a fine-grained ASBA tasks that are proposed recently. Cai et al. introduced two new datasets annotating sentimental quadruples [3]. Zhang et al. proposed a Paraphrase modeling strategy, which transformed the target sequence into the obtained natural language sequence by combining the pre-established template with the labeled ASQP quadruples [21]. Bao et al. introduced a opinion tree generation model to reveal a comprehensive attribute-level sentiment structure [1].

**Datasets:** As shown in Table 1: The series of SemEval sentiment analysis datasets [12–14] have been widely used and pushed forward related research. Compiling user reviews from e-commerce websites, they are small size but qualified domain sentiment analysis datasets (SA) datasets. The restaurant subset includes 5 aspect categories (i.e., Food, Service, Price, Ambience and Anecdotes/Miscellaneous) and 4 polarity labels (i.e., Positive, Negative, Conflict and Neutral). MAMS [7] tailors SE-ABSA14 to make it more challenging, in which each sentence contains at least two aspects with different sentiment polarities. TOWE [5], ASC-QA [18] are high-quality datasets respectively oriented at aspect opinion term extraction and Q&A form ABSA tasks; The ASAP [2] dataset is the largest Chinese ABSA dataset consisted of 46370 restaurant comments, with 18 aspect categories and aspect-based sentiment polarity annotated; The ACOS [3] and the ASBA-QUAD [21] datasets are proposed to solve ASQP tasks.

## 3 Dataset Collection and Analysis

The following facts may be helpful for understanding the dataset construction.

### 3.1 Data Source and Domain

CCD-ASQP dataset comes from JD.com, one of China's mainstream e-commerce platforms, which allows users to post coarse-grained reviews on restaurant items they have purchased. Monitoring customer feedback can be automated with Aspect Level Sentiment Classification (ALSC) which allows us to analyse specific aspects of the products in reviews.

In order to test domain difference in terms of sentimental information, we selected 6 domain data with low correlation and covering common categories to construct a cross-category dataset. They are the *car domain*, *beauty makeup and skin care domain*, *food domain*, *sports domain*, *jewelry domain* and *smart home domain*. Then, for each domain we chose no less than 1 product to represent the domain. That is, there are 10 sub-domains in total, namely *bicycles* for car, *lipsticks and perfumes*for beauty and skin care, *milk* for food, *skipping rope* for sports,*ear stud* and *ear clip* for jewelry, *cooking machine*, *smart switch*, *smart doorbell* and *sweeping robot* for smart home. To facilitate later processing, data cleaning was performed before data annotation.

### 3.2 Sentiment Quadruple

To fulfill ASQP task, the complete 4 sentimental elements need to be annotated manually from the above source domain, namely:

*(1) aspect term:* the entity or entity aspect mentioned explicitly or implicitly in the text. *(2) aspect category:* the category to which the aspect items mentioned in the article belong. *(3) opinion term:* the opinion description of the aspect term mentioned in the text. *(4) sentiment polarity:* the emotional description of the aspect items mentioned in the article.

Their language representations are illustrated in Fig. 1:

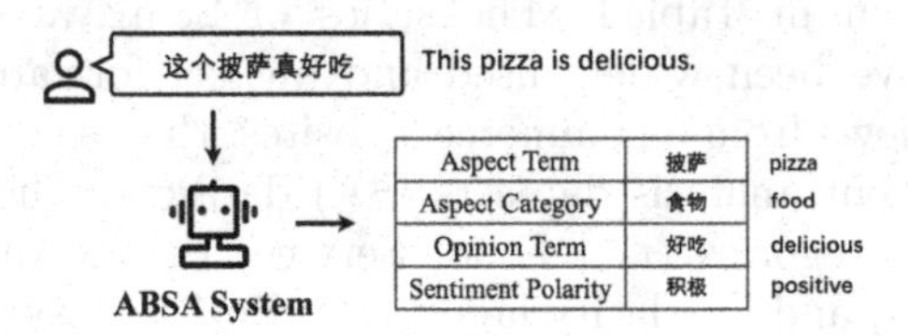


**Fig. 1.** The ABSA sentiment quadruple

### 3.3 Aspect Categories

In CCD-ASQP, we extract 14 common aspects for the 10 subdomains, namely *price*, *merchant service*, *appearance design*, *functional effect*, *quality*, *material*, *capacity*, *component function*, *installation and use*, *sound*, *safety*, *taste*, *production date*, *portability and user experience*. It should be pointed out that since it is a cross-domain dataset, even the same aspect category may represent different information in different domains. *Taste* refers to whether it tastes good in the milk category, but it refers to whether it smells good in the perfume category, as shown in Table 2.

**Table 2.** Different semantics of *Taste* aspect across domains. The meaning of "更香浓的牛奶味" is "have rich milk taste", and the sentiment quadruple is (taste, taste, positive, rich), the meaning of 这味道真是一点都不好闻 is "it smells really bad", and the sentimen quadruple is (smell, taste, negative, bad)

| Domain | Example | Sentiment quadruple |
|---|---|---|
| Milk | 更香浓的牛奶味 | (牛奶味,味道,positive,香浓) |
| Perfume | 这味道真是一点都不好闻 | (味道,味道,negative,不好闻) |

### 3.4 Sentiment Data Ratio Control

In real life, the distribution of positive and negative reviews in different domains may be inconsistent with the distribution of existing datasets. In order to test the adaptive ability of known data of current approaches on different domains, the ratio of positive and negative in the verification set and test set is manually controlled to different proportions. The details of the positive and negative reviews in the 10 domains are shown in Table 3.

**Table 3.** The details of the sentiment polarity in the 10 domains.

| | Sweeping Robot | Bicycle | Lipstick | Jumping Rope | Smart Switch |
|---|---|---|---|---|---|
| Pos vs. Neg | none | 144:156 | 57:243 | 193:106 | 193:106 |
| Ratio | none | 1:1 | 2:8 | 6:4 | 6:4 |
| | Smart Doorbell | Milk | Perfume | Cooking Machine | Ear Stud |
| Pos vs. Neg | 205:95 | 190:110 | 24:176 | 71:229 | 89:211 |
| Ratio | 7:3 | 6:4 | 1:9 | 2:8 | 3:7 |

### 3.5 Annotation Guidelines and Process

2 annotators need to consider the predefined 14 aspect categories and annotate the sentiment quadruple from cleaned row data. Only positive and negative are considered when labelling the sentiment polarity. The work flow of the labeling process is shown in Fig. 2. The F1 value between quadruples marked by the two annotators is 75.1%, which shows that there is a substantial agreement between the annotators in the extraction of sentimental quadruples.

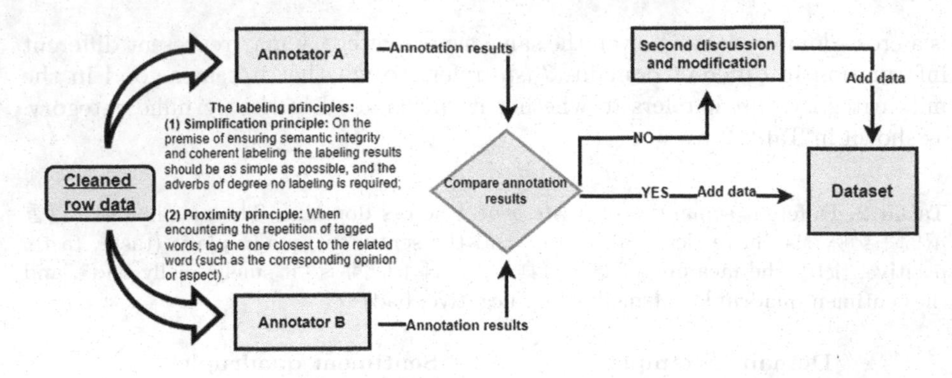


**Fig. 2.** Labeling process, an overview

### 3.6 Dataset Analysis

CCD-ASQP consists of 3,700 real user reviews out of 7,440 sentence, with 15,878 manually-labelled sentiment quadruples. Table 4 shows the number of quadruples in the 10 subdomains and average quadruple distribution of CCD-ASQP.

**Table 4.** The statistics and label distribution of CCD-ASQP.

| | Sweeping robot | Bicycle | Lipstick | Jumping rope | Smart switch |
|---|---|---|---|---|---|
| Average quadruple per review | 4.54 | 4.59 | 4.94 | 5.64 | 3.63 |
| Average tokens per review | 90.62 | 91.65 | 74.11 | 79.77 | 78.77 |
| Pos-aspect: Neg-aspect | 723:196 | 722:338 | 258:1137 | 441:832 | 486:585 |
| Number of quadruples | 4543 | 1379 | 1483 | 1692 | 1090 |
| | Smart doorbell | Milk | Perfume | Cooking machine | Ear stud |
| Average quadruple per review | 3.06 | 3.53 | 4.65 | 4.15 | 3.57 |
| Average tokens per review | 77.70 | 70.44 | 88.20 | 87.02 | 60.01 |
| Pos-aspect: Neg-aspect | 723:196 | 722:338 | 258:1137 | 441:832 | 486:585 |
| Number of quadruples | 919 | 1060 | 1395 | 1246 | 1071 |

The average number of aspects per review in the 10 subdomains are all around 5, the highest is 5.64 for the skipping rope subdomain, and the lowest is 3.06 for the smart doorbell subdomain. Except for ear stud and ear clips, the average tokens per review of each other subdomains is about 70%–90%, it means CCD-ASQP has a large amount of data, which alleviates the problem of data sparseness. As shown in Table 3, the ratio of positive aspects and negative ones among all domains is deliberately set as different.

In a nutshell, CCD-ASQP is based on concrete resource with dense sentiment information. And its sentiment polarity ratio have been carefully conditioned to monitor different real-world situation.

# 4 Experiments

## 4.1 Experimental Setup

**Dataset Split.** The entire dataset is divided into 10 subdomains. We use models fine-tuned on the sweeping robot subdomain to predict the other 9 subdomains, which consist in out-of-distribution data. Data split can be seen from Table 5. Training set, validation set and test set refer to data of the sweeping robot subdomain.

**Table 5.** Detailed data split. The OOD represents data from the other 9 subdomains, respectively bicycles, skipping ropes, perfumes, lipsticks, milk, cooking machines, ear studs and ear clips, smart switches and smart doorbells. Each consists of 300 data.

| Train | Dev | Test | OOD Dev | OOD Test |
|---|---|---|---|---|
| 700 | 200 | 100 | 200 | 100 |

**Evaluation Metrics.** We adopt the F1 score as the main evaluation metric. A sentiment quadruple prediction is correct if and only if all predicted elements are exactly the same as the gold label. At the same time, the precision and recall scores of the ASQP task are also used as evaluation indicators.

**Experiment Details.** The model parameters are optimized by Adam with a learning rate of 3e−4. The batch size is 16. Our experiments are carried out with an Nvidia RTX 3090 GPU. The experimental results are obtained by averaging 20 runs with random initialization.

**Experiment on ChatGPT.** We conducted experiments using the gpt-3.5-turbo model by OpenAI, with temperature being the default value of 1. Instructions are as follows: *You are a useful assistant for extracting emotional quads, understanding and remembering the following concepts: The Concept of Emotional Quads and specific examples. Please identify the emotional quads contained in the following sentence.* Due to the unique format of the task, we conducted experiments based on three settings: one-shot, three-shot, and five-shot. The experimental results were calculated using a matching method.[1]

## 4.2 Baselines

Since the ASQP task has not been explored previously, in addition to choosing two models for experiments on the English dataset, we also selected two models with better effect on the extraction of triplets, and made some modifications to them and used them on the ASQP task.

[1] The "aspect category" and "sentient polarity" items were accurately matched, while the "aspect term" and "opinion" items were fuzzily matched, that is, they are correct as long as they contain key information.

- **TASO-BERT** [17]. TAS is proposed to extract (ac, at, sp) triplets, it reduces the problem of joint detection to two sub-problems of text classification and sequence labeling question. By changing the labeling scheme in sequence labeling, the model can predict aspect items and opinion items at the same time.
- **GAS** [22]. It is the first work to deal aspect level sentiment analysis with generative method. Its basic model is the pre trained language model T5. We improve it by changing its base model to mT5 and taking the emotion quadruple sequence directly as the target sequence.
- **Extract-Classify** [3]. It implements quadruple extraction in two steps. The first step is to extract aspect-opinion binary groups in comments, and then obtain aspect category and sentiment polarity based on aspect-opinion classification.
- **Paraphrase** [21]. It is also a generation-based method. It proposes a new paraphrase modeling paradigm, transforming the ASQP task into a paraphrase generation problem, and transforms the sentiment quadruple q = (c, a, o, p) through the projection function "*Pc(c) is Pp(p) because Pa(a) is Po (o)*" is linearly transformed into a natural language sentence. Both input and output are natural language sentences, which can more naturally utilize the rich knowledge in the pre-trained generative model.

**Table 6.** The F1 value for ASQP task for baseline and ChatGPT in 10 subdomains

| | Sweeping robot | Bicycle | Lipstick | Jumping rope | Smart switch |
|---|---|---|---|---|---|
| TSAO-BERT | 42.67 | 22.10 | 18.68 | 22.34 | 29.16 |
| GAS | 45.69 | 16.53 | 2.13 | 24.04 | 19.35 |
| Extract-Classify | 17.89 | 15.03 | 10.96 | 16.44 | 16.91 |
| Paraphrase | **67.14** | 22.24 | 10.37 | 28.57 | **35.47** |
| ChatGPT-oneshot | 42.39 | **28.53** | 18.12 | 22.27 | 22.70 |
| ChatGPT-threeeshot | 45.80 | 27.34 | **21.17** | **31.33** | 28.40 |
| ChatGPT-fiveshot | 44.54 | 27.48 | 19.73 | 26.95 | 23.68 |
| | Smart doorbell | Milk | Perfume | Cooking machine | Ear stud |
| TSAO-BERT | 36.23 | 13.24 | 24.07 | 22.34 | 18.45 |
| GAS | 33.33 | 10.37 | 2.45 | 8.26 | 13.41 |
| Extract-Classify | 17.28 | 16.87 | 12.82 | 12.44 | 17.95 |
| Paraphrase | **45.21** | 16.36 | 10.76 | 16.01 | 26.43 |
| ChatGPT-oneshot | 28.57 | 19.27 | 30.23 | **29.20** | 23.81 |
| ChatGPT-threeeshot | 35.47 | 19.97 | **34.72** | 24.75 | **29.55** |
| ChatGPT-fiveshot | 31.33 | **20.92** | 31.82 | 26.12 | 25.38 |

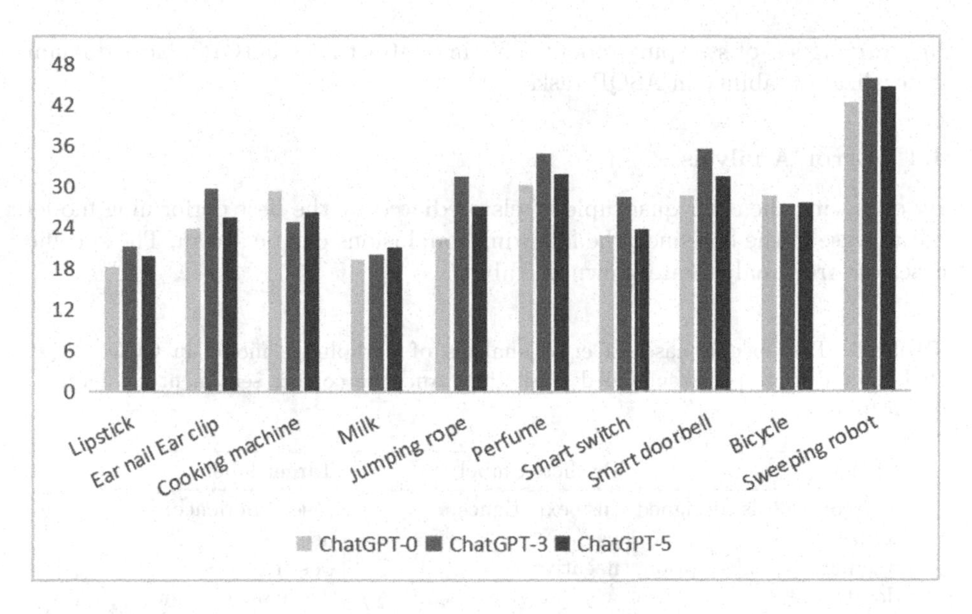


**Fig. 3.** The F1 value for ASQP task based on ChatGPT (%). ChatGPT-1/3/5 represents providing corresponding cases for each category

### 4.3 Main Results

As shown in Table 6 and Fig. 3, all models achieve the best f1 on the test set of sweeping robots, and the change into other domain will affect the extraction of sentiment quadruples. The unsatisfactory results in the other 9 domains show that the migration ability of the model on OOD dataset is not strong. Second, ChatGPT perform the best 7 of 10 times, and for current approaches Paraphrase performs slightly better.

Further analyse on domain change, we find model's performance on sentiment analysis is affected by the ratio of sentiment polarity. For example, *milk* and *jumping rope* subsets belong to different domain, but their F1 values are similar, while *smart doorbell* and *smart switch* don't show the situation. As shown in Table 3, *milk* and *jumping rope* share the similar sentiment polarity distribution.

T5-based generative model (i.e., GAS and Paraphrase) performs slightly better than Bert-based models. Which shows that encoding natural language tags into the target output through a unified generative architecture can make full use of label semantics.

When more context is given in the few-shot setting (one-shot to three-shot), ChatGPT learned this task form from the example and the effect is improved. However, as the context further increases (from three-shot to five-shot), the effectiveness of ChatGPT fluctuates slightly, which may be due to existing cases interfering with ChatGPT's understanding of the task. ChatGPT performs best on the test set of sweeping robots, as the example of instruction comes from the training set of sweeping robots. This indicates that ChatGPT lacks domain generalization ability in ASQP tasks.

### 4.4 Error Analysis

By observing the error quadruple labels predicted by the best performing model paraphrase in the baselines, the following conclusions can be drawn. The specific cases of error analysis are shown in Table 7.

**Table 7.** The specific cases of error analysis of paraphrase model in CCD-ASQP. Red colored fonts represent failed predictions and the correct sentiment elements in quadruples.

| Review | Predicted label | Target label |
|---|---|---|
| The overall is designed with magnetic absorption, don't worry about the use of a long time will not fasten | (use experience, use, negative, will not fasten securely) | (use experience, use, positive, don't worry about not fasten securely) |
| This smell is very special, a little heady, others don't know, but I can not accept it | (taste, smell, positive, special) | (taste, smell, negative, special) |
| The map construction is fast, but a little incomplete, but the efficiency of the map construction is still high | (functional effect, build, positive, incomplete) | (functional effect, build, negative, incomplete) |

**Double Negation Affects Polarity Judgment.** In Chinese, double negation affects the judgment of sentiment polarity in that double negation means the contrary. The model fails to recognize the first negation and only recognize the second. Therefore, in the judgment of polarity of emotion, the positive emotion will be judged as the negative emotion. For example, as in review 1, there are other contents mixed with *used for a long time* between the first negative *don't worry* and the second negative *not fasten*, and the model only recognizes the second negative *not fasten*. That is why the emotional polarity judgment given by the model is *negative*. But in fact, it means positive *it can be fastened after a long time.*

**The Context Influences the Judgment of Sentiment Polarity.** In Chinese, situational context also affects the judgment of sentiment polarity. It is hard to judge the correct polarity simply by word semantics. However, if the context is complex or obscure, the model may not be accurate in the recognition of the sentiment, which will lead to the wrong prediction. For example, in review 2, the adjective *special* refers to *unusual and different*, if we judge emotion polarity only by its word meaning, it is easy to be misjudged as positive, but in Chinese cultural context, *a little heady* is a more subtle way of saying *I'm not comfortable with the smell* while *I'm not comfortable with it anyway* is a more obvious negative emotion. Therefore, judging from the context, *special* here expresses a negative emotion.

**The Predictive Express Different Sentiment Connotation with Different Noun Head.** In Chinese, the same predictive has both positive and negative meaning, depending on the noun head. Therefore, the sentiment polarity of the former evaluation has an impact on the latter. For example, in review 3, the words related to *map construction* include *fast*, *incomplete* and *high (efficiency)*. For *map construction*, *fast* and *high (efficiency)* are positive evaluation, while *incomplete* is negative evaluation. However, because *incomplete* is sandwiched between *fast and high (efficiency)*, the model is influenced by the positive emotions before and after the recognition, so *incomplete* is also judged as a positive emotion.

## 5 Conclusions and Limitation

ASQP task has become a popular task in sentiment analysis of NLP, but there is still no studies on Chinese datasets. We built a Chinese cross-domain sentiment dataset, and conducted sentiment quadruple extraction task on it. Comparison has been done between current ASQP approaches and popular LLM, ChatGPT. Together with error analysis from a linguistic perspective, the unsatisfactory results suggest poor domain adaptation for current methods and call for deeper investigation on this dataset.

As far as we are considered, the limitation of our research may cover (1) domain richness, this article attempts to test cross-domain generalization ability of current approaches, however, the entities included are not comprehensive enough, and the conclusion may not be reusable when replaced with other domains. (2) The sentiment analysis of Chinese dataset may be strongly affected by unique Chinese rhetorical techniques and social context also, which we have not considered in our dataset construction. Beyond that, performing ASQP tasks on Chinese remains a challenging problem that deserves further exploration.

**Acknowledgement.** This research project is supported by Science Foundation of Beijing Language and Culture University (supported by "the Fundamental Research Funds for the Central Universities") (23YJ080009).

## References

1. Bao, X., Zhongqing, W., Jiang, X., Xiao, R., Li, S.: Aspect-based sentiment analysis with opinion tree generation. In: Raedt, L.D. (ed.) Proceedings of the Thirty-First International Joint Conference on Artificial Intelligence, IJCAI-22, pp. 4044–4050. International Joint Conferences on Artificial Intelligence Organization (2022). https://doi.org/10.24963/ijcai.2022/561. Main Track
2. Bu, J., et al.: ASAP: a Chinese review dataset towards aspect category sentiment analysis and rating prediction. In: Proceedings of the 2021 Conference of the North American Chapter of the Association for Computational Linguistics: Human Language Technologies, pp. 2069–2079. Association for Computational Linguistics, Online, June 2021. https://doi.org/10.18653/v1/2021.naacl-main.167. https://aclanthology.org/2021.naacl-main.167
3. Cai, H., Xia, R., Yu, J.: Aspect-category-opinion-sentiment quadruple extraction with implicit aspects and opinions. In: Proceedings of the 59th Annual Meeting of the Association for Computational Linguistics and the 11th International Joint Conference on Natural Language Processing (Volume 1: Long Papers), pp. 340–350. Association for Computational Linguistics, Online, August 2021. https://doi.org/10.18653/v1/2021.acl-long.29. https://aclanthology.org/2021.acl-long.29
4. Chen, X., et al.: Aspect sentiment classification with document-level sentiment preference modeling. In: Proceedings of the 58th Annual Meeting of the Association for Computational Linguistics, pp. 3667–3677 (2020)
5. Fan, Z., Wu, Z., Dai, X., Huang, S., Chen, J.: Target-oriented opinion words extraction with target-fused neural sequence labeling. In: Proceedings of the 2019 Conference of the North American Chapter of the Association for Computational Linguistics: Human Language Technologies, Volume 1 (Long and Short Papers), pp. 2509–2518 (2019)
6. Hu, M., Liu, B.: Mining and summarizing customer reviews. In: Proceedings of the tenth ACM SIGKDD International Conference on Knowledge Discovery and Data Mining, pp. 168–177 (2004)
7. Jiang, Y.Z., et al.: Genomic and transcriptomic landscape of triple-negative breast cancers: subtypes and treatment strategies. Cancer Cell **35**(3), 428–440 (2019)
8. Liu, B.: Sentiment Analysis and Opinion Mining. Springer, Cham (2022). https://doi.org/10.1007/978-3-031-02145-9
9. Liu, P., Joty, S., Meng, H.: Fine-grained opinion mining with recurrent neural networks and word embeddings. In: Proceedings of the 2015 Conference on Empirical Methods in Natural Language Processing, pp. 1433–1443. Association for Computational Linguistics, Lisbon, Portugal, September 2015. https://doi.org/10.18653/v1/D15-1168. https://aclanthology.org/D15-1168
10. Pang, B., Lee, L., Vaithyanathan, S.: Thumbs up? Sentiment classification using machine learning techniques. In: Proceedings of the 2002 Conference on Empirical Methods in Natural Language Processing (EMNLP 2002), pp. 79–86. Association for Computational Linguistics, July 2002. https://doi.org/10.3115/1118693.1118704. https://aclanthology.org/W02-1011
11. Peng, H., Xu, L., Bing, L., Huang, F., Lu, W., Si, L.: Knowing what, how and why: a near complete solution for aspect-based sentiment analysis. In: Proceedings of the AAAI Conference on Artificial Intelligence, vol. 34, pp. 8600–8607 (2020)
12. Pontiki, M., et al.: SemEval-2016 task 5: aspect based sentiment analysis. In: Proceedings of the 10th International Workshop on Semantic Evaluation (SemEval-2016), pp. 19–30. Association for Computational Linguistics, San Diego, California,

June 2016. https://doi.org/10.18653/v1/S16-1002. https://aclanthology.org/S16-1002

13. Pontiki, M., Galanis, D., Papageorgiou, H., Manandhar, S., Androutsopoulos, I.: SemEval-2015 task 12: aspect based sentiment analysis. In: Proceedings of the 9th International Workshop on Semantic Evaluation (SemEval 2015), pp. 486–495. Association for Computational Linguistics, Denver, Colorado, June 2015. https://doi.org/10.18653/v1/S15-2082. https://aclanthology.org/S15-2082
14. Pontiki, M., Galanis, D., Pavlopoulos, J., Papageorgiou, H., Androutsopoulos, I., Manandhar, S.: SemEval-2014 task 4: aspect based sentiment analysis. In: Proceedings of the 8th International Workshop on Semantic Evaluation (SemEval 2014), pp. 27–35. Association for Computational Linguistics, Dublin, Ireland, August 2014. https://doi.org/10.3115/v1/S14-2004. https://aclanthology.org/S14-2004
15. Qiu, G., Liu, B., Bu, J., Chen, C.: Opinion word expansion and target extraction through double propagation. Comput. Linguist. **37**(1), 9–27 (2011)
16. Tuveri, F., Angioni, M.: Sentiment analysis and opinion mining (2012)
17. Wan, H., Yang, Y., Du, J., Liu, Y., Qi, K., Pan, J.Z.: Target-aspect-sentiment joint detection for aspect-based sentiment analysis. In: AAAI Conference on Artificial Intelligence (2020)
18. Wang, J., et al.: Aspect sentiment classification towards question-answering with reinforced bidirectional attention network. In: Proceedings of the 57th Annual Meeting of the Association for Computational Linguistics, pp. 3548–3557. Association for Computational Linguistics, Florence, Italy, July 2019. https://doi.org/10.18653/v1/P19-1345. https://aclanthology.org/P19-1345
19. Wang, W., Pan, S.J., Dahlmeier, D., Xiao, X.: Recursive neural conditional random fields for aspect-based sentiment analysis. arXiv preprint arXiv:1603.06679 (2016)
20. Wang, W., Pan, S.J., Dahlmeier, D., Xiao, X.: Coupled multi-layer attentions for co-extraction of aspect and opinion terms. In: Proceedings of the AAAI Conference on Artificial Intelligence, vol. 31 (2017)
21. Zhang, W., Deng, Y., Li, X., Yuan, Y., Bing, L., Lam, W.: Aspect sentiment quad prediction as paraphrase generation. In: Proceedings of the 2021 Conference on Empirical Methods in Natural Language Processing, pp. 9209–9219. Association for Computational Linguistics, Online and Punta Cana, Dominican Republic, November 2021. https://doi.org/10.18653/v1/2021.emnlp-main.726. https://aclanthology.org/2021.emnlp-main.726
22. Zhang, W., Li, X., Deng, Y., Bing, L., Lam, W.: Towards generative aspect-based sentiment analysis. In: Proceedings of the 59th Annual Meeting of the Association for Computational Linguistics and the 11th International Joint Conference on Natural Language Processing (Volume 2: Short Papers), pp. 504–510. Association for Computational Linguistics, Online, August 2021. https://doi.org/10.18653/v1/2021.acl-short.64. https://aclanthology.org/2021.acl-short.64

# Move Structure Recognition in Scientific Papers with Saliency Attribution

Jinkun Lin[1], Hongzheng Li[2], Chong Feng[1(✉)], Fang Liu[2(✉)], Ge Shi[3], Lei Lei[2], Xing Lv[2], Ruojin Wang[2], Yangguang Mei[2], and Lingnan Xu[1]

[1] School of Computer Science and Technology, Beijing Institute of Technology, Beijing 100081, China
{ljkun,fengchong,xln}@bit.edu.cn

[2] School of Foreign Languages, Beijing Institute of Technology, Beijing 100081, China
{lihongzheng,liufang,leilei,lvxing,wangruojin,meiyangguang}@bit.edu.cn

[3] Faculty of Information Technology, Beijing University of Technology, Beijing 100124, China
shige@bjut.edu.cn

**Abstract.** Move analysis is a primary research topic in computational linguistics that relates to pragmatics. It plays a crucial role in analyzing the intent and coherence of the text. This paper introduces a innovative exploration of move analysis to scientific papers and presents a novel task - move structure recognition in scientific papers. Existing datasets are inadequate to support this task. Thus, we manually annotated a dataset called Scientific Abstract Moves Dataset (SAMD). The implicit mixture and counterfactual reasoning in the move structure's content has led to poor performance in move recognition. This research examines the issue in depth and presents a new concept of move saliency attribution, which can illuminate the contribution of words to specific move structures. On this foundation, we design a new move recognition training mechanism, which fully consider the context information of the move to achieve promising performance on SAMD and NLPContributionGraph shared task dataset (NCG). This is the first attempt at interpretability of move recognition, giving us the possibility to understand how the model makes decisions and identify potential biases or errors in the model.

**Keywords:** move structure recognition · saliency attribution · scientific papers

## 1 Introduction

Move analysis, proposed by Swales et al. in 1981, has gained significant attention in the field of academic discourse research [14]. Moves refer to rhetorical units that serve as complete communicative expressions, representing segmented discourse fragments with consistent linguistic orientation [15]. This paper aims to apply move analysis to scientific papers and introduces a new task called move structure recognition in scientific papers.


H. Wang et al. (Eds.): CCKS 2023, CCIS 1923, pp. 246–258, 2023.
https://doi.org/10.1007/978-981-99-7224-1_19

The goal of this task is to employ automated techniques for the recognition of move structures in scientific papers at the sentence level. Move structure recognition is essentially a text classification problem. An example diagram is shown in Fig. 1. This research lays the foundation for subject knowledge mining, and also provides the possibility to give more accurate feedback and evaluation on the move structure and language characteristics of scientific writing.

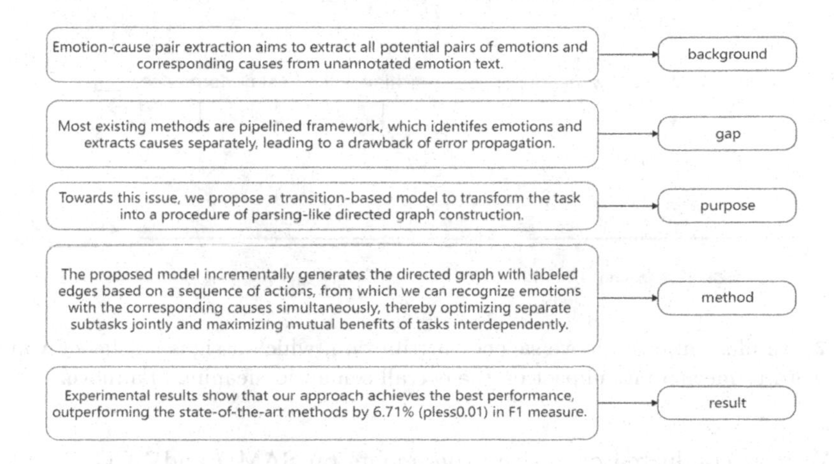


**Fig. 1.** An example of move recognition in a scientific paper abstract, each sentence corresponds to one or more move labels.

The existing datasets only focus on the abstract of structural rules, which means they have limited domain coverage. The move shows different characteristics depending on the field. Therefore, it is necessary to develop a large-scale dataset with manual labeling to accelerate research on scientific move recognition. We construct a high-quality moves corpus for abstracts of scientific papers (SAMD). This corpus marked the move structure labels of the paper, such as background, gap, method, purpose, result, conclusion, contribution and implication.

In irregularly structured abstracts of scientific papers, a common challenge arises where the content encompassing move structures is implicitly intertwined. This often results in the blending of multiple move elements and counterfactual reasoning within a single sentence. For example, consider the following sentence: "We propose neural models that simulate the distributions of these response offsets, taking into account the response turn as well as the preceding turn [10]" While the sentence structure aligns with the purpose move, its actual meaning is more closely associated with the method move. This phenomenon leads to blurring of the boundaries of move determination, which brings difficulties to the automatic move recognition of scientific papers.

To solve the above problems, this paper introduced the concept of move saliency attribution. Drawing inspiration from feature attribution methods [11, 13], we treat each word as a feature and compute its contribution (saliency value) towards predicting the move label, Fig. 2 illustrates the key idea. Then, this paper design a training mechanism based on saliency attribution. The proposed strategy fully considers the contextual information and forcing the model to focus on the most significant text related to the label [12], which can be seen as a more stringent version of attention.

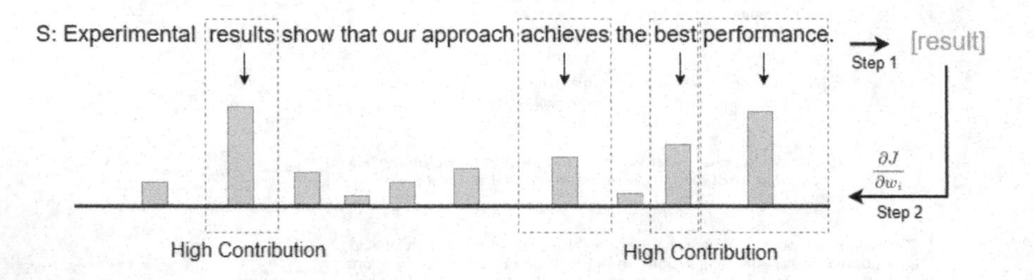


**Fig. 2.** An illustration of move saliency attribution, which assigns a saliency value to each word to measure its impact on the overall semantic meaning of a move.

We have conducted extensive experiments on SAMD and NCG (NLPContributionGraph shared task[1]). According to the results: contextual information effectively assists the judgment of the move function of the current sentence, the saliency attribution method captures the underlying model and explains the model prediction well. On the two tasks, the F1 gains are 1.8% and 1.9% compared to the baseline method.

Our contributions can be summarized: we propose a novel task, step structure recognition in scientific papers, and manually labeled a new dataset (SAMD). We propose a move structure recognition method for abstracts of scientific papers with saliency attribution, which uses sentence position characteristics and adjacent preceding and following sentences as contextual information, and uses saliency as evidence to enhance learning. Experimental results show that our method is significantly better than the baseline method in many aspects and achieves better performance.

## 2 Related Work

### 2.1 Move Structure Recognition Dateset

Only a few datasets have been built for move structure recognition. Wang et al. constructed a standard abstract dataset selected from the Applied Linguistics [17]; Ding et al. constructed a structured dataset of abstracts of scientific papers in various disciplines including natural sciences, social sciences, and

[1] https://ncg-task.github.io/.

humanities [4]; Dayrell et al. constructed a corpus of abstracts in the fields of Physical Sciences and Engineering, Life and Health Sciences [3]; Liakata et al. constructed the ART Corpus, which consists of the full text of papers in the fields of physical chemistry and biochemistry [6].

### 2.2 Feature Attribution

The aim of feature attribution (FA) is to identify the features or inputs in a model that have the most significant impact on its output or prediction [8]. It is important for people to understand how the model makes decisions and identify potential biases or errors in the model. Formally, suppose we have an input vector $x = (x_1, x_2, ..., x_n) \in R^n$ and a function $\mathcal{F} : R^N \rightarrow (0, 1)$ that represents a deep network. The attribution value of $x$, with respect to the output $\mathcal{F}(x)$, is defined as a vector $A_{\mathcal{F}(x)} = (a_1, a_2, ..., a_n) \in R^n$, where $a_i$ measures the contribution of $x_i$ to $\mathcal{F}(x)$ [13]. There are two technical routes for feature attribution. The first approach involves computing the gradient of the output for the correct class, with respect to the input vector of a given model. This gradient is then used as an attribution value map for the masked input, enabling the identification of the input features that contribute the most to the model's output [1]. Another research approach is to use a locally additive model to approximate the interpretable model. This helps in explaining the variation in the model's output compared to a predetermined "reference" output, based on the difference between the input and the corresponding "reference" input [2]. FA have been used to interpret model predictions in applications including image classification [11], machine translation [5], text classification [2], event detection [8] and others. We made the first attempt on the interpretability of move structure recognition.

Integrated Gradients is a specific attribution method, which regards the feature dependent value as the cumulative gradient between the model input $x$ and the baseline $x'$ (for text models the baseline could be the zero embedding vector) [13]. Compared with other FA methods, the integral gradient method has high computational efficiency and is very effective in dealing with a wide range of text-based tasks.

## 3 Move Structure Recognition with Saliency Attribution

This paper proposes a method for recognizing the move structure in scientific paper abstracts with saliency attribution. Specifically, each sentence is first assigned a semantic move label that expresses the overall semantics. Then, each word is considered as a feature and its contribution (saliency value) to a specific move type is calculated. Finally, saliency values are used as evidence to enhance learning for move structure recognition.

This paper call a complete model that fully considers contextual information and uses saliency values as embedding enhancement learning as Saliency_SR. The framework is illustrated in Fig. 3, and technical specifics are presented below. We model the move structure recognition task as a single-sentence-based multi-label classification problem, for each sentence $s = [w_1, w_2, ..., w_n]$, one or more move labels $y$ are assigned, where $y$ belongs to $\mathcal{T}$, $\mathcal{T}$ is a set that includes all the predefined move types.

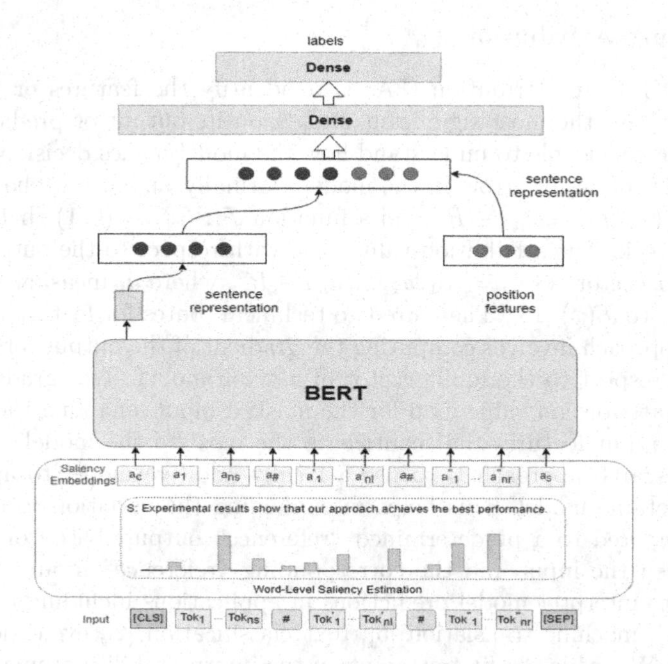


**Fig. 3.** The overview of our move structure recognition model (Saliency_SR). The backbone of the model is BERT. Based on the full consideration of contextual information, the model uses word saliency embeddings to enhance learning.

## 3.1 Sentence-Level Move Classification

First, each sentence $s$ is assigned a sentence-level move label $\mathcal{G}_S$ that conveys the overall semantic meaning of the move. Let the label be $\mathcal{G}_s = [g_1, g_2, ..., g_{|\mathcal{T}|}] \in R^{|\mathcal{T}|}$, where $g_i \in \{0, 1\}$ indicates whether $s$ can be $(g_i = 1)$ or not $(g_i = 0)$ assigned a move label of the $i^{th}$ move type. Next, a sentence-level move classifier is created to learn the mapping from $s$ to $\mathcal{G}_S$.

Notably, abstracts of scientific papers usually use some common structures to organize the information. Inspired by the encoding of positional information in words, we design a BERT-based sentence classifier that considers adjacent preceding and following sentences and positional information. Specifically, (1) each sentence is spliced with its adjacent preceding and following sentences and input into bert for encoding to obtain a text representation. (2) Use a combination of numerical features (the offset of the sentence relative to the chapter, the offset divided by the number of sentences in the chapter) to characterize the position of each sentence in the document. (3) Concatenate the text representation with the location feature to obtain the sentence representation and input it into two dense layers. The classification function uses sigmoid and the loss function uses multi-label binary cross-entropy loss:

$$\mathcal{L}\left(\mathcal{G}_s; X_s\right) = -\frac{1}{|\mathcal{T}|}\sum_{i=1}^{|\mathcal{T}|} g_i \cdot \log\left(o_i^s\right) + \left(1 - g_i\right) \cdot \log\left(1 - o_i^s\right) \tag{1}$$

where $X_s$ represents the input embedding of $s$ in BERT, $o^s \in R^{|\mathcal{T}|}$ is the logits vector computed by the classifier, and $o_i^s$ refers to the $i^{th}$ element of $o^s$.

### 3.2 Word-Level Saliency with Move

Based on the sentence-level move classifier, we employ Integrated Gradient [13] to compute the contribution (saliency value) of each word to the prediction. The integrated gradient defines the attribution of the $i^{th}$ input feature as the path integral of the straight line path from the baseline ${x_i}'$ to the input $x_i$ :

$$IntegratedGrads_i(x) ::= (x_i - {x_i}') \times \int_{a=0}^{1} \frac{\partial \mathcal{F}(x' + a \times (x - x'))}{\partial x_i} da \tag{2}$$

where $\mathcal{F} : R^N \rightarrow (0, 1)$ stands for a neural network and $\frac{\partial \mathcal{F}(x)}{\partial x_i}$ is the gradient of $\mathcal{F}(x)$ in the $i^{th}$ dimension [9].

To compute the saliency of each word $w_i$, more precisely, its BERT representation $x_i \in X_s$, we utilize the loss function as our desired model and evaluate it accordingly [16]:

$$\alpha_{w_i} = (x_i - x_i') \times \int_{\alpha=0}^{1} \frac{\partial \mathcal{L}\left(\mathcal{G}_s; X' + \alpha \times (X_s - X')\right)}{\partial x_i} d\alpha \tag{3}$$

where $X'$ is a sequence of all-zero vectors (serving as a reference input), and ${X_i}'$ denotes the $i^{th}$ element in $X'$. We then normalize $\alpha_{w_i}$ as a scalar value $a_{w_i}$ with a sentence-wise normalization:

$$a_{w_i} = e^{\|\alpha_{w_i}\|_2} / \sum_{n=1}^{N} e^{\|\alpha_{w_n}\|_2} \tag{4}$$

where $\|\|$ denotes the $L_2$ norm.

In actuality, we may not be interested in a word's saliency to the general move semantic $\mathcal{G}_S$, but instead in a specific move type $T \in \mathcal{T}$. Therefore, we replace $\mathcal{G}_S$ with the one-hot representation of $T$ in Eq. 3 for evaluation. Finally, we represent the word-level saliency of $w_i$ with respect to the move type $T$ by $a_{w_i}$.

### 3.3 Saliency Enhanced Move Structure Recognition

Using move saliency attribution, we develop a novel training approach for move structure recognition.

Recognizing that each word may have a different impact on predicted move labels, we incorporate a mechanism called word saliency embeddings (WSEs) into our model to capture such patterns. Specifically, we quantified the importance of each word to the move label as a value from 0 to 1 based on $a_{w_i}$, and then combined the saliency values into the model using a single embedding vector. This is similar to word embedding to increase the focus on words that are more important when recognizing the move structure. As shown in Fig. 4, the final input of the BERT is summed by four embedding vectors, namely Token Embeddings, Segment Embeddings, Position Embeddings and Saliency Embeddings.

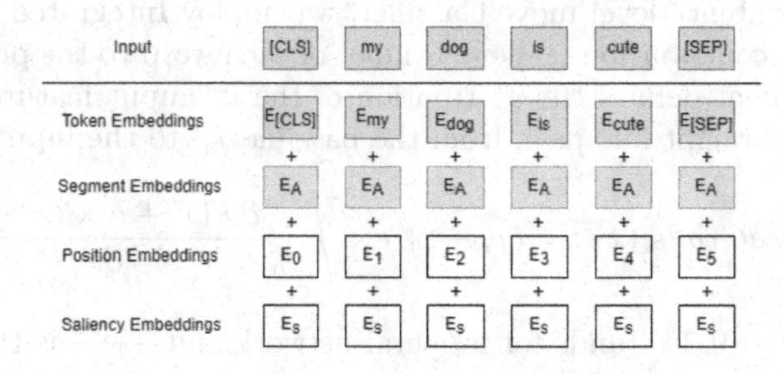


**Fig. 4.** The final input of the BERT.

## 4 Experimental Setup

### 4.1 Datasets

**Scientific Abstract Moves Dataset.** We have created a new dataset called the Scientific Abstract Moves Dataset (SAMD) by manually annotating abstracts of conference papers in the field of NLP. The dataset consists of 1,500 abstracts and 8,934 sentences, sourced from conferences such as ACL, EMNLP, and NAACL, which are part of the ACL Anthology[2].

The manual annotation process strictly followed the move analysis method, which is a widely recognized approach in the field of English for Specific Purposes (ESP) research. In this process, each sentence was considered as the smallest annotation unit, and a combination of top-down and bottom-up annotation methods was employed to determine the move label for each sentence. The annotation task was carried out by six professional annotators from the School of Foreign Languages, and the inter-annotator agreement was measured using the

[2] https://aclanthology.org/.

Kappa coefficient. The resulting Kappa coefficient was calculated to be 0.785, indicating substantial agreement among the annotators. This level of agreement implies that approximately 93.375% of the corpus was independently annotated consistently. The annotation scheme used in this study consists of an eight-move structure, which is as follows:

1) Background: The background section highlights the content of the literature review, background information, and relevant research.
2) GAP: The gap section identifies the knowledge gaps in previous research and illustrates aspects that were not covered in earlier studies, which is the motivation for conducting this research.
3) Method: The method section explains the methods, tools, and procedures used in the study, as well as the experimental design on which the study is based.
4) Purpose: The purpose section outlines the research objectives and research questions.
5) Result: The result section presents the study's findings and analysis, as well as the implications of the results.
6) Conclusion: The conclusion section indicates the contribution of the research to the discipline and summarizes the research's importance to the field.
7) Contribution: The contribution section emphasizes the research's impact on the discipline and is used to summarize the research's significance.
8) Implication: The implication section highlights the potential implications of the research results for future studies or their impact on practice.

**Table 1.** Statistics of the dataset for each move structure class, where BAC, GAP, MTD, PUR, RST, CLN, CTN and IMP represent Background, GAP, Method, Purpose, Result, Conclusion, Contribution and Implication respectively.

| Statistical | BAC | GAP | MTD | PUR | RST | CLN | CTN | IMP |
|---|---|---|---|---|---|---|---|---|
| #Sen | 1,773 | 1,026 | 2,441 | 1,447 | 1,110 | 748 | 321 | 78 |
| Percentage(%) | 19.8 | 11.5 | 27.3 | 16.2 | 12.3 | 8.4 | 3.6 | 0.9 |

The data volume statistics for each move in SAMD are shown in Table 1. Due to the characteristics of the writing style of nlp scientific papers, the data volume of each move label is unbalanced. The number of sentences introducing "method" and "purpose" is high, and the number of sentences introducing "contribution" and "implication" is relatively small.

**NLPContributionGraph Dataset.** In addition, in order to further verify the effectiveness of our method, relevant experiments were performed on the NLPContributionGraph Shared Task Dataset (NCG)[3]. NCG is a public dataset covering 24 NLP domain tasks, which defines 12 more fine-grained contribution sentence types. Table 2 presents the data statistics for both datasets in detail.

[3] https://ncg-task.github.io/.

**Table 2.** Statistics of SAMD and NCG, where #Type, #Doc, and #Sen indicate the number of move types, documents, and sentences respectively.

| Dataset | #Type | #Doc | Split | #Sen |
|---|---|---|---|---|
| SAMD | 8 | 1,500 | Training | 7,147 |
| | | | Test | 1,787 |
| NCG | 12 | 287 | Training | 4,200 |
| | | | Test | 1,051 |

### 4.2 Evaluation Metrics

This paper uses the following metrics to evaluate model performance: (i) Precision (P), Recall (R), and (Micro) F1, which are commonly used to assess move structure recognition models. (ii) (Macro) F1, which represents the average of class-wise F1 scores, and is lower for models that perform poorly on rare types compared to common types.

### 4.3 Implementations

Our move saliency attribution method employs a sentence-level classifier built on BERT-base with a batch size of 16 and a learning rate of $1e^{-5}$. For the move structure recognition model, we use BERT-base architectures with a batch size of 16. The word saliency embeddings are set to a dimension of 100 based on empirical analysis. Our dataset and code are made available at https://github.com/ljk1228/Saliency_SR.

## 5 Results and Analysis

### 5.1 Overall Results

To evaluate the effectiveness of Saliency_SR, the paper conducted comparisons with other models on two different datasets. (1) For SAMD, this model compare with BERT and BERT+Context, where the BERT+Context uses position information and the preceding and following sentences as additional contextual features. (2) For NCG, this model compare with UIUC_BioNLP [7], which uses BERT to encode the original sentence and its title in a parameter-sharing manner, and incorporates positional features of the sentence. This is also the best performing method in the NLPContributionGraph (NCG) shared task. In the variant models, +WSE means to supplement UIUC_BioNLP with word saliency embeddings. For the model and its variationswe conduct three runs, and calculate the average of their performance and standard deviations. Table 3 displays performances of different models.

Overall, we make the following observations: (1) Saliency_SR achieves the best Micro F1 score (78.53%) on SAMD, with an improvement of 1.81% over

**Table 3.** Results on SAMD and NCG, where P △, R △, and F1 △ indicate Precision, Recall, and Micro F1 respectively.

| Dataset | Method | P△ | R△ | F1△ |
|---|---|---|---|---|
| **SAMD** | BERT | 74.06 | 79.58 | 76.72 |
| | BERT+Context | 74.55 | 81.23 | 77.60 |
| | Saliency_SR | **75.01** | **82.34** | **78.53** |
| **NCG** | UIUC_BioNLP | 69.31 | 68.68 | 68.99 |
| | UIUC_BioNLP+WSE | **69.80** | **72.08** | **70.92** |

BERT and 0.93% over BERT+Context. We believe that the model not only better considers the order constraint relationship between context information and move labels, but also forces the model to focus on the more important part of the recognition results of specific move types. These two points effectively solve the problem of implicit combination of content corresponding to move structure. (2) UIUC_BioNLP+WSE shows good results on the NCG dataset, with improvements of 0.49%, 3.4%, and 1.93% in P, R, and Micro F1, respectively. These further prove that the method has good robustness and generalization capabilities.

### 5.2 Ablation Study

We undertake an ablation study in Table 4. In the variant models, +WSE and +Context denote supplementing BERT with word saliency embeddings and contextual information, respectively.

**Table 4.** Ablation study of different components. P △, R △, and F1 △ indicate Precision, Recall, and Micro F1 respectively; F1 ▽ denotes Macro F1.

| Method | P△ | R△ | F1△ | F1▽ |
|---|---|---|---|---|
| BERT | 73.78 | 80.22 | 76.81 | 68.17 |
| BERT+WSE | 74.07 | 80.41 | 77.11 | 69.98 |
| BERT+Context | 74.55 | 81.23 | 77.60 | 69.73 |
| BERT+Con+WSE | **75.01** | **82.34** | **78.53** | **70.56** |

The results showed that: (1) combining word saliency embeddings can improve the performance (P+0.29, R+0.19, Micro F1+0.30, and Macro F1+1.81). Saliency attribution can capture the importance of individual words or phrases in a sentence or text and help the model learn to identify key structural elements and relationships in a piece of text. This is especially important for tasks that heavily rely on understanding the order and organization of information. (2) Furthermore, context information is also helpful (P+0.77, R+1.01,

Micro F1+0.79, and Macro F1+1.56). It can provide background knowledge of the context and help the model better understand the meaning and function of words or phrases in a sentence. (3) When the two are combined, the best results are achieved. These two factors help the model learn the complex relationship between syntactic structure and contextual information, and ultimately enhance the robustness and generalization ability of the move recognition model.

### 5.3 Camparison with ChatGPT

In addition, we selected 30 abstracts of scientific papers and applied our model and ChatGPT for the recognition of the move structure. When calling the gpt-3.5-turbo API for experimentation, we first need to define a function for batch processing. The function is able to receive multiple sentences as input and return the result of move recognition for each sentence. At the beginning of the conversation, we provided the following instructions to gpt-3.5-turbo to ensure professionalism and accuracy of the experiment, as shown in the Table 5 below.

**Table 5.** The requested text content is provided in the form of a dictionary. "system" is used to give ChatGPT a statement at the beginning of the session, and "user" represents the content of the user's question. Inputinfo represents a sentence.

| Role | Content |
|---|---|
| "system" | The move structure of a scientific paper refers to the categorical composition of the linguistic rhetorical components of the academic discourse in the paper. Move recognition is essentially a classification problem in sentences. Now the moves are background, gap, method, purpose, result, conclusion, contribution, implication. Here are a few examples of move recognition: Detecting emotion in text allows social and computational scientists to study how people behave and react to online events. [background] However, developing these tools for different languages requires data that is not always available. [gap] This paper collects the available motion detection datasets across 19 languages. [purpose] We train a multilingual emotion prediction model for social media data, XLM-EMO. [method] The model shows competitive performance in a zero-shot setting, suggesting it is helpful in the context of low-resource languages. [results] We release our model to the community so that interested researchers can directly use it. [contribution] Below I will give you some sentences, these sentences are from scientific papers, please complete the step recognition. |
| "user" | "Inputinfo" + What is the move of this sentence? |

We also asked domain experts to evaluate the results without knowing which model produced them, using a percentage-based scoring system. Our model received a score of 80, while ChatGPT received a score of 65. These results indicate that Saliency_SR is better able to learn the relationships between syntactic structures and contextual information in text. Furthermore, fine-tuning the model for specific tasks enables it to be optimized specifically for that task, which results in better performance compared to more general pre-trained models like ChatGPT.

## 6 Conclusion

In this study, we proposed a new task – move structure recognition in scientific papers, and manually constructed a dataset SAMD. In addition, we analyze the causes of poor move recognition and build a Saliency_SR model. Experiments are conducted on two datasets and the results show that the proposed method achieves significant improvements and proves its effectiveness and generality. In future work, we will consider combining move structure recognition and review generation to automatically evaluate scientific paper abstracts, such as giving writing suggestions and providing reference sentences to help study, etc.

## References

1. Chen, J.: Towards Interpretability and Robustness of Machine Learning Models. University of California, Berkeley (2019)
2. Chen, J., Song, L., Wainwright, M., Jordan, M.: Learning to explain: an information-theoretic perspective on model interpretation. In: International Conference on Machine Learning, pp. 883–892. PMLR (2018)
3. Dayrell, C., et al.: Rhetorical move detection in English abstracts: multi-label sentence classifiers and their annotated corpora. In: LREC, pp. 1604–1609 (2012)
4. Ding, L., Zhang, Z., Liu, H.: Factors affecting rhetorical move recognition with SVM model. Data Anal. Knowl. Disc. **3**(11), 16–23 (2019)
5. Ding, Y., Liu, Y., Luan, H., Sun, M.: Visualizing and understanding neural machine translation. In: Proceedings of the 55th Annual Meeting of the Association for Computational Linguistics (Volume 1: Long Papers), pp. 1150–1159 (2017)
6. Liakata, M., Saha, S., Dobnik, S., Batchelor, C., Rebholz-Schuhmann, D.: Automatic recognition of conceptualization zones in scientific articles and two life science applications. Bioinformatics **28**(7), 991–1000 (2012)
7. Liu, H., Sarol, M.J., Kilicoglu, H.: Uiuc_bionlp at semeval-2021 task 11: a cascade of neural models for structuring scholarly NLP contributions. arXiv preprint arXiv:2105.05435 (2021)
8. Liu, J., Chen, Y., Xu, J.: Saliency as evidence: event detection with trigger saliency attribution. In: Proceedings of the 60th Annual Meeting of the Association for Computational Linguistics (Volume 1: Long Papers), pp. 4573–4585 (2022)
9. McCloskey, K., Taly, A., Monti, F., Brenner, M.P., Colwell, L.J.: Using attribution to decode binding mechanism in neural network models for chemistry. Proc. Natl. Acad. Sci. **116**(24), 11624–11629 (2019)

10. Roddy, M., Harte, N.: Neural generation of dialogue response timings. arXiv preprint arXiv:2005.09128 (2020)
11. Simonyan, K., Vedaldi, A., Zisserman, A.: Deep inside convolutional networks: visualising image classification models and saliency maps. arXiv preprint arXiv:1312.6034 (2013)
12. Song, W., Song, Z., Liu, L., Fu, R.: Hierarchical multi-task learning for organization evaluation of argumentative student essays. In: IJCAI, pp. 3875–3881 (2020)
13. Sundararajan, M., Taly, A., Yan, Q.: Axiomatic attribution for deep networks. In: International Conference on Machine Learning, pp. 3319–3328. PMLR (2017)
14. Swales, J.M.: Aspects of article introductions. No. 1, University of Michigan Press (2011)
15. Upton, T.A., Cohen, M.A.: An approach to corpus-based discourse analysis: the move analysis as example. Discourse Stud. **11**(5), 585–605 (2009)
16. Wallace, E., Tuyls, J., Wang, J., Subramanian, S., Gardner, M., Singh, S.: AllenNLP interpret: a framework for explaining predictions of NLP models. arXiv preprint arXiv:1909.09251 (2019)
17. Wang, L., Liu, X.: Constructing a model for the automatic identification of move structure in English research article abstracts. TEFLE, pp. 45–50+ (2017)

# Moral Essential Elements: MEE-A Dataset for Moral Judgement

Ying Wang[1], Xiaomeng Du[1], Pengyuan Liu[1,2], and Dong Yu[1](✉)

[1] Beijing Language and Culture University, Beijing, China
yudong@blcu.edu.cn

[2] National Language Resources Monitoring and Research Center for Print Media, Beijing, China

**Abstract.** Moral judgments, including moral polarity judgments, moral intensity judgments, and moral type judgments, are important in helping people to understand the moral characteristics of behaviors. Moral judgments do not rely on all elements in behavior, but on those strongly related to them, which are named as moral essential elements in this paper. By conducting research on the moral essential elements in machine morality, it can help machines adapt to downstream tasks such as eliminating moral bias while making existing artificial intelligence more ethical. Existing research on machine morality knowledge contains moral vocabulary, moral sentences, etc. This information has been able to help machines achieve relatively good results when making moral judgments, but there are still problems such as incomplete injection of information and multiple actions present in a sentence with different moral judgments. To address the above problems, this paper summarizes and designs a moral judgments system, proposes the concept of moral essential elements, constructs a moral essential elements dataset containing more than 10,000 behaviors, identifies behavioral words and their moral essential elements in sentences using the traditional model and the large model ChatGPT respectively, then makes moral judgments.

**Keywords:** artificial intelligence ethics · moral judgment · moral essential elements

## 1 Introduction

With the development of AI technology, the ethical issues of machines have received more and more attention, and at the same time become an urgent issue


H. Wang et al. (Eds.): CCKS 2023, CCIS 1923, pp. 259–269, 2023.
https://doi.org/10.1007/978-981-99-7224-1_20

of AI ethics. So what kind of ethics is needed for the moral judgments made by machines, and what information does it need to circumvent immorality so as to avoid negative impacts on human beings and society, while breaking the restrictions on the development of AI?

Currently, in previous work, such as the Moral Foundations Dictionary (MFD), [5] used the classification system of moral foundations theory and polarity labels to label moral words. These datasets contain much information. However, we found that moral judgments do not depend on all elements in the sentences, but only on some of them, as shown in Fig. 1. And different elements of behavior lead to various moral judgments. We name some of the specific elements as moral essential elements on which these moral judgments depend.

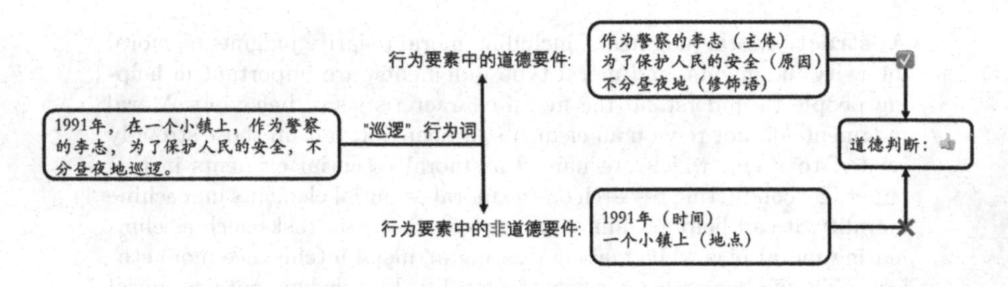


**Fig. 1.** The essential elements of moral judgment.

To address the above issues, this paper summarizes and designs a moral judgment system, and proposes the concept of moral essential elements, which refers to the elements of behavior that are strongly related to moral judgment. It includes morally relevant subjects, places, and other elements of the behavior, which are important information in the task of machine moral judgment. Meanwhile, we constructed a machine moral judgment essential elements dataset MEE (moral essential elements dataset) with more than 10,000 behaviors, aiming to investigate the identification of the essential elements of behavioral moral judgment. The dataset is shown in Fig. 2. Then, we identify the words indicating behavior in the sentences, the moral essential elements, and the moral judgments they form on the traditional and large language models. Our data can be found at https://github.com/blcunlp/MEE.

源文本：

趁老人熟睡的时候，小偷盗窃了老人的东西，警察抓住了小偷。

行为1&道德判断&道德要件：

| 行为词：抓住 |
|---|
| 道德属性：正向（道德倾向）；道德（道德强度）；爱岗敬业（道德类型） |
| 警察（主体）；小偷（客体） |

行为2&道德属性&道德要件：

| 行为词：盗窃 |
|---|
| 道德属性：负向（道德倾向）；不道德（道德强度）；遵纪守法（道德类型） |
| 论元：趁老人熟睡的时候（背景）；小偷（主体）；老人的东西（客体） |

**Fig. 2.** An example of the dataset.

## 2 Related Work

In recent years, textual morality has gradually come to the attention of researchers in the field of natural language processing research. Currently, [4] attempted to learn a priori knowledge of moral value orientations using comic pictures. [14] proposed a moral judgment system including the moral word extraction stage and moral judgment stage. [1] classified moral judgments into two valence classes: positive and negative valence. [8] introduced the description of moral dilemmas as utility functions. Later [3] introduce moral stories.

Moral-related datasets can help researchers gain insight into human moral behaviors and decisions. The moral corpus collected by [15] consists of a set of sentences labeled as "positive" or "negative" from a moral perspective. [9] followed the principle of consequentialism and other principles [11], constructed sentence-level moral sentences. [6] present a corpus called "MFTC", which collects 35,108 tweets on Twitter. [12] collected 16,123 comments on Reddit and proposed a Reddit corpus called "MFRC". [13] constructed a Chinese moral dictionary that uses universal social norms to classify words by type and label separately.

There are also a few existing studies on moral essential elements such as comprehensive and non-redundant moral information that complement moral judgments, for which we construct a relevant dataset for a more in-depth investigation.

# 3 Theory

## 3.1 Moral Judgment System

The moral judgment system covers three moral judgments: moral polarity, moral type, and moral intensity. Moral polarity refers to the conformity of behavior to moral norms and can be divided into positive moral polarity and negative moral polarity. We also focus here on these two moral polarities, namely positive morality and negative morality.

Moral intensity specifically refers to the moral pressure or urgency involved in the ethical issue itself, [7] divides moral intensity into six dimensions, severity of consequences, social consistency, and so on. This paper divides moral intensity into four dimensions: very immoral, immoral, moral, and very moral.

Moral types are the different types of behavior involved and can form a variety of classifications based on different criteria. The moral types discussed in this paper are divided into three categories according to the sphere of influence, responsible for one's own behavior, responsible for the behavior of others, and responsible for the behavior of society, i.e., person and self, person and others, person and society, and referring to [13]'s delineation of ethical behavior at the level of the scenario environment, then subdivided. In this paper, the moral judgment system is designed inductively to distinguish the different impacts of different moral behaviors. The moral judgment system designed in this paper is summarized as shown in Table 1. In particular, for example, the moral type of "遵纪守法的行为" and "不遵纪守法的行为" are marked as "遵纪守法". For we have made a distinction between them in moral polarity.

**Table 1.** Classification of moral judgment systems.

| 道德判断体系 | 道德判断体系的分类 |
|---|---|
| 道德极性 | 正向，负向 |
| 道德强度 | 非常不道德，不道德，道德，非常道德 |
| 道德类型 | 人与自己：诚实守信，文明礼貌，爱岗敬业，奉献 |
| | 人与他人：助人为乐，尊老，爱幼 |
| | 人与社会：遵纪守法，办事公正，保护环境，爱护公物 |

## 3.2 Definition of Moral Essential Elements

Moral essential elements refer to those elements of the behavior elements that are strongly related to morality and that will have an impact on the moral judgment of the behavior. We define moral essential elements mathematically: for a sentence $S$, the set of words indicating behaviors is $E = \{e_1, e_2...\}$, then for each act $e \in E$, the set of argument elements is denoted as $W = \{w_1, w_2...\}$, then the moral essential elements, denoted as the set $A = \{a_1, a_2...\}$, $A \subseteq W$. The textual definition of moral essential elements is shown in Table 2.

**Table 2.** The defination of moral essential elements.

| 道德要件 | 道德要件的定义 |
|---|---|
| 主体要件 | 与道德判断相关的施事 |
| 客体要件 | 与道德判断相关的受事 |
| 时间要件 | 与道德判断相关的时间 |
| 地点要件 | 与道德判断相关的地点 |
| 原因要件 | 与道德判断相关的原因 |
| 结果要件 | 与道德判断相关的后果 |
| 方式要件 | 与道德判断相关，发出行为的方式 |
| 工具要件 | 与道德判断相关，发出行为使用的工具 |
| 背景要件 | 与道德判断相关，发出行为时所处状态 |
| 修饰语要件 | 与道德判断相关，修饰行为词的词语，如“虚假诊疗”中的“虚假”，大多为副词 |
| 频率要件 | 与道德判断相关，行为产生的次数 |

# 4 Dataset Construction

## 4.1 Data Sources

The dataset of machine moral judgment was intercepted from the part of CMOS dataset [9] labeled with positive and negative polarity. In addition, since the source of the CMOS dataset could not cover all the moral types we designed, we additionally expanded part of the dataset by collecting news from various sections of websites such as China Good People Website and Southern Website through crawling techniques.

Secondly, the format of collected data by crawling websites could not be directly labeled for format and other problems, so we pre-selected and cleaned the collected dataset, removing sentences with incomplete and unknown meanings, such as “台湾商人到西安参，瓶可乐醒来400万珠宝没了”, and “黑出租车计价器跳得，子上车就没手机信号”, they were both removed to reduce the propagation of errors in the subsequent manual annotation.

## 4.2 Data Annotation

**Annotating Rules.** We recruited 13 annotators with linguistic backgrounds for the annotation. They should mark five areas: moral polarity, moral type, moral intensity, words indicating behaviors, and moral essential elements. For some issues that arise in the annotation process, we make the following provisions:

**Special Sentences Ignore Syntactic Structure.** This includes special sentences such as “把字句” and “被字句”. For example, “她正在读医学院的22岁女儿，在健身房锻炼时，“因被误认成黑帮成员女友被枪杀”.

**Avoid the Problem of Disambiguation.** The denotational phenomenon can avoid the problems of bloated and redundant utterances caused by the repeated occurrence of the same words [2]. In order to reduce the impact of the problem, we stipulate that the subject and object elements are marked to the most specific words, such as “刘启荣在东升金海岸大门边的街道上散步，突然走上来一对老夫妇握着他的手，动情地说：刘书记，感谢你”. “刘启荣” and ”刘秘书” refer to the same person, so we mark “刘启荣” as the subject essential element.

### 4.3 Annotating Results

**Moral Judgement.** Statistically, the distribution of the number of moral judgments is shown in Table 3, Table 4, and Table 5. Among all moral sentences, sentences containing multiple behaviors account for 15% of them. Among the multi-behavior sentences, each sentence containing different moral judgments accounted for 87% of them. Because the collected corpus belongs to the sentence level, the percentage of sentences containing multiple behaviors is relatively small, but it can be seen from the moral types varying in a sentence that the probability of different moral judgments in a multi-behavior sentence is much greater than the probability of the same.

Table 3 indicates that in the MEE dataset, the amount of negative data is about 1.4 times more than positive, and the data distribution is fairly balanced in this dimension. And in the statistics of moral intensity, the largest proportion of moral behaviors and a very small proportion of very moral behaviors indicate the trend of moral judgment in news reports. From Table 4, it can be seen that the moral type parent category of “人与社会” has the largest range of moral influence; among the subcategories of moral types, the type of “遵纪守法” accounts for about 50%, and the data are extremely unbalanced. The reasons for this are the one-sidedness of the collected data and the fact that the news reports in real society mostly focus on the moral type of “遵纪守法” while reflecting the moral orientation of human beings concerned with social life.

**Table 3.** The number of each moral tendency.

| 道德倾向 | 负向 | 正向 |
|---|---|---|
| 数量 | 9738 | 6795 |

**Table 4.** The number of each moral intensity.

| 道德强度 | 不道德 | 道德 | 非常不道德 | 非常道德 |
|---|---|---|---|---|
| 数量 | 9738 | 6795 | 2111 | 282 |

**Table 5.** The number of each moral type.

| 道德类型 | 道德类型父类 | | | 道德类型子类 | | | | | | | | | | |
|---|---|---|---|---|---|---|---|---|---|---|---|---|---|---|
| | 人与社会 | 人与他人 | 人与自己 | 遵纪守法 | 助人为乐 | 爱岗敬业 | 奉献 | 文明礼貌 | 尊老 | 诚实守信 | 办事公正 | 爱幼 | 保护环境 | 爱护公物 |
| 数量 | 10269 | 5307 | 3350 | 9859 | 2541 | 2076 | 1367 | 1354 | 610 | 510 | 199 | 199 | 145 | 66 |

**Moral Essential Elements.** From Table 6, we conclude that for moral judgment, in addition to "主体要件" and "客体要件", "方式要件" and "修饰语要件", etc. are all important influencing factors for moral judgment.

**Table 6.** The number of each moral essential elements.

| 要件分类 | 主体要件 | 客体要件 | 方式要件 | 修饰语要件 | 原因要件 | 结果要件 | 状态要件 | 地点要件 | 工具要件 | 频率要件 | 时间要件 |
|---|---|---|---|---|---|---|---|---|---|---|---|
| 数量 | 18040 | 16998 | 2614 | 2525 | 2192 | 2104 | 1788 | 1503 | 1168 | 985 | 782 |

## 5 Experiments and Analysis

In this paper, we follow the event extraction model Casee [10] and use the large language model ChatGPT to identify words indicating behaviors and moral essential elements， determine the corresponding moral parent type, moral subtype, moral tendency, and moral intensity in zero-shot and few-shot scenarios. In the experiment of ChatGPT, we give sentences, and definitions of moral judgments, then let the model generate answers from the original text, such as "句子A。道德要件是指对道德判断产生影响的行为要素。道德极性有正向和负向。用原文回答句子A中有哪些行为词？这些行为词各自的道德极性是什么？用原文回答影响每个行为词道德极性的道德要件有哪些？" and for the few-shot scenario, we give three standard test samples that have been labeled.

The results of the conventional model under different subtasks, words indicating behaviors and moral essential element identification and moral judgment, and the results of the large model in the zero-Shot, few-shot scenarios, respectively, are shown in Table 7, 8, 9 and 10.

From the above table, we can see that the traditional model Casee does not solve the task of moral judgment and identification of moral essential elements very well. The reasons for this are, first, the data problem, whether it is the moral tendency, moral type, or moral intensity, our data are unbalanced, so the recall of the model is low. Second, we use a cascade decoding model, and the pipeline propagation will lead to cascade errors. Third, the complexity of the Chinese language leads to the fact that the real span of marked words indicating behaviors and moral essential elements is more difficult to identify.

**Table 7.** The performance of models in words indicating behavior and moral essential elements identification and moral tendency.

| Models | 行为词识别 | | | 道德极性判断 | | | 道德要件识别 | | |
|---|---|---|---|---|---|---|---|---|---|
| | P | R | F1 | P | R | F1 | P | R | F1 |
| casee | 50.9 | **68.0** | 58.2 | 49.4 | 67.3 | 56.9 | 55.9 | 50.2 | 52.9 |
| ChatGPT-zero shot | 63.2 | 67.5 | 65.6 | 64.3 | **70.7** | 67.1 | 60.4 | 58.3 | 59.1 |
| ChatGPT-few shot | **67.3** | 65.2 | **66.2** | **70.8** | 67.7 | **68.9** | **67.2** | **65.5** | **67.3** |

**Table 8.** The performance of models in words indicating behavior and moral essential elements identification and moral intensity.

| Models | 行为词识别 | | | 道德强度判断 | | | 道德要件识别 | | |
|---|---|---|---|---|---|---|---|---|---|
| | P | R | F1 | P | R | F1 | P | R | F1 |
| casee | 51.4 | 59.8 | 55.3 | 35.4 | **48.4** | 40.9 | 38.6 | 44.1 | 41.2 |
| ChatGPT-zero shot | 61.2 | **67.3** | 64.8 | 43.2 | 42.7 | 42.9 | 45.2 | **58.3** | **51.2** |
| ChatGPT-few shot | **68.3** | 65.2 | **67.0** | **47.1** | 43.7 | **45.1** | **50.1** | 50.0 | 49.9 |

**Table 9.** The performance of models in words indicating behavior and moral essential elements identification and moral parent type.

| Models | 行为词识别 | | | 道德父类判断 | | | 道德要件识别 | | |
|---|---|---|---|---|---|---|---|---|---|
| | P | R | F1 | P | R | F1 | P | R | F1 |
| casee | 49.7 | 63.1 | 55.6 | 38.7 | 54.2 | 45.2 | 43.3 | 45.0 | 44.2 |
| ChatGPT-zero shot | 55.3 | 62.1 | 58.4 | 45.6 | 57.8 | 50.6 | 45.2 | **58.3** | 51.1 |
| ChatGPT-few shot | **63.3** | **67.2** | **65.5** | **54.9** | **66.4** | **60.2** | **53.4** | 57.8 | **55.1** |

**Table 10.** The performance of models in words indicating behavior and moral essential elements identification and moral subtype.

| Models | 行为词识别 | | | 道德子类判断 | | | 道德要件识别 | | |
|---|---|---|---|---|---|---|---|---|---|
| | P | R | F1 | P | R | F1 | P | R | F1 |
| casee | 74.0 | 50.7 | 60.1 | 56.2 | 43.5 | 49.1 | 58.8 | 30.9 | 40.5 |
| ChatGPT-zero shot | 72.4 | 58.3 | 65.2 | 58.7 | 46.5 | 47.9 | 62.3 | 40.1 | 50.2 |
| ChatGPT-few shot | **78.7** | **62.4** | **75.5** | **63.4** | **51.2** | **56.9** | **69.8** | **45.2** | **54.5** |

Figure 3 is a response about moral subtype from ChatGPT in a few-shot scenario. We can see that its response is incomplete. But in terms of overall performance, ChatGPT improves in most metrics in both zero-shot and few-shot scenarios. In zero-shot scenarios, there are problems in that ChatGPT's responses do not match the original meaning. For example, in the sentence "从私家车的车窗里，不停地有橘子皮被扔出来......", the result essential element of ChatGPT feedback is "私家车被扔了很多橘子皮" which is totally different from the meaning of "私家车中扔出橘子皮". In the few-shot scenario, ChatGPT showed some shortcomings, but also strong learning ability. After inputting the sentence "4名男子专抢滴滴司机" and the prompt of the question, ChatGPT answered "其道德要件包括：主体要件："4名男子"，客体要件："滴滴司机"，方式要件："专抢"，时间要件：未明确说明", a response that demonstrates ChatGPT's strong natural language understanding.

| 输入（Input） |
|---|
| 中南部克拉伦登区一名警员前一天试图阻止劫匪在街头抢劫两名中国公民，中弹殉职。<br>A police officer in the south-central Clarendon district was shot and killed in the line of duty the day before when he tried to stop a robber who was robbing two Chinese citizens on the street. |
| 输出（Output） |
| 行为词：试图阻止 \|\| 抢劫<br>word indicating behavior：tried to stop \|\| robbing<br>道德子类型："办事公正" 或 "爱岗敬业" \|\| 非常不道德<br>moral subtype："Fairness" or "dedication to the job" \|\| very immoral<br>道德要件：主体要件："一名警员"；客体要件："劫匪在街头抢劫两名中国公民"；原因要件："试图阻止" \|\| 客体要件："两名中国公民"<br>moral essential elements：subject essential element："A police officer"；object essential element："a robber who was robbing two Chinese citizens on the street"；reason essential element："tried to stop" \|\| object essential element:："two Chinese citizens"<br>与所问问题无关的额外回答：需要注意的是，道德类型、倾向和强度的判断都是基于常识和社会价值观的，并非严格的定义。不同人可能会有不同的判断。同时，道德要件的分类也可以根据不同的分类标准进行，以上分类只是其中一种可能。<br>additional response not related to the question asked：It is important to note that judgments of moral type, tendency, and intensity are based on common sense and social values, and are not strictly defined. Different people may have different judgments. Also, the classification of moral elements can be based on different classification criteria, and the above classification is just one of the possibilities. |

**Fig. 3.** A response of ChatGPT.

# 6 Summary and Future Work

The main work of this paper is to construct a behavior-oriented moral judgment imperative dataset with tens of thousands of levels and identify moral judgments and moral essential elements of words indicating behaviors. In the experimental part, we use the traditional model and the large model to complete the identification task and get the preliminary results, respectively. As we can see, we are not yet able to perform the recognition effectively because of the cascading errors of the model and the complexity of Chinese. In the future, we will make full use of the constructed dataset to optimize the model approach and design high-performance recognition models. Also, try to identify the essential elements of machine moral judgment more efficiently with better prompt learning and other methods.

**Acknowledgements.** This work is funded by the Humanity and Social Science Youth foundation of Ministry of Education (19YJCZH230) and the Fundamental Research Funds for the Central Universities in BLCU (No. 21PT04).

## References

1. Botzer, N., Gu, S., Weninger, T.: Analysis of Moral Judgement on Reddit (Jan 2021). http://arxiv.org/abs/2101.07664,arXiv:2101.07664
2. Clark, K., Manning, C.D.: Improving coreference resolution by learning entity-level distributed representations (2016)
3. Emelin, D., Bras, R.L., Hwang, J.D., Forbes, M., Choi, Y.: Moral Stories: Situated Reasoning about Norms, Intents, Actions, and their Consequences (Dec 2020). http://arxiv.org/abs/2012.15738,arXiv:2012.15738
4. Frazier, S., Nahian, M.S.A., Riedl, M., Harrison, B.: Learning Norms from Stories: A Prior for Value Aligned Agents (Dec 2019). http://arxiv.org/abs/1912.03553,arXiv:1912.03553
5. Garten, J., Boghrati, R., Hoover, J., Johnson, K.M., Dehghani, M.: Morality Between the Lines: Detecting Moral Sentiment In Text
6. Hoover, J., et al.: Moral Foundations Twitter Corpus: A collection of 35k tweets annotated for moral sentiment. preprint, PsyArXiv (Apr 2019). https://osf.io/w4f72
7. Jones, T.: Ethical decision making by individuals in organizations: an issue-contingent model. Acad. Manag. Rev. **16**, 366–395 (1991)
8. Kim, R., et al.: A Computational Model of Commonsense Moral Decision Making. In: Proceedings of the 2018 AAAI/ACM Conference on AI, Ethics, and Society, AIES 2018, pp. 197–203. Association for Computing Machinery, New York (Dec 2018)
9. Peng, S., Liu, C., Deng, Y., Yu, D.: (morality between the lines: Research on identification of Chinese moral sentence). In: Proceedings of the 20th Chinese National Conference on Computational Linguistics, pp. 537–548. Chinese Information Processing Society of China, Huhhot, China (Aug 2021), https://aclanthology.org/2021.ccl-1.49
10. Sheng, J., et al.: Casee: a joint learning framework with cascade decoding for overlapping event extraction (2021)
11. Thiroux, J.P.: Ethics: Theory and Practice. Pearson Prentice Hall, Boston (2008)
12. Trager, J., et al.: The Moral Foundations Reddit Corpus (Aug 2022). arXiv:2208.05545
13. Wang, H., Liu, C., Yu, D.: (Construction of a Chinese Moral Dictionary for Artificial Intelligence Ethical Computing). In: Proceedings of the 19th Chinese National Conference on Computational Linguistics, pp. 539–549. Chinese Information Processing Society of China (2020), https://aclanthology.org/2020.ccl-1.50, event-place: Haikou, China

14. Yamamoto, M., Hagiwara, M.: A moral judgment system using evaluation expressions. Trans. Japan Soc. Kansei Eng. **15**(1), 153–161 (2016)
15. Yamamoto, M., Hagiwara, M.: A moral judgment system using attention-based distributed representation and co-occurrence information. Inter. J. Affect. Eng. **17**(2), 137–145 (2018)

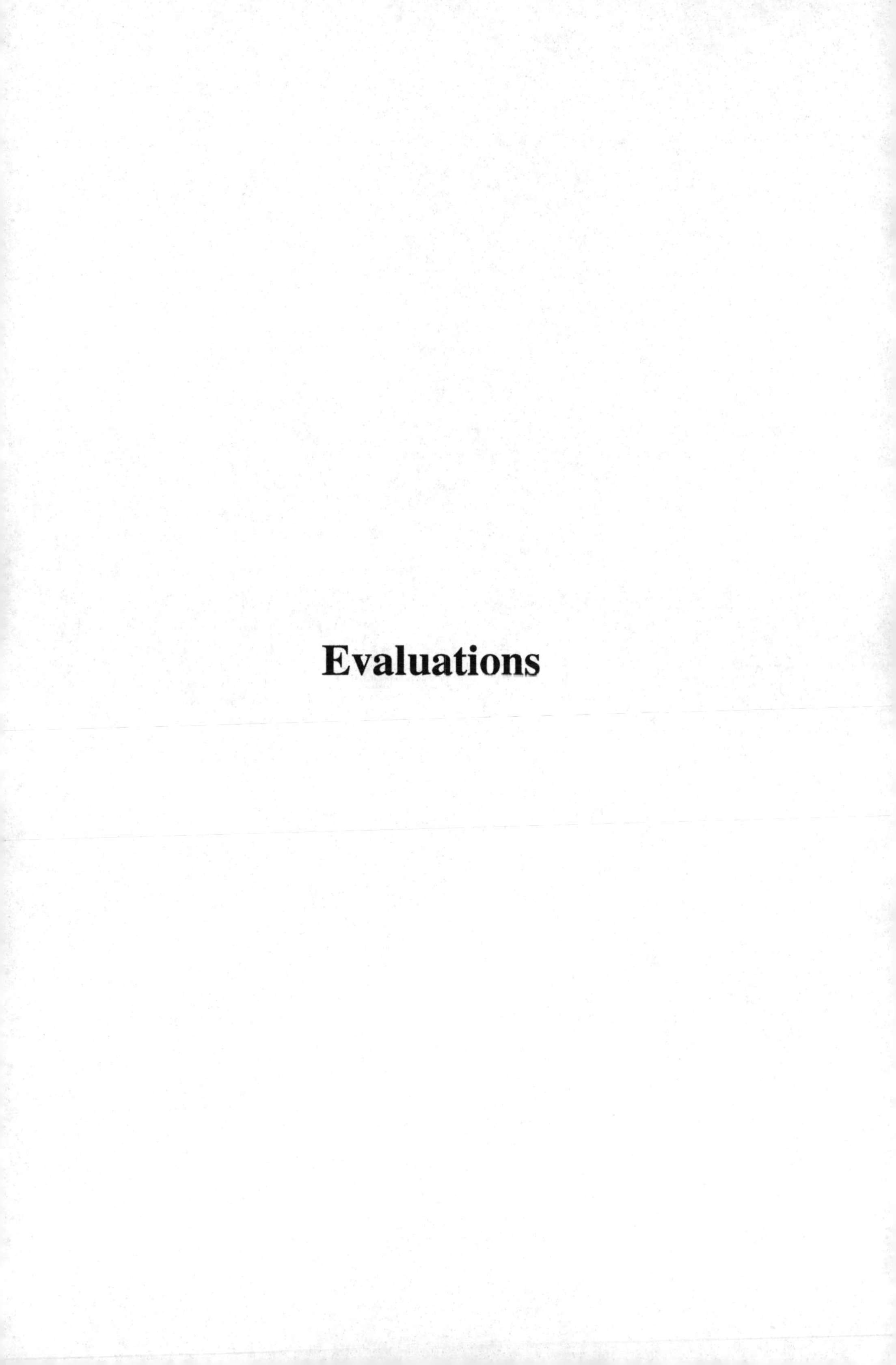

# Evaluations

# Improving Adaptive Knowledge Graph Construction via Large Language Models with Multiple Views

Yilong Chen[1,2], Shiyao Cui[1(✉)], Kun Huang[1,2], Shicheng Wang[1,2], Chuanyu Tang[1,2], Tingwen Liu[1,2(✉)], and Binxing Fang[1]

[1] Institute of Information Engineering, Chinese Academy of Sciences, Beijing, China
{chenyilong,cuishiyao,huangkun,wangshicheng,tangchuanyu, liutingwen}@iie.ac.cn

[2] School of Cyber Security, University of Chinese Academy of Sciences, Beijing, China

**Abstract.** Knowledge graph construction (KGC) aims to build the semantic network which expresses the relationship between named entities. Despite the success of prior studies, it is struggling to accommodate existing KGC models with evolving entity-relation knowledge schema. In this paper, we propose a schema-adaptive KGC method driven by the instruction-tuning large language models (LLM). We fine-tune a LLM with tailored KGC corpus, through which the generalization ability of LLMs are transfered for KGC with evolving schema. To alleviate the bias of a single LLM, we integrate the superiority of several expert models to derive credible results from multiple perspectives. We further boost KGC performances via an elaborately designed schema-constrained decoding strategy and a LLM-guided correction module. Experimental results validate the advantages of our proposed method. Besides, our method achieved the first place in the first task of CCKS-2023 Knowledge Graph Construction.

**Keywords:** Large language models · Knowledge graph construction

## 1 Introduction

Knowledge Graph (KG) [12] is a semantic network which expresses the semantic relationship between named entities. Since KG facilitates various applications [15], KG Construction [12] has gain great research attention in recent years.

Despite of the success of prior studies [2,7,11], it is struggling to achieve flexible knowledge updates across different knowledge schema, since these KGC models are developed via once-and-for-all training. Nevertheless, as the volume of data increases across scenarios and domains, it is significant to accommodate KGC models with evolving knowledge schema. To address the above challenge, it is intuitive to incrementally re-train the KGC models with new-class data, namely incremental learning [3]. However, they are subject to the catastrophic


H. Wang et al. (Eds.): CCKS 2023, CCIS 1923, pp. 273–284, 2023.
https://doi.org/10.1007/978-981-99-7224-1_21

forgetting [10] problem, as a cost of adaption to new schema. Fortunately, recent studies show that large language models (LLMs) maintain a great generalization ability in various downstream tasks, which can perform adaptively across scenarios with instruction-tuning.

Motivated by this, we intend to fine-tune a fundamental LLM (e.g. LLaMA) model with elaborately designed KGC corpus and instructions, to construct an Instruction-Driven Adaptive Knowledge Graph. However, there are extra issues in relational triplets across schema. For instance, the requirements for prior knowledge in RTE differ with the schema. Besides, the interdependence of the elements in a relation triplet varies with the schema. Hence, it is crucial to aggregate relational triplets predicted by different models and prompts to facilitate the KGC procedure.

In this paper, we propose a method towards Adaptive Knowledge Graph Construction with evolving schema. Our method consists of three main components. First, we fine-tune KGC-tailed LLMs with elaborately designed corpus for adaptable KGC. Second, inspired by the superiority of ensemble learning, we enhance the prediction results with parametric knowledge utilizing distinct LLMs. Finally, we design a schema-constrained decoding strategy with automated correction, in order to mitigate the general errors and further improve the prediction quality. We conduct extensive experiments on the InstructKGC task. The experimental results demonstrate that our method exhibits desirable generalization, accuracy, and completeness in low-resource, cross-task scenarios. Our approach achieves 5.28% improvement comprehensively over the baseline method, with 7.22% improvement in terms of F1 score and 4.23% improvement in terms of Rouge-2 score. Our method won the first place with a significant advantage in the CCKS2023 instruction-driven adaptive knowledge graph construction task.

Our contributions are summarized as follows:

- We develop a KGC-tailored large language model with strong generalization capabilities across knowledge schema.
- We introduce a Multi-view Experts (MvE) framework, which integrates different expert models from multiple views, to mitigate the generative bias for KGC.
- We propose the GPT-guided Supervision and Schema-constrained Decoding method which can effectively correct general errors in reasoning results.
- Experimental results demonstrate that our method achieves highly competitive performance and secures the first-place position in the competition.

## 2 Task Definition

Instruction-Driven Adaptive Knowledge Graph Construction (InstructionKGC) is a task that updates and optimizes the structure of a knowledge graph based on user instructions. The goal of this task is to renew the content of the knowledge graph according to user requirements, achieving more accurate and efficient information retrieval and reasoning.

In this task, we mainly refer to the objectives defined in the CCKS2023 competition. The model extracts entities and relationships of specific types based on the instructions provided by the user, in order to construct a knowledge graph. Unlike conventional triplet extraction tasks, the specified entities and categories are more diverse and may not be fully present in the training set. Additionally, 10% of the data in the test set contains some missing information, which requires the model to excavate own reasoning abilities for inference as well as completion. The competition provides a comprehensive definition of the task and evaluation metrics, along with a training set of 5000 samples and a test set of 1000 samples without ground truth. The detailed descriptions are displayed here[1].

## 3 Method

In this section, we propose a novel method to tackle the above-mentioned challenges, which mainly consists of three essential components. Specifically, first, we collect diverse instruction data and then fine-tune a large language model in an efficient way, which enables to adapt generalization ability to downstream KGC task. Next, we design a MvE (Multi-view Experts) module, which combines multiple models to promote performance utilizing their respective strengths. In this way, we alleviate the accumulation errors and instability during the generation process. To address the general errors existed in distinct models, we employ schema-constrained decoding strategy and devise a supervised model for correction, achieving a significant improvement on results.

### 3.1 KGC-Tailored Large Language Models

Since large language models, such as ChatGPT, are mainly devised for dialogue scenarios, apparently it is inconsistent with downstream information extraction tasks, leading to low accuracy and completeness. Taking the powerful generalization ability of LLMs into account, we tend to fine-tune a specific large language model for knowledge graph construction with various information extraction tasks. Specifically, Fig. 1, the overall procedure can be divided into three stages: data pool construction, instruction pool construction, and multi-task instruction fine-tuning. We elaborate the details as follows.

**Data Pool Construction.** In this sub-module, we aim to collect a substantial amount of data for information extraction tasks, which includes: the collection of 32 publicly available datasets, covering three types of IE tasks [1]; the CCKS Instruction KGC competition [5], which provide around 5000 knowledge graph data samples as training sets. During the dataset construction process, we aim to ensure balance in task types and domains. We design dozens of prompt templates to transform the datasets into a unified instruction set following a consistent standard.

**Instruction Pool Construction.** It is well-known that improving the quality of instructions, especially the diversity characteristics, is beneficial to gather

[1] https://tianchi.aliyun.com/competition/entrance/532080/information.

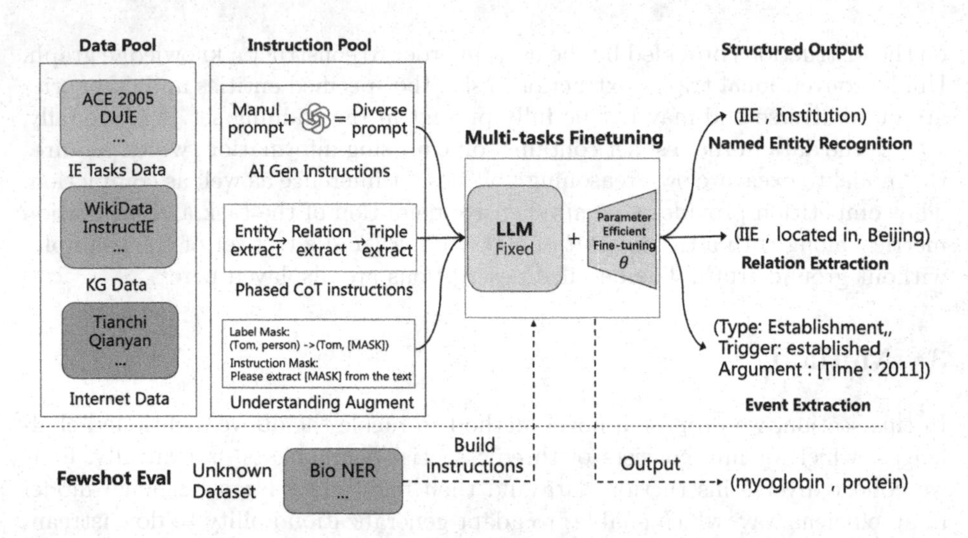


**Fig. 1.** The overall process of fine-tuning large models for knowledge graph construction tasks.

high-quality data and further promote the downstream tasks. To be specific, our instructions consist of three key components: task descriptions, constraints, and inputs. The task descriptions provide concrete information on how to perform extraction process. Constraints define the valid output labels and guide the mapping from predicted outputs to semantic concepts and the inputs are the original texts. Notably, we first construct instructions manually as seeds and then utilize Prompt GPT3.5 to obtain enhanced instruction prompts. Finally we filter the results manually and select the appropriate instructions to ensure the quality of instructions.

Considering that the model is not explicitly trained with structured information generation tasks during pretraining. We decompose complex tasks into sub-tasks. For instance, in the Relation Extraction task, we introduce entity pair extraction and entity pair relation recognition. The introduction of these subtasks provides the model with more diverse training signals, thereby improving the model's performance and generalization ability.

**Multi-task Instruction Fine-Tuning.** The instruction fine-tuning framework consists of two components: the base model and fine-tuning methods. The base model is a large-scale, open-source language model pre-trained on extensive datasets. We thoroughly investigated existing open-source base models and curated a GitHub repository[2]. After extensive research, we selected KnowLM-ZhiXi [9] and ChatGLM [13] as the base models. KnowLM-ZhiXi was a further pre-trained LLaMA model on Chinese corpora to boost its Chinese comprehension. ChatGLM was selected as the base model due to its advantages in deployment, sequence length, and training capabilities.

[2] https://github.com/Longyichen/Alpaca-family-library.

To adapt the pre-trained models to downstream tasks, we employ LoRA [6] technique to fine-tune the fundamental model, which is an efficient manner to reduce computation costs and facilitate practicability. Especially, LoRA freezes the pre-trained parameters and introduces trainable layers (low-rank decomposition matrices) to Transformer blocks. The implementation details are described in Sect. 4.2. Through fine-tuning, we achieved the following improvements:

- Augmented the model's proficiency in adhering to multifaceted instructions.
- Elevated the model's capacity to synthesize structured information.
- Amplified the comprehensiveness of the information generated.
- Enhanced the efficacy in the execution of specialized tasks.

### 3.2 Multi-view Expert Models Ensembling

We observed that the model's performance is highly sensitive to different instructions [4]. Existing generative methods for triple extraction suffer from sequential generation and overlook interdependencies among triple elements, making them susceptible to autoregressive noise. Specifically, the following problems exist:

- Paradigm differences: Triple extraction is not a sequence generation task. Modeling structured information as sequence generation can lead to issues like incomplete or erroneous generation.
- Element importance: The significance and importance of elements in extracted triples vary greatly and are strongly correlated with the corresponding relationships. However, this correlation is challenging to learn through instruction, demonstration, and sequence generation fine-tuning.
- Instability: Language models are highly sensitive to different prompts. Thus, some error generation can be attributed to noise introduced by prompts.
- Error accumulation: Predictive errors from autoregressive models accumulate and affect subsequent predictions. This manifests as decreased accuracy in subsequent predictions and issues like repeated generation

Drawing inspiration from how humans cooperate from different perspectives and roles to solve challenging problems [14], we use both KnowLM-ZhiXi and ChatGPT models and introduce instruction methods based on different prompts and decoding methods to control the generation order of head and tail entities. By permuting elements, we eliminate the error accumulation and instability in the generation process, allowing the extraction model to capture common understanding from the input. We find that KnowLM-ZhiXi and ChatGPT generate results that are highly complementary. KnowLM-ZhiXi provides more complete results, while ChatGPT demonstrates stronger capabilities in understanding challenging samples and achieving higher accuracy. By integrating the diverse output results from both models and filtering out individual errors, we strike a balance between accuracy and comprehensiveness (Fig. 2).

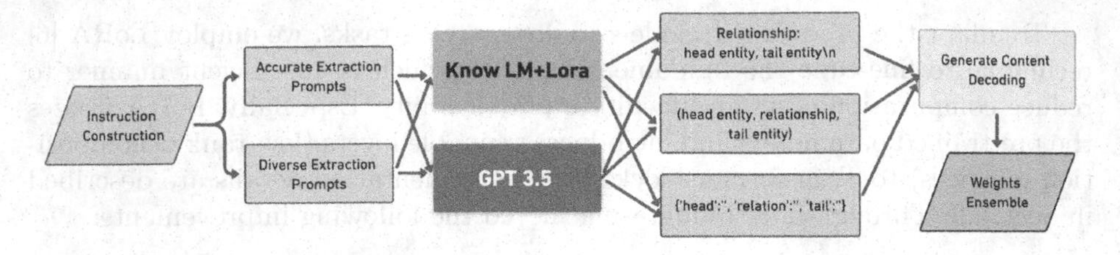


**Fig. 2.** We designed various instruction prompt models to extract text content using multiple methods. During inference, we employed the Multi-view of Experts method to integrate common features from different expert models' outputs, obtaining more accurate answers.

**Prompts Construction for Multi-view Experts Generation.** To enable the extraction of triplets from multiple perspectives, MvE introduces a generative structural prompting mechanism. Specifically, we design the task objective as multiple generation tasks with different structures and orders, and prompt the model to complete these tasks. For instance, given a standard triplet array [[A,B,C],[D,E,F]], we transform the task objective into generating specific formatted content such as $B : A, C; E : D, F$ and $\{`head' : A, 'relation' : B, `tail' : C\}$. We construct 10 different instructions and 8 distinct generation tasks. These are randomly combined to form a training set of 20,000 pairs. Subsequently, we utilize the prompting mechanism to guide the model's attention towards different views of the input during the inference process. Our prompts, along with the model trained on our dataset, effectively adhere to the instructions and accomplish the task objectives across multiple perspectives.

**Multi-view Inference of Experts Models.** For reasoning, the MvE method guides models to accurately and consistently generate multiple triples following a predefined structure and order. For pre-trained models, instructions are sufficient. With ChatGPT, we select the three most relevant instances for each input and construct demonstrations. Through context-based learning, ChatGPT proficiently generates multiple tuples with precision. Ultimately, we aggregate all outcomes to obtain the most rational tuples.

**Weighted Voting for Ensemble of Multiple Output Results.** Due to our guidance for each expert model to predict tuples from multiple task views, we first aggregate all the prediction results. Subsequently, after evaluating the performance of each task, we weight the predicted tuples of the model based on the relative task performance. We set a threshold and consider tuples with votes exceeding a specific threshold as the final result. Overall, for an input sentence $x$, assuming we prompt $m$ experts models to generate from a selected set of $n$ permutations, the predicted tuple set for each permutation $p_i$ is denoted as $T_{p_i}$, which may contain one or multiple sentiment tuples. The voting weight for this set is denoted as $\alpha_i$. We set the voting weight as $\theta$, and thus we obtain the final aggregated result $T'_{\text{MvE}}$.

$$T'_{\text{MvE}} = \left\{ t \mid t \in \bigcup_{i=1}^{m*n} T'_{p_i} \text{ and } \left( \sum_{i=1}^{m*n} \alpha_{\exists T'_{p_i}}(t) \geq \theta \right) \right\}$$

## 3.3 Automatic Correction with Schema Constraints and GPT Guidance

Due to the presence of noisy training data, task complexity, and the difficulty for the model to fully understand instructions and data patterns, the integrated results still contain common errors from the models. To address this, we leveraged the relation schema mined from the training set and an "observer" model to constrain and effectively correct the model's output. This led to better performance.

As shown in the Fig. 3, the overall implementation architecture can be divided into four parts: training set relation-library construction, context demonstration recommendation, and instruction building, diverse instruction prediction result integration, and GPT-guided post-processing.

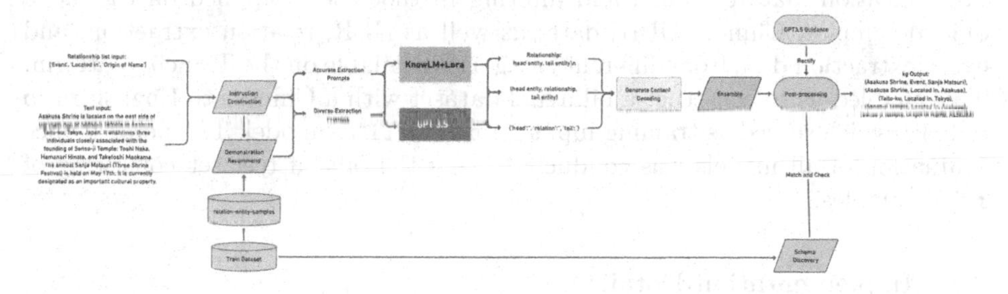


**Fig. 3.** Method Implementation Overall Architecture. The method implementation can be divided into four parts: training set relation library construction, contextual demonstration recommendation and instruction construction, diverse instruction prediction result integration, and GPT-guided post-processing.

**Recommending Contextual Demonstrations.** We processed each sample in the training dataset and constructed the relation library. Each sample contains a list of relations. For each relation, we extracted matching triples and sample information were added to the relation library. Based on the type and relations of each input, we recommend corresponding prompts.

**Schema Constrained Decoding** Despite guiding the model to generate the target from different perspectives, the generated results may not conform to the target pattern format. Therefore, we have designed a pattern-based constraint decoding. First, we ensure that the generated elements are within the respective vocabulary sets. Secondly, we check whether the elements in the triple satisfy the pattern constraints for specific patterns, such as the order of head and tail entities. Third, we determine the plausibility by comparing them against relationship patterns from the training set.

**GPT-Guided Error Correction.** For errors that are difficult to handle using rule-based methods, we first filter them out. Then, we design a "checkprompt" approach to convert them into single-choice questions in the form of reading comprehension. We use ChatGPT to complete the checkprompt task and correct the results. The corrected results from ChatGPT are considered as the final results.

# 4 Experiment

## 4.1 Datasets

The chatGLM (LORA) [13] model was fine-tuned using 5,000 official provided data samples. We augmented the official dataset to 25,000 samples by employing diverse instructions and combined it with Chinese DUIE data for further refinement of the chatGLM (LORA) [13] model. Various techniques such as instruction diversification, data cleaning, and filtering methods were applied on the 5,000 official samples, Chinese DUIE data, as well as NER, relation extraction, and event extraction data from different categories available on the Tianchi platform. This enabled us to construct a bilingual dataset with a Chinese to English ratio of 2.3:1 which served as training input for our DITING model. The performance evaluation of all models was conducted using the official test set consisting of 1,000 samples.

## 4.2 Implementation Details

We trained the model on two A100 80G GPUs. We utilized the AdamW optimizer to tune the model. During the training process, the DITING model was trained for 5 epochs with the learning rate of 1e-4 and LORA rank of 16. After the DITING model was trained for 4,000 steps, we achieved an eval loss of 0.031, and the final learning rate was decayed to 20% of the peak learning rate. The ChatGLM model was trained for 50,000 steps with learning rate of 2e-5 and LORA rank of 32.

During the inference process, our KnowLM-ZhiXi(MvE) model utilized the results from chatGPT with five instructions and KnowLM-ZhiXi with four instructions for voting. During inference, we set the temperature of KnowLM-ZhiXi to 0 and the number of beams to 5. For ChatGPT, we set the temperature to 0.2. To balance the votes between chatGPT and KnowLM-ZhiXi, we assigned a weight of 2 to the KnowLM-ZhiXi votes that used JSON instructions and set the voting threshold to 3.

### 4.3 Evaluation Metrics

We employ ROUGE-2 and F1 as evaluation metrics, utilizing F1 to gauge the model precision in generating entity triplets, while serializing the triplet results for assessment using ROUGE-2. Ultimately, we compute the average of both metrics to derive the final Score.

### 4.4 Baseline

In our research, we compared our model with the following models:

- ChatGLM [13]: A bilingual language model with 6.2 billion parameters based on the GLM architecture. It has undergone extensive pretraining on both Chinese and English corpora, along with longer sequence length capabilities.
- KnowLM-ZhiXi [9]: A model based on LLaMA-13B, further pretrained on Chinese and English corpora to enhance the model's Chinese language capabilities while retaining its English language capabilities.

### 4.5 Main Result

**Table 1.** Main Result:'LORA' means that the model is fine-tuned using the LORA method [6], 'p-tuning' means that the model is fine-tuned using the p-tuning method [8]

| model | F1 | ROUGE-2 | Score |
|---|---|---|---|
| ChatGLM-6B(LORA) [13] | 32.97 | 77.69 | 55.33 |
| ChatGLM-6B(p-tuning) [13] | 34.12 | 78.23 | 56.18 |
| KnowLM-ZhiXi(LORA) [9] | 52.32 | 79.33 | 65.83 |
| DITING(LORA) | 56.47 | 79.61 | 68.04 |
| KnowLM-ZhiXi(MvE) | **59.54** | **82.68** | **71.11** |

As presented in Table 1, our KnowLM-ZhiXi(MvE) model demonstrates significant enhancements in F1, ROUGE-2, and overall score compared to the baseline KnowLM-ZhiXi model, with improvements of 13.51, 3.35, and 5.28 respectively. The notable improvement in F1 highlights the effectiveness of our MvE approach in mitigating biases arising from diverse instructions and models. Moreover, our DITING model surpasses KnowLM-ZhiXi with a score improvement of 2.21, showcasing the efficacy of incorporating additional training corpora. It is important to note that fine-tuning ChatGLM solely using the available data falls short when compared to KnowLM-ZhiXi due to its extensive training on tens of thousands of data as opposed to only a few thousand instances used for ChatGLM's fine-tuning process. The increase in data volume significantly contributes to both capabilities and performance enhancement.

**Table 2.** Abalation Study: Gradually removing each module.

| model | F1 | ROUGE-2 | score |
|---|---|---|---|
| KnowLM-ZhiXi(MvE) | **59.54** | **82.68** | **71.11** |
| - Schema | 58.78 | 82.46 | 70.62 |
| - gpt fixed | 57.08 | 82.54 | 69.81 |
| - MvE | 54.66 | 79.00 | 66.83 |
| - chatGpt | 52.32 | 79.33 | 65.83 |

### 4.6 Abalation Study

To further evaluate the effectiveness of each component in our proposed method, we conduct the ablation experiments and report the results in Table 2. After removing Schema Constrained Decoding, F1 is reduced by 0.76, and ROUGE2 is reduced by 0.2, which proves that it can effectively repair common errors in model results. After removing the GPT correction module, F1 dropped by 1.7, indicating that the design hints that GPT can be a good supervisor in KG construction. After removing the MVE method, F1 decreased by 2.42, and ROUGE2 decreased by 3.54, which shows that it can effectively extract the correct consensus of the model under different perspectives, and also shows that our method can effectively remove the noise in the generation. After removing ChatGPT, F1 dropped by 2.34, indicating that using different models can effectively supplement the results and improve the accuracy of predictions.

### 4.7 Case Study

According to Table 3, it can be observed that in Example 1, KnowLM-ZhiXi exhibited confusion between the head and tail entities of the triples; however, this issue was resolved upon implementing MvE. Furthermore, both Example 1 and Example 2 demonstrate that KnowLM-ZhiXi(MvE) generated a higher number of accurate triples compared to the individual KnowLM-ZhiXi model. This significant improvement in performance can be attributed to the incorporation of the MvE module. Nevertheless, despite employing multiple expert models for voting in Example 1, it still introduced some erroneous triples. This occurrence arises from certain expert models struggling with precise determination of entity boundaries, resulting in partial errors. Addressing this concern remains an important aspect for future research.

**Table 3.** Case study: ✓ means the tirplet is right, ✗ means the triplet is wrong.

| Exp 1: People's Square, located in Haigang District of Qinhuangdao City, Hebei Province, is the landmark building of Qinhuangdao City. | |
|---|---|
| KnowLM-ZhiXi | (People's Square, is located in, Haigang District)✓<br>(Hebei Province, is located in, Qinhuangdao City)✗<br>(Qinhuangdao City, is located in, Haigang District)✗<br>(People's Square, achievement, landmark building of Qinhuangdao City)✓ |
| KnowLM-ZhiXi(MvE) | **(People's Square, is located in, Qinhuangdao City)**✓<br>(People's Square, is located in, Haigang District)✓<br>(People's Square, achievement, landmark buildings in Qinhuangdao City)✓<br>(Qinhuangdao City, is located in, Hebei Province)✓<br>**(Haigang District, is located in, Qinhuangdao City)**✓<br>(People's Square, achievement, landmark buildings)✗ |
| Exp2: On January 7, 1987, Zhao Yazhi and Huang Jinshen's son, Huang Kaijie, was born. | |
| KnowLM-ZhiXi | (Zhao Yazhi, spouse, Huang Jinshen)✓<br>(Huang Kaijie, parent, Zhao Yazhi)✓<br>(Huang Kaijie, parent, Huang Jinshen)✓<br>(Huang Kaijie, was born on, January 7, 1987)✓ |
| KnowLM-ZhiXi(MvE) | (Zhao Yazhi, spouse, Huang Jinshen)✓<br>**(Huang Jinshen, spouse, Zhao Yazhi)**✓<br>(Huang Kaijie, parent, Zhao Yazhi)✓<br>(Huang Kaijie, parent, Huang Jinshen)✓<br>(Huang Kaijie, was born on, January 7, 1987)✓ |

## 5 Conclusion

In this paper, we propose a method which performs knowledge graph construction adaptively with evolving knowledge schema. Three key ingredients contribute to the success of our method. First, a KGC-tailored large language model (LLM) is built, where the generalization ability of LLM is transferred for KGC. Second, a multi-model integration strategy is designed to mitigate the bias of a single LLM. Finally, schema-constrained decoding and LLM-guided correction improve KGC performances via the automatic error corrections. Experiments show the superiority of our method. In the future, we would like to extend our methods to related tasks including knowledge graph completion.

**Acknowledge.** This work was supported by the Strategic Priority Research Program of Chinese Academy of Sciences under Grant No.XDC02040400 , the National Key Research and Development Program of China under Grant No.2021YFB3100600, the Youth Innovation Promotion Association of CAS under Grant No. 2021153.

## References

1. InstructUIE: Multi-task Instruction Tuning for Unified Information Extraction. https://doi.org/10.48550/arXiv.2304.08085

2. Chen, X., et al.: Relation extraction as open-book examination: Retrieval-enhanced prompt tuning. In: Amigó, E., Castells, P., Gonzalo, J., Carterette, B., Culpepper, J.S., Kazai, G. (eds.) SIGIR 2022: The 45th International ACM SIGIR Conference on Research and Development in Information Retrieval, Madrid, Spain, 11–15 July 2022, pp. 2443–2448. ACM (2022). https://doi.org/10.1145/3477495.3531746
3. Cui, L., Yang, D., Yu, J., Hu, C., Cheng, J., Yi, J., Xiao, Y.: Refining sample embeddings with relation prototypes to enhance continual relation extraction. In: Proceedings of the 59th Annual Meeting of the Association for Computational Linguistics and the 11th International Joint Conference on Natural Language Processing (Volume 1: Long Papers), pp. 232–243. Association for Computational Linguistics, Online (Aug 2021). https://doi.org/10.18653/v1/2021.acl-long.20
4. Gou, Z., Guo, Q., Yang, Y.: MvP: Multi-view Prompting Improves Aspect Sentiment Tuple Prediction (May 2023). http://arxiv.org/abs/2305.12627
5. Gui, H., Zhang, J., Ye, H., Zhang, N.: InstructIE: A Chinese Instruction-based Information Extraction Dataset (May 2023)
6. Hu, E.J., et al.: Lora: Low-rank adaptation of large language models. arXiv preprint arXiv:2106.09685 (2021)
7. Liu, K., et al.: Noisy-labeled NER with confidence estimation. In: Proceedings of the 2021 Conference of the North American Chapter of the Association for Computational Linguistics: Human Language Technologies, NAACL-HLT 2021, Online, 6–11 June 2021, pp. 3437–3445. Association for Computational Linguistics (2021). https://doi.org/10.18653/v1/2021.naacl-main.269
8. Liu, Xet al.: P-tuning: prompt tuning can be comparable to fine-tuning across scales and tasks. In: Proceedings of the 60th Annual Meeting of the Association for Computational Linguistics (Volume 2: Short Papers), pp. 61–68 (2022)
9. Ningyu Zhang, Jintian Zhang, X.W., et al.: Deepke-llm: A large language model based knowledge extraction toolkit. GitHub repository (2023)
10. Thrun, S.: Lifelong learning algorithms. In: Thrun, S., Pratt, L.Y. (eds.) Learning to Learn, pp. 181–209. Springer (1998). https://doi.org/10.1007/978-1-4615-5529-2_8
11. Wang, X., et al.: Automated concatenation of embeddings for structured prediction. In: Zong, C., Xia, F., Li, W., Navigli, R. (eds.) Proceedings of the 59th Annual Meeting of the Association for Computational Linguistics and the 11th International Joint Conference on Natural Language Processing, ACL/IJCNLP 2021, (Volume 1: Long Papers), Virtual Event, 1–6 August 2021, pp. 2643–2660. Association for Computational Linguistics (2021). https://doi.org/10.18653/v1/2021.acl-long.206
12. Yao, Y., et al.: Schema-aware reference as prompt improves data-efficient knowledge graph construction. In: Chen, H., Duh, W.E., Huang, H., Kato, M.P., Mothe, J., Poblete, B. (eds.) Proceedings of the 46th International ACM SIGIR Conference on Research and Development in Information Retrieval, SIGIR 2023, Taipei, Taiwan, 23–27 July 2023, pp. 911–921. ACM (2023). https://doi.org/10.1145/3539618.3591763
13. Zeng, A., et al.: Glm-130b: An open bilingual pre-trained model. arXiv preprint arXiv:2210.02414 (2022)
14. Zhang, X., et al.: Wider and Deeper LLM Networks are Fairer LLM Evaluators (Aug 2023). http://arxiv.org/abs/2308.01862
15. Zhu, Y., et al.: Llms for knowledge graph construction and reasoning: Recent capabilities and future opportunities. ArXiv abs/ arXiv: 2305.13168 (2023). https://api.semanticscholar.org/CorpusID:258833039

# Single Source Path-Based Graph Neural Network for Inductive Knowledge Graph Reasoning

Deqiang Huang, Guoqing Zhao, Tianyu Liu, Tong Xu(✉), and Enhong Chen

School of Data Science, University of Science and Technology of China, Hefei, China
{deqianghuang,gqzhao,tianyu_liu}@mail.ustc.edu.cn,
{tongxu,cheneh}@ustc.edu.cn

**Abstract.** In this paper, we introduce our solution for the inductive knowledge graph reasoning task organized by the 2023 China Conference on Knowledge Graph and Semantic Computing (CCKS 2023). Specifically, this inductive knowledge graph reasoning task has two main challenges, namely 1) How to predict the entities that are not in the training set, and 2) How to train and reason more efficiently. To deal with these challenges, we adapt the Neural Bellman-Ford Networks (NBFNet) with grid search strategy for achieving the best performance. Along this line, we further refine this solution with ensemble learning and post-processing. Extensive experiments have demonstrated the effectiveness of our solution, which won the first place in the competition. Code is publicly available at https://github.com/smart-lty/CCKS-2023-Task2

**Keywords:** Inductive Knowledge Graph Reasoning · Representation Learning · Graph Neural Network

## 1 Introduction

Recent years have witnessed the rapid development of Knowledge Graph (KG) techniques, which stores quantities of structured human knowledge in the form of factual triples, and further benefits various downstream domains, like natural language processing [1], recommendation Systems [2], and question answering [3]. In the context of real-world knowledge graphs, continuous emergence of new entities, such as users and products in e-commerce, leads to the persistent incompleteness of KG. For instance, even widely-used large-scale knowledge graphs, like Freebase [4], Wikidata [5], and Yago3 [6] suffer from this problem. To that end, several knowledge graph completion models, including RotatE [7] and R-GCN [8], have been developed. Unfortunately, they encounter difficulties in dealing with emerging entities due to their reliance on observing test entities during training. Consequently, there is a growing interest in the field of inductive link prediction, which aims to address this challenge by predicting missing links in evolving knowledge graphs [9].


H. Wang et al. (Eds.): CCKS 2023, CCIS 1923, pp. 285–296, 2023.
https://doi.org/10.1007/978-981-99-7224-1_22

Indeed, the fundamental concept underlying inductive relation prediction on knowledge graphs revolves around the learning of logical rules, which capture co-occurrence patterns between relations in an entity-independent manner, enabling seamless generalization to previously unseen entities [10,11]. Although existing models like AMIE+ [12] and Neural LP [13] explicitly reveal logical rules, thereby offering interpretability [13], their performance is constrained by the extensive search space and discrete optimization requirements [14,15]. Recently, a promising alternative is represented by subgraph-based methods, exemplified by GraIL [10], TACT [11], and CoMPILE [16], which introduce implicit rule mining through reasoning over the subgraph induced from the target link. These innovative approaches leverage the power of Graph Neural Networks (GNNs) [8] to learn subgraph representations, effectively encoding all mined rules. However, there are still some irrelevant rules [17] within the subgraph.

At the same tine, we realize that NBFNet, a path-based Method, can solve this problem. NBFNet drawing inspiration from conventional path-based methodologies, proposes to represent a pair of nodes by employing a generalized sum of all path representations between them. Each path representation is defined as the generalized product of the edge representations along that particular path. Intriguingly, various well-known link prediction methods, including Katz index [18], personalized PageRank [19], graph distance [20], and classical graph theory algorithms such as widest path [21] and most reliable path [21], can be viewed as special instances of our path formulation, distinguished by the specific summation and multiplication operators employed. Motivated by the efficiency of the polynomial-time algorithm for the shortest path problem [22], they demonstrate that our formulation lends itself to an efficient solution via the generalized Bellman-Ford algorithm [21], even under mild conditions, making it highly scalable for large graphs.

To that end, we adapt NBFNet framework with improved grid-search-based training strategies, and further refine this solution with weighted ensemble learning and post-processing. Extensive experiments have demonstrated the effectiveness of our solution, which won the first place in the inductive knowledge graph reasoning competition of CCKS 2023.

## 2 Related Works

In recent academic explorations, link prediction within knowledge graphs (KGs) has emerged as a focal area of research, largely due to the inherent incompleteness in many existing graphs. Generally, this task can be categorized into two primary classifications: transductive link prediction and inductive link prediction. At present, four predominant paradigms for KG link prediction have been identified: rule-based approaches, path-based approaches, embedding-based techniques, and graph neural networks.

### 2.1 Rule-Based Approaches

Rule-based approaches predominantly focus on the formulation of rules to reason and predict new triples. The AIME model [23] is illustrative of this, as it predicts rules for each relation through the introduction of both dangling and closing atoms, thereby extending the rule set of the relation. Despite its commendable interpretability and capability to automatically discern reasoning rules, its vast search space constrains performance. The DRUM methodology [24], an end-to-end differentiable approach, offers another perspective by learning weights of specific paths through probabilistic logical rule sets, leveraging the relation's adjacency matrix for single-hop graph calculations.

### 2.2 Path-Based Approaches

Historical path-based techniques such as Katz [18] and PageRank [19] have employed path features for link prediction purposes. Within the domain of knowledge graphs, models can utilize paths between entity pairs to ascertain and predict direct 1-hop relations. Path-RNN [25], for instance, encodes and amalgamates paths using recurrent neural networks to subsequently predict relations. Similarly, the Path Ranking Algorithm (PRA) [26] conceptualizes paths as features analogous to specific Horn clauses, allowing for logical rule-based inferences. Notably, the NBFNet model [27], which forms the foundation for our research, is another path-centric technique leveraging the Bellman-Ford algorithm for path determinations.

### 2.3 Embedding Techniques

Embedding techniques aim to learn distributed representations of KG entities and relations, employing these representations to compute triple scores. Contemporary knowledge graph embedding methodologies like TransE [28], DistMult [29], and RotatE [7] have demonstrated considerable efficacy in link prediction, even in expansive graphs. Nonetheless, one salient limitation of these embedding techniques is their transductive nature, rendering them ineffective with previously unseen entities and relations.

### 2.4 Graph Neural Networks (GNNs)

Several GNN-inspired knowledge graph embedding techniques have been developed to address link prediction challenges. GNNs inherently capture the topological intricacies of graphs, translating entities into embeddings. Decoders then employ these embeddings to predict the interrelations between nodes. SEAL [30], as an exemplar, extracts a k-hop subgraph surrounding each target link and subsequently assigns integer labels to each node as supplementary features. A graph neural network is then deployed for link existence predictions. It is also worth noting that GraIL [10], which stands as the benchmark model for this competition, is rooted in GNN methodologies.

# 3 Methodology

In the pursuit of optimal results, this study adopts NBFNet as its foundational model and strategically employs a weighted ensemble technique.

## 3.1 Insight into NBFNet

NBFNet serves as the primary modeling framework for our methodology. With its roots in path-based link prediction techniques and the sophisticated Generalized Bellman-Ford algorithm, NBFNet incorporates a single-source Graph Neural Network (GNN) for the inductive inference of links. This discourse delves into the theoretical underpinnings of the Generalized Bellman-Ford algorithm and elucidates the structural intricacies of NBFNet.

**Theoretical Framework of the Generalized Bellman-Ford Algorithm.** Consider a knowledge graph represented by $\mathcal{G} = (\mathcal{V}, \mathcal{E}, \mathcal{R})$. Within this framework, $\mathcal{V}$ delineates entities, $\mathcal{E}$ typifies relations within the Knowledge Graph (KG), and $\mathcal{R}$ signifies the diverse relation types present in the graph. For the purpose of this discussion, let $\mathcal{N}(u)$ signify the neighbors of entity $u$ and $\mathcal{E}(u)$ denote the relations associated with $u$.

*Construct of Path Definition.* Within the realm of this research, the objective of link prediction can be stated as follows: given a primary entity, denoted by $u$, and a specific query relation $q$, the aim is to ascertain the tail entities, symbolized by $v$, that are congruent with the established facts in the KG. For methodologies rooted in path-based techniques and GNNs, discerning the subgraph structure interlinking $u$ and $v$ is paramount for deducing their relational dynamics. The representation of this subgraph structure, given a query relation $q$, is designated as $h_q(u, v)$. Conventional practice leverages the product operator for deducing the representation of a singular path bridging $u$ and $v$, while employing the sum operator to amalgamate representations from diverse paths.

Define $\mathcal{P}_{uv}$ as set of paths from $u$ to $v$ and $w_q(e_i)$ as the representation of edge $e_i$. The representation of subgraph structure between $u$ and $v$ is:

$$\boldsymbol{h}_q(u,v) = \boldsymbol{h}_q(P_1) \oplus \boldsymbol{h}_q(P_2) \oplus \ldots \oplus \boldsymbol{h}_q\left(P_{|\mathcal{P}_{uv}|}\right)\Big|_{P_i \in \mathcal{P}_{uv}} \triangleq \bigoplus_{P \in \mathcal{P}_{uv}} \boldsymbol{h}_q(P) \tag{1}$$

And the representation of a single path is:

$$\boldsymbol{h}_q\left(P = \left(e_1, e_2, \ldots, e_{|P|}\right)\right) = \boldsymbol{w}_q(e_1) \otimes \boldsymbol{w}_q(e_2) \otimes \ldots \otimes \boldsymbol{w}_q\left(e_{|P|}\right) \triangleq \bigotimes_{i=1}^{|P|} \boldsymbol{w}_q(e_i) \tag{2}$$

This process is similar to Depth-First Search (DFS), where it explores all possible paths and combines their representations. As a result, it is transductive and can be applied to various graphs. $\oplus$ and $\otimes$ is a general sum and multiplication that different algorithms could replace.

*Generalized Bellman-Ford Algorithm.* Bellman-Ford algorithm is a famous algorithm for finding the shortest paths of a single source using the idea of dynamic programming. Every node aggregates all the information from its neighbors and updates its information. Since the number of paths grows exponentially with the path length, it's expensive to calculate representations path by path. NBFNet uses a more scalable solution. Generalized Bellman-Ford algorithm is defined as follows:

$$\boldsymbol{h}_q^{(0)}(u,v) \leftarrow \mathbb{1}_q(u=v) \tag{3}$$

$$\boldsymbol{h}_q^{(t)}(u,v) \leftarrow \left( \bigoplus_{(x,r,v)\in\mathcal{E}(v)} \boldsymbol{h}_q^{(t-1)}(u,x) \otimes \boldsymbol{w}_q(x,r,v) \right) \oplus \boldsymbol{h}_q^{(0)}(u,v) \tag{4}$$

$\mathbb{1}_q(u=v)$ is the indicator function which judges if $u=v$. It's used for boundary conditions. The current representation of the path depends on prior neighbors' representation and representation of edge $e=(x,r,v)$. Given entity $u$ and query relation $q$, the algorithm can compute pair representation $\boldsymbol{h}_q(u,v)$ for all $v \in \mathcal{V}$ in parallel.

### 3.2 Single Source Path-Based GNN

If we have a source node $u$, we can determine the relationships between the nodes. In this competition, if we're given source node $u$ and query relation $q$, we'll calculate the representations of all nodes. Then, we'll calculate the likelihood of the node $v$ having relation $q$ with $u$. When working with $u$ and $q$, the probability of $v$ is determined by the equation $p(v|u,q) = \sigma(f(hq(u,v)))$. Here, $\sigma(.)$ represents the sigmoid function, and $f()$ refers to a feed-forward neural network.

The training for this model needs to be done carefully.

$$\boldsymbol{h}_v^{(0)} \leftarrow \text{INDICATOR}(u,v,q) \tag{5}$$

$$\boldsymbol{h}_v^{(t)} \leftarrow \textsc{Aggragate}\left(\left\{\textsc{Message}\left(\boldsymbol{h}_x^{(t-1)}, \boldsymbol{w}_q(x,r,v)\right) \mid (x,r,v)\in\mathcal{E}(v)\right\}\right) \tag{6}$$

In the knowledge graph, *MESSAGE* could be the translation and scaling used in TransE and DistMult, respectively. Translation is summation of $\boldsymbol{h}_x^{(t-1)}$ and $\boldsymbol{w}_q(x,r,v)$ and scaling is multiplication. Consider different edges have different contribution to different query so edge representations for $\boldsymbol{w}_q(x,r,v)$ is defined as $\boldsymbol{w}_q(x,r,v) = \boldsymbol{W}_r\boldsymbol{q} + \boldsymbol{b}_r$.

We design the *AGGREGATE* operator using methods in GNNs such as sum, mean, or max. Then a transformation and activation are used to produce the final state.

The *INDICATOR* operator needs to give a representation of source node $u$. To make the initial state of queries different, we learn a query embedding $\boldsymbol{q}$ and define the operator as $\mathbb{1}(u=v) * \boldsymbol{q}$

After defining all the operators used in the method, we can convert the reasoning task to a single source message-passing process. And every node will learn its representation based on source node $u$, then the relations between them will be determined.

For this contest, when given source node $u$ and query relation $q$, we calculate the representations of all nodes. Then the conditional likelihood of the node $v$ has relation $q$ with $u$ is $p(v \mid u, q) = \sigma(f(\boldsymbol{h}_q(u, v)))$, where $\sigma()$ is the sigmoid function and $f()$ is a feed-forward neural network.

The training process minimizes the negative log-likelihood of positive and negative triples. The negative triples are generated using Partial Completeness Assumption (PCA) [23]. And the loss is defined as follows:

$$\mathcal{L}_{KG} = -\log p(u, q, v) - \sum_{i=1}^{n} \frac{1}{n} \log\left(1 - p\left(u_i', q, v_i'\right)\right) \tag{7}$$

# 4 Insights and Methodology

## 4.1 NBFNet Parameter Investigation

The NBFNet framework, characterized by its singular source initialization methodology, serves as a general message-passing framework. In our pursuit of optimizing performance within the context of the CCKS competition, an empirical exploration of specific hyperparameters within the NBFNet architecture has been undertaken.

**Exclusion of Trivial Edges.** In the course of training, a pivotal operation that significantly enhances model performance involves the temporary exclusion of trivial edges within the training graph. In essence, given a graph denoted as G, alongside a target link represented by $\tau = (u, r_q, v)$, the link $\tau$ is temporarily removed from G. To illustrate, when faced with a scenario where the training graph G encompasses three links, namely $e_1, e_2$, and $e_3$, yielding $G = (e_1, e_2, e_3)$, the endeavor to predict the link $e_1$ necessitates the temporary removal of $e_1$ from G. Thus, the modified input transforms from $(G = (e_1, e_2, e_3), e_1)$ to $(G = (e_2, e_3), e_1)$.

This parameterization serves the purpose of facilitating the unobstructed propagation of information along reasoning paths. Failing to eliminate facile edges during the model's training phase results in a suboptimal learning outcome.

**Utilization of Layer Normalization.** Layer Normalization (LayerNorm), a normalization technique frequently deployed in the realm of Graph Neural Networks (GNNs), assumes significance. The employment of normalization techniques, such as LayerNorm, is pivotal in stabilizing the learning trajectory and expediting the process of convergence.

Within the ambit of GNNs, LayerNorm can be applied iteratively to node features at each layer of the network. This normalization procedure is conducted across the feature dimension for individual nodes, as opposed to a collective normalization across nodes, reminiscent of techniques like Batch Normalization.

In the context of NBFNet, LayerNorm is implemented post the execution of each AGGREGATE function, thereby augmenting the training velocity of the NBFNet model. Its role is to harmonize the dimensional disparities among diverse samples, while simultaneously preserving the relative dimensional relationships between distinct features.

**Incorporation of Shortcut Connections.** Shortcut connections, recognized alternately as residual connections or skip connections, constitute a technique initially introduced within the architecture of ResNet, tailored to convolutional neural networks. NBFNet adopts this mechanism to expedite model training.

Given a transformation denoted by $F(x)$, the layer output is expressed as $F(x)+x$, diverging from the solitary transformation $F(x)$. This construct, termed a "ortcut,"forges a direct pathway from input to output, circumventing intervening transformations.

Within the realm of GNNs, shortcut connections can be seamlessly integrated at each network layer. They prove instrumental in alleviating the quandary of vanishing gradients that may beset deep network training. Additionally, they facilitate facile learning of identity functions for node transformations and the preservation of input feature-derived information.

**Negative Sampling Strategy.** The incorporation of negative samples within NBFNet adheres to the principles of Partial Completeness Assumption (PCA). This entails generating negative samples through the substitution of one entity within a ground truth triple. The overarching training objective involves maximizing the discrepancy between the loss incurred by positive and negative triples.

## 4.2 Grid Search Procedure

For the fine-tuning of parameters within the NBFNet framework, a systematic grid search methodology is employed to identify the optimal model configuration, thereby corroborating the findings outlined in Sect. 4.1.

Experimental endeavors reveal that the exclusion of trivial edges and the application of layer normalization yield discernible enhancements in model performance. Concurrently, the integration of shortcut connections accelerates training. As regards negative sampling, the dimensionality of 32 is ascertained as optimal.

In light of these determinations, the aforementioned parameters are held constant, while the broader spectrum of parameters undergoes examination via grid search. The ensuing parameters subjected to this investigation are presented in Table 1.

**Table 1.** Parameters under Grid Search Analysis

| Model ID | Dependency | Removal of One-Hop | Adversarial Temperature | Message Function |
|---|---|---|---|---|
| 23 | True | False | 1 | Distmult |
| 30 | False | False | 0.5 | Distmult |
| 54 | True | False | 0.5 | RotatE |
| 58 | False | True | 0.5 | RotatE |
| 59 | False | True | 1 | RotatE |
| 62 | False | False | 0.5 | RotatE |

### 4.3 Model Ensemble Strategy

To harness the complementary attributes inherent to individual models and enhance the overall predictive prowess, the imperative of model ensemble emerges. The chosen strategy for this purpose is labeled as "ingle model weighted voting." Within this paradigm, models operating with distinct parameter configurations yield distinct score files, encapsulating predictive scores for edge establishment. Subsequent to their generation, these scores are subjected to diverse weights prior to aggregation, culminating in the synthesis of the ultimate output.

## 5 Experiments

### 5.1 Dataset

Our experimentation is conducted upon the CCKS 2023 inductive knowledge graph reasoning competition dataset. The initial training set encompasses 26,874 triples, involving 46 relationships and 4,050 entities. Each triplet $(head, relation, tail)$ is numerically represented. Similarly, the preliminary test set contains 3,533 entities and 43 relationships. Within this, the support set consists of 11,960 triples, while the query set comprises 2,110 triples.

The support set encapsulates triplets $(head, relation, tail)$, wherein the numerical identifiers correspond to entities and relationships. While the relationship identifiers align with the training set, the entity identifiers differ. Entities within the test set represent novel instances not encountered during training. The support set configuration closely mirrors that of the training set.

The query set is composed of tuples $(head, relation)$ along with a list of candidate tail entities. The entity numbering is coherent with the support set structure. The primary task involves ranking the candidate tail entities based on their likelihood of being true.

### 5.2 Data Pre-processing

Preliminary data pre-processing involves constructing entity dictionaries and relational dictionaries for all training set and support set triples. This entails encoding entities and relationships, appending previously unrecorded entities and relationships to the dictionaries. Consequently, a new knowledge graph is formed featuring encoded nodes, facilitating the segregation of training and validation sets.

### 5.3 Experimental Configuration

The NBFNet model serves as the foundational architecture for our experiments. To enhance the model's capacity to learn relation representations, we augment the number of triples. For each $(h, r, t)$ triplet, a corresponding $(t, r^{-1}, h)$ triplet is introduced, thereby incorporating an inverse edge. Subsequently, when predicting the score of a given edge $(h, r, t)$ within the input graph G, the edge's presence in G's training set necessitates temporary removal. Residual connections are employed to mitigate gradient vanishing. Parameter settings involve a random seed of 1024, PNA as the aggregate function, a hidden dimension of $[32, 32, 32, 32, 32, 32]$, input dimension of 32, Adam optimizer, learning rate of 0.005, and further details are found in Table 1. An ensemble approach is adopted to ascertain the optimal model.

### 5.4 Experimental Outcomes

To ensure the identification of the optimal model, grid search is employed. The principal outcomes derived from the validation set are documented in Table 2.

**Table 2.** Key Results on the Validation Set and Final Test Set

| Model ID | Hits@3 | Hits@10 | Validation MRR | Final Score |
|---|---|---|---|---|
| 23 | 0.7697 | 0.8498 | 0.7065 | 0.6320 |
| 30 | 0.7313 | 0.8641 | 0.6558 | 0.6480 |
| 54 | 0.7892 | 0.8857 | 0.7204 | 0.6410 |
| 58 | 0.7280 | 0.8589 | 0.6558 | 0.6240 |
| 59 | 0.7301 | 0.8844 | 0.6602 | 0.6230 |
| 62 | 0.7959 | 0.9028 | 0.7139 | 0.6400 |

Subsequently, employing a weighted ensemble, we determine the optimal output, we use the model that performs best on the validation set for the ensemble. Our most successful submission combines Model No. 30 with a weight of 3, Model No. 54 with a weight of 1, and Model No. 62 with a weight of 1. This ensemble achieves a score of 0.6510 in the final test set, consequently securing the first prize in the competition.

## 6 Conclusion

In this paper, we improved NBFNet framework with utilizing a single source path-based graph neural network to complete the inductive knowledge graph reasoning competition of CCKS2023. Specifically, with integrating a weighted ensemble strategy, we achieved the score as 0.6510 and won the first prize.

## References

1. Zhang, Z., et al.: ERNIE: enhanced language representation with informative entities. In: Proceedings of the 57th Annual Meeting of the Association for Computational Linguistics, Florence, Italy, July 2019, pp. 1441–1451. Association for Computational Linguistics. https://doi.org/10.18653/v1/P19-1139. https://aclanthology.org/P19-1139
2. Wang, H., et al.: RippleNet: propagating user preferences on the knowledge graph for recommender systems. In: Proceedings of the 27th ACM International Conference on Information and Knowledge Management, CIKM 2018, Torino, Italy, pp. 417–426. Association for Computing Machinery (2018). ISBN 9781450360142. https://doi.org/10.1145/3269206.3271739
3. Huang, X., et al.: Knowledge graph embedding based question answering. In: Proceedings of the Twelfth ACM International Conference on Web Search and Data Mining, WSDM 2019, Melbourne, VIC, Australia, pp. 105–113. Association for Computing Machinery (2019). ISBN 9781450359405. https://doi.org/10.1145/3289600.3290956
4. Bollacker, K., et al.: Freebase: a collaboratively created graph database for structuring human knowledge. In: Proceedings of the 2008 ACM SIGMOD International Conference on Management of Data, SIGMOD 2008, Vancouver, Canada, pp. 1247–1250. Association for Computing Machinery (2008). ISBN 9781605581026. https://doi.org/10.1145/1376616.1376746
5. Vrandečić, D., Krötzsch, M.: Wikidata: a free collaborative knowledgebase. Commun. ACM **57**(10), 78–85 (2014). ISSN 0001-0782. https://doi.org/10.1145/2629489
6. Mahdisoltani, F., Biega, J.A., Suchanek, F.M.: YAGO3: a knowledge base from multilingual Wikipedias. In: Conference on Innovative Data Systems Research (2015). https://api.semanticscholar.org/CorpusID:6611164
7. Sun, Z., et al.: RotatE: knowledge graph embedding by relational rotation in complex space. arXiv:1902.10197 [cs.LG] (2019)
8. Schlichtkrull, M., Kipf, T.N., Bloem, P., van den Berg, R., Titov, I., Welling, M.: Modeling relational data with graph convolutional networks. In: Gangemi, A., et al. (eds.) ESWC 2018. LNCS, vol. 10843, pp. 593–607. Springer, Cham (2018). https://doi.org/10.1007/978-3-319-93417-4_38
9. Galkin, M., Berrendorf, M., Hoyt, C.T.: An open challenge for inductive link prediction on knowledge graphs (2022)
10. Teru, K.K., Denis, E.G., Hamilton, W.L.: Inductive relation prediction by subgraph reasoning. In: Proceedings of the 37th International Conference on Machine Learning, ICML 2020. JMLR.org (2020)
11. Chen, J., et al.: Topology-aware correlations between relations for inductive link prediction in knowledge graphs. Proc. AAAI Conf. Artif. Intell. **35**(7), 6271–6278 (2021). https://doi.org/10.1609/aaai.v35i7.16779. https://ojs.aaai.org/index.php/AAAI/article/view/16779.

12. Galárraga, L., et al.: Fast rule mining in ontological knowledge bases with AMIE+. VLDB J. **24**(6), 707–730 (2015). ISSN 1066-8888. https://doi.org/10.1007/s00778-015-0394-1
13. Yang, F., Yang, Z., Cohen, W.W.: Differentiable learning of logical rules for knowledge base reasoning. In: Proceedings of the 31st International Conference on Neural Information Processing Systems, NIPS 2017, Long Beach, California, USA, pp. 2316–2325. Curran Associates Inc. (2017). ISBN 9781510860964
14. Kok, S., Domingos, P.: Statistical predicate invention. In: Proceedings of the 24th International Conference on Machine Learning, ICML 2007, Corvalis, Oregon, USA, pp. 433–440. Association for Computing Machinery (2007). ISBN 9781595937933. https://doi.org/10.1145/1273496.1273551
15. Wang, W.Y., Mazaitis, K., Cohen, W.W.: Structure learning via parameter learning. In: Proceedings of the 23rd ACM International Conference on Conference on Information and Knowledge Management, CIKM 2014, Shanghai, China, pp. 1199–1208. Association for Computing Machinery (2014). ISBN 9781450325981. https://doi.org/10.1145/2661829.2662022
16. Mai, S., et al.: Communicative message passing for inductive relation reasoning. arXiv:2012.08911 [cs.AI] (2021)
17. Rameshkumar, K., Sambath, M., Ravi, S.: Relevant association rule mining from medical dataset using new irrelevant rule elimination technique. In: 2013 International Conference on Information Communication and Embedded Systems (ICICES), ICICES 2013, pp. 300–304 (2013). https://doi.org/10.1109/ICICES.2013.6508351
18. Katz, L.: A new status index derived from sociometric analysis. Psychometrika **18**(1), 39–43 (1953)
19. Brin, S.: The PageRank citation ranking: bringing order to the web. In: Proceedings of ASIS 1998, vol. 98, pp. 161–172 (1998)
20. Liben-Nowell, D., Kleinberg, J.: The link prediction problem for social networks. In: Proceedings of the Twelfth International Conference on Information and Knowledge Management, pp. 556–559 (2003)
21. Baras, J., Theodorakopoulos, G.: Path Problems in Networks. Springer, Cham (2022). https://doi.org/10.1007/978-3-031-79983-9
22. Bellman, R.: On a routing problem. Q. Appl. Math. **16**(1), 87–90 (1958)
23. Galárraga, L.A., et al.: AMIE: association rule mining under incomplete evidence in ontological knowledge bases. In: Proceedings of the 22nd International Conference on World Wide Web, pp. 413–422 (2013)
24. Sadeghian, A., et al.: DRUM: end-to-end differentiable rule mining on knowledge graphs. arXiv:1911.00055 [cs.LG] (2019)
25. Das, R., et al.: Chains of reasoning over entities, relations, and text using recurrent neural networks. arXiv:1607.01426 [cs.CL] (2017)
26. Lao, N., et al.: Efficient relational learning with hidden variable detection. In: Proceedings of the 23rd International Conference on Neural Information Processing Systems - Volume 1, NIPS 2010, Vancouver, British Columbia, Canada, pp. 1234–1242. Curran Associates Inc. (2010)
27. Zhu, Z., et al.: Neural Bellman-Ford networks: a general graph neural network framework for link prediction. arXiv:2106.06935 [cs.LG] (2022)
28. Bordes, A., et al.: Translating embeddings for modeling multi-relational data. In: Proceedings of the 26th International Conference on Neural Information Processing Systems - Volume 2, NIPS 2013, Lake Tahoe, Nevada, pp. 2787–2795. Curran Associates Inc. (2013)

29. Yang, B., et al.: Embedding entities and relations for learning and inference in knowledge bases. arXiv:1412.6575 [cs.CL] (2015)
30. Zhang, M., Chen, Y.: Link prediction based on graph neural networks. arXiv:1802.09691 [cs.LG] (2018)

# A Graph Learning Based Method for Inductive Knowledge Graph Relation Prediction

Shen Jikun(✉), Lu TaoYu, Zhang Xiang, Luo Hong, and Xu Yun

China Mobile (Hangzhou) Information Technology Co., Ltd., Hangzhou 311100, Zhejiang, China
shenjikun@cmhi.chinamobile.com

**Abstract.** In recent years, with the continuous progress of knowledge graph technologies, lots of large-scale knowledge graph have been applied to a variety of downstream tasks, including question answering and information retrieval. However, even the largest knowledge graph at present is still incomplete. Specially, in an open environment, the knowledge graph itself is dynamic, and new knowledge graph are constantly being constructed. Therefore, it is necessary to complete the knowledge graph with relation prediction methods. However, since the transductive relation prediction method assumes that the entities in the testing samples have appeared in the training samples, it is impossible to perform relation prediction on unseen entities. This paper propose a new inductive knowledge graph completion method called EGraIL, in which the entity embeddings are initialized by thc surrounding relation embeddings and then calculate the possibility of triples through conventional knowledge graph embedding score function. The model is trained on the triple classification and relation prediction task at the same time. Finally, the results are significantly better than baseline method GraIL, and achieve good rank in the Inductive Knowledge Graph Relation Prediction Task in CCKS2023.

**Keywords:** knowledge graph · knowledge graph complement · relation prediction

## 1 Introduction

With the deepening of research on knowledge graphs, even the state-of-the-art knowledge graphs suffer from incompleteness and incorrectness issue: 1)Incompleteness issue, such as persons in graphs have no career information, which may be caused by data missing or emergence of new entities.; 2)Incorrectness issue, like incorrect attribute or relation between entities, which may be caused by information extraction.

Relation prediction aims to predict the missing part in known triples, and triple classification aims to judge whether a given triplet is correct or not. Both of them are important tasks in knowledge graph completion. Compare to transductive methods, inductive methods have been probed to have higher practical value.

The-state-of-art inductive methods for link prediction and triple classification are a series of graph neural networks (GNNs) based methods. GraIL [1] and its following


H. Wang et al. (Eds.): CCKS 2023, CCIS 1923, pp. 297–303, 2023.
https://doi.org/10.1007/978-981-99-7224-1_23

works construct enclosing subgraph from source knowledge graph and learn to predict relation from the subgraph structure around candidate relation.

However, in the Inductive Knowledge Graph Relation Prediction Task in CCKS2023, GralL suffer from over fitting, due to the too small amount of source KGs in training. To alleviate it, we propose several modifications based on GralL, and build a new inductive graph learning method, termed Enhanced GralL (EGralL). In EGralL, we apply a new entity node initialization method inspired by MorsE [2], to replace the GNN scoring with knowledge graph embedding(KGE) model, and joint strategy for training. Experiments show that EGralL outperforms GralL baseline method and achieve 3rd rank in the finals of this task.

## 2 Background: GralL

For inductive relation prediction, GralL is the first method proposed to model enclosing subgraph structure around the target triple based on GNNs. The overall task is to score a triple $(h, r, t)$ to predict the likelihood of a possible relation $r$ between a head entity node $h$ and tail entity node $t$ in a KG. The framework mainly consists of three sub-tasks: 1) Subgraph Extraction, extracting the enclosing subgraph around the target nodes; 2) Entity Node Initialization, initializing the entity nodes in the extracted subgraph; 3) GNN Scoring, obtaining representation of enclosing subgraph by GNN.

### 2.1 Subgraph Extraction

Given target triple $(h, r, t)$, the enclosing subgraph is defined as the graph generated by all the nodes that occur on a path between $h$ and $t$. There are three steps for subgraph extraction in GralL. Firstly, getting the node sets of k-step neighborhood, $\mathcal{N}_k(h)$ and $\mathcal{N}_k(t)$ for node $h$ and $t$ respectively. Secondly, taking intersection between $\mathcal{N}_k(h)$ and $\mathcal{N}_k(t)$ as the enclosing subgraph. Finally, filtering out isolated nodes or nodes that are more than k away from any target node.

### 2.2 Entity Node Initialization

GNNs require a node feature matrix X as input to initialize the neural message passing algorithm. GralL adopt the double radius vertex labeling scheme. Firstly, caculating the shortest distance from $h$ and $t$, $d(i, h)$ and $d(i, t)$, for each node i in the subgraph around nodes $h$ and $t$. While h and t are represented as (0,1) and (1,0) respectively, the final representation vector of node i is concatenated by one-hot representation, like [one-hot($d(i, h)$) $\oplus$ one-hot($d(i, t)$)], where $\oplus$ denotes concatenation. However, the initialization method probably suffer from sparsity, due to one-hot.

### 2.3 GNN Modulation and Scoring

The Given the extracted and labeled subgraph around the target nodes set $G(h, t, r)$, to score the likelihood of triple. GralL adopt a GNN based general message-passing scheme work, in the scheme, a entity node representation is updated iteratively by combination with the aggregation of neighbors representation. The $k$-th layer of GNN:

$$a_t^k = AGGREGATE^k\left(\left\{h_s^{k-1} : s \in \mathcal{N}(t)\right\}, h_t^{k-1}\right)$$

$$h_t^k = COMBINE^k(\{h_t^{k-1}, a_t^k\})$$

where $a_t^k$ is the aggregated message from the neighbors, $h_t^k$ is the hidden representation of entity node $t$ in the $k$-th layer, $\mathcal{N}(t)$ is the set of immediate neighbors of node $t$. For any node i, $h_i^0$ is the initialization for the node features. The GNN architectures consist of AGGREGATE and COMBINE.

AGGREGATE for modulation is defined as:

$$AGGREGATE = \sum_{r=1}^{R} \sum_{s \in \mathcal{N}_r(t)} \alpha_{rr_t st}^k W_r^k h_s^{k-1}$$

COMBINE for scoring is defined as:

$$\text{COMBINE} = ReLU\left(W_{self}^k h_t^{k-1} + a_t^k\right)$$

where $R$ is the relations number of knowledge graph, $\mathcal{N}_r(t)$ denotes the immediate outgoing neighbors set of node $t$ with relation $r$, $W_r^k$ is the transformation matrix to propagate messages in the $k$-th layer over relation $r$; $\alpha_{rr_t st}^k$ is the edge attention weight in the $k$-th layer corresponding to the relation $r$ edge connecting nodes $s$ and $t$. This attention weight fuction is defined as:

$$\alpha_{rr_t st}^k = \upsilon\left(A_2^k s + b_2^k\right)$$

$$\text{s} = \text{ReLU}\left(A_1^k\left[h_s^{k-1} \oplus h_t^{k-1} e_r^a e_{r_t}^a\right] + b_1^k\right)$$

where $h_s^k$ and $h_t^k$ denote the hidden node representation of the nodes in the $k$-th layer of the GNN, $e_r^a$ and $e_{r_t}^a$ denote trainable attention embeddings of the relations.

## 3 Proposed Enhanced GraIL

We propose a new inductive graph learning framework, EGraIL, which is the enhanced modifications based on GraIL. In this framework, we replace entity node initialization, remove GNN scoring and combine with KGE model to calculate triple score, and apply joint strategy for training.

### 3.1 Entity Node Initialization

Relation representation consists of two part: $\boldsymbol{r}^{in}$ and $\boldsymbol{r}^{out}$, from in-direction and out-direction. We initialize entity embeddings with the surrounding relations.

$$\boldsymbol{H}_e = \frac{\sum_{i=1}^{M} \boldsymbol{r}^{in} + \sum_{j=1}^{N} \boldsymbol{r}^{\text{out}}}{M + N}$$

where M is the number of in-direction relation around entity, N is the number of out-direction relation around entity.

### 3.2 GNN Modulation

Compare to GraIL, we remove GNN Scoring to obtain the final entity embedding for each entity node directly, which will be used to feed into KGE.

### 3.3 Ensemble KGE

Following the GNN Modulation, we feed entity embedding into KGEs to calculate triple score instead of scoring in GraIL. Given a triple (*h, r, t*), we use three KGE models in this paper, including TransE [3], TransR [4], and RotatE [5] as below:

**TransE**, a simple and effective method for knowledge graph embedding, which encodes entities and relations of a knowledge graph into a low-dimensional embedding vector space. TransE takes a relation as translation from head entity to tail entity. For a relation triplet (*h, r, t*), TransE requires the embedding of entities $\boldsymbol{h}, \boldsymbol{t}$ and the embedding of relation $\boldsymbol{r}$ meet the formula $\boldsymbol{h} + \boldsymbol{r} \approx \boldsymbol{t}$. During training, the score function is defined as

$$f(\boldsymbol{h}, \boldsymbol{r}, \boldsymbol{t}) = \|\boldsymbol{h} + \boldsymbol{r} - \boldsymbol{t}\|_2^2$$

The score is low when (h,r,t) is a golden triplet, and high otherwise. TransE achieives outstanding performance in 1-to-1 relations. However, when dealing with 1-to-N, N-to-1, N-to-N relations, its performance is not so good as expected.

**TransR**, a modification based on TransE, which sets a mapping matrix for every relation $\boldsymbol{r}$ to help project entities from entity space to relation space.

$$\boldsymbol{h}_r = \boldsymbol{h}\boldsymbol{M}_r, \boldsymbol{t}_\mathrm{r} = \boldsymbol{t}\boldsymbol{M}_\mathrm{r}$$

Then the score function is defined as

$$f(\boldsymbol{h}, \boldsymbol{r}, \boldsymbol{t}) = \|\boldsymbol{h}_\mathrm{r} + \boldsymbol{r} - \boldsymbol{t}_\mathrm{r}\|_2^2$$

**RotatE**, which is inspired by Euler's identity, maps the head and tail entities h, t to the complex embeddings, and considers each relation as a rotation from the head entity to the tail entity in the complex vector space.

$$\boldsymbol{t} = \boldsymbol{h} \circ \boldsymbol{r}, \text{ where } |r_i| = 1$$

Then the score function is defined as

$$f(\boldsymbol{h}, \boldsymbol{r}, \boldsymbol{t}) = \|\boldsymbol{h} \circ \boldsymbol{r} - \boldsymbol{t}\|$$

**Ensemble KGEs**, in testing, given entity and relation embeddings from GNN modulation, we expect significant gains to be obtained by ensembling different KGE model on reciprocal rank metrics, called late fusion.

$$mrr_{\text{ensemble}} = \frac{1}{N} \sum_{i=1}^{N} \sum_{j=1}^{M} \frac{1}{\text{rank}_j^k}$$

where N is the size of the query set in testing, M is the number of KGE models, $rank_j^k$ is the rank of correct entity in the ranking of the predicted candidate entities list for test triple *i*, with KGE model *j*.

### 3.4 Joint Training

We joint triple classification and relation prediction tasks for training, while apply binary cross entropy loss function for triple classification task and self-adversarial negative sampling loss function referred to RotatE on query triples for relation prediction task.

The loss function of triple classification task is defined as:

$$\mathcal{L}_{\text{triple classification}} = -\sum_{(h,r,t)\in T} \log \sigma(f(h,r,t))$$

where $\sigma$ denotes sigmoid function, $T$ denotes triples including positives and negatives.

The loss function of relation prediction task is defined as:

$$\mathcal{L}_i = \sum_{(h,r,t)\in Q_i} -\log \sigma(\gamma + f(h,r,t)) - \sum_{i=1}^{k} p(h_i', r, t_i') \log \sigma(-\gamma - f(h_i', r, t_i'))$$

$$p\left(h_j', r, t_j'\right) = \frac{\exp \beta f\left(h_j', r, t_j'\right)}{\sum_i \exp \beta f(h_i', r, t_i')}$$

where $\sigma$ denotes sigmoid function; $\gamma$ is a fixed margin; $k$ is the number of negative samples for each triple; $(h_i', r, t_i')$ is the $i$-th negative triple by corrupting head or tail entity; $\beta$ is the temperature of sampling.

The final learning objective of our work is defined as the combination of the above two:

$$\mathcal{L} = \alpha \mathcal{L}_{\text{triple classification}} + (1-\alpha)\mathcal{L}_{\text{relation prediction}}$$

where $\alpha$ controls the contribution of loss of the above two task. With this joint training strategy, our model is capable of modeling subgraph with complete relations while capturing neighboring relations aware of both local and global structural information.

## 4 Experiments

### 4.1 Data Sets

In this task, we evaluate our system with the Inductive Knowledge Graph Relation Prediction Task dataset in CCKS2023, and Table 1 lists some statics of the dataset.

**Table 1.** Statics for the datasets used

| DataSet | #Rel | #Source Ent | #Support Ent |
|---|---|---|---|
| CCKS2023 | 46 | 4050 | 3533 |

### 4.2 Training Details

In the experiment, we select learning rate for SGD among {0.05, 0.01, 0.005},the margin $\gamma$ among {4, 8, 10, 16}, the dimensions of entity embedding $k$ and relation embedding $d$ among {32, 64, 128}, the batch size $B$ among {32, 64, 128}. The best experimental setting obtained by valid set are: $\lambda = 0.01$, $\gamma = 10$, k = 32, d = 32, B = 128. All experiments are implemented by PyTorch and run on Nvidia Tesla V100 16GB.

To compare fairly with the prior methods, we use Mean Reciprocal Rank(MRR) metrics as evaluation protocol, which is defined as:

$$MRR = \frac{1}{N} \sum_{i=1}^{N} \frac{1}{rank_i}$$

where N is the size of the query set in testing, $rank_i$ is the rank of correct tail entity in the ranking of the predicted candidate tail entities.

### 4.3 Main Results

**Table 2.** The test result in primary and final dataset

| Method | Test dataset | |
|---|---|---|
| | Primary | Final |
| GraIL | 0.828 | 0.562 |
| EGraIL(TransE) | 0.835 | 0.576 |
| EGraIL(TransR) | **0.851** | **0.613** |
| EGraIL(RotatE) | 0.845 | 0.587 |

As can be seen from Table 2, combining GNN modulation with KGE model TransE, TransR and Rotate, all perform better than GraIL. EGraIL(TransR) obtain the best result both in test primary dataset which is ranked 4th on the leaderboard, and achieve rank 3rd in test final.

## 5 Conclusion and Future Work

In the Inductive Knowledge Graph Relation Prediction Task, we propose a new inductive graph learning framework, EGraIL, which is the enhanced modifications based on GraIL. In this framework, we replace entity node initialization, remove GNN scoring and combine with KGE model to calculate triple score, and apply joint strategy for training. It capable of modeling subgraph with complete relations while capturing neighboring relations aware of both local and global structural information.

Experiments demonstrate that this method has good performance in inductive relation inference scenarios, and achieve good ranks both in test primary section and test finals.

We think the following issues should be further studied as bellow: First, all the entities in different relations share the same mapping matrix in TransR.But as known, the types and attributes of the entities are distinct. We think it would be difficult for one universial mapping matrix to character these different attributions hidden in different entities. Second, when the amount of relations become very large, the amount of multiplication computations between Matrix or Vector in KGEs would become much larger, thus it would take lots of time for training the whole model. Both of these two issues should be well studied in the future.

## References

1. Teru, K.K., Denis, E., Hamilton, W.: Inductive relation prediction by subgraph reasoning. In: ICML (2020)
2. Chen, M.-Y., et al.: Meta-knowledge transfer for inductive knowledge graph embedding. In: SIGIR (2022)
3. Bordes, A., Usunier, N., Garcia-Duran, A.: Translating embeddings for modeling multi-relationaldata. In: Proceedings of NIPS, pp. 2787–2795 (2013)
4. Lin, Y., Zhang, J., Liu, Z., Sun, M., Liu, Y., Zhu, X.: Learning entity and relation embeddings forknowledge graph completion. In: Proceedings of AAAI (2015)
5. Sun, Z., Deng, Z.-H., Nie, J.-Y., Tang, J.: RotatE: knowledge graph embedding by relational rotation in complex space. In: ICLR (2019)

# LLM-Based SPARQL Generation with Selected Schema from Large Scale Knowledge Base

Shuangtao Yang(✉), Mao Teng, Xiaozheng Dong, and Fu Bo

Lenovo Knowdee (Beijing) Intelligent Technology, Haidian District, Beijing, China
{yangst,Maoteng,dongxz}@knowdee.com

**Abstract.** Knowledge base question answering (KBQA) aims to answer natural language questions using structured knowledge bases. Common approaches include semantic parsing-based approaches and retrieval-based approaches. However, both approaches have some limitations. Retrieval-based methods struggle with complex reasoning requirements. Semantic parsing approaches have a complex reasoning process and cannot tolerate errors in earlier steps when generating the final logical form. In this paper, we proposed a large language model (LLM)-based SPARQL generation model, which accepts multiple candidate entities and relations as inputs, reducing the reliance on mention extraction and entity linking performance, and we found an entity combination strategy based on mentions, which can produce multiple SPARQL queries for a single question to boost the chances of finding the correct answer. Finally, our model achieves state-of-the-art performance in the CCKS2023 CKBQA competition, F1 score is 75.63%.

**Keywords:** KBQA · Large Language Model · SPARQL Generation

## 1 Introduction

Knowledge based question answering (KBQA) [1] has recently gained research interest, as it provides an intuitive way to access factual knowledge. The KBQA system makes use of structured knowledge bases such as Freebase, Wikidata, and DBpedia, which have logically organized entities and relations. A knowledge base typically contains a large number of triples, which can be represented as *(head, relation, tail)*, the head refers to main entity, the tail refers to another entity or a literal value, and the relation is a directed relationship between head and tail [2]. KBQA systems can infer answers to questions by matching relevant entities and relations.

The existing KBQA approaches can be divided into two main categories: retrieval-based methods [3–9] and semantic parsing-based methods [10–23]. Retrieval-based methods directly represent and rank entities parsed from the input question. Among them, some methods first extract a subgraph containing only question-relevant entities from the knowledge base before performing reasoning. By narrowing the focus to a subset of KB, these methods can reduce the space for reasoning and be more efficient, while still struggling with complex questions. In contrast, semantic parsing-based (SP-based) methods parse a question into a logic form like SPARQL [10], Lambda-DCS [11], and


H. Wang et al. (Eds.): CCKS 2023, CCIS 1923, pp. 304–316, 2023.
https://doi.org/10.1007/978-981-99-7224-1_24

KoPL [12] that can be executed against the KB. However, these methods rely heavily on expensive annotations of intermediate logic forms and tend to be limited to narrow domains. With the advance of pre-trained language models (PLMs), many works have reformulated the semantic parsing task as a sequence-to-sequence (Seq2Seq) logical expression generation problem, which directly translate natural language queries into logical forms.

More recently, Large language models(LLM) have made significant advancements in natural language processing (NLP), such as GPT-3 [24], PaLM [25], LLaMA [26], which has proven to be an effective technique for improving performance on a wide range of language tasks [27]. Considering the large scale of the knowledge graph to process, containing 66,630,393 triplets, 11,327,935 entities, and 408,794 relations, we adopt a semantic parsing-based method with LLM as CKBQA solution. Like traditional semantics-based approaches, our method adopts a staged pipeline architecture. Traditional semantic parsing pipeline comprises mention extraction, entity linking, and SPARQL generation. However, for extremely large knowledge graphs, SPARQL generation performance by traditional semantic parsing pipeline often decreases substantially due to error propagation across pipeline. Due to the outstanding capabilities of large language models (LLMs) [20], we proposed large language model (LLM)-based SPARQL generation model that accepts multiple candidate entities and relations as inputs, which helps to reduce the reliance on mention extraction and entity linking performance. We incorporate an entity relation selection model into the pipeline to prune noisy inputs for the generation model. Additionally, we implement an entity combination strategy based on mentions, which can produce multiple SPARQL queries for a single question to boost the chances of finding the correct answer.

The main contributions of this paper are summarized below:

- This work represents the first attempt at leveraging large pre-trained language models (LLM) for SPARQL generation to address Chinese knowledge graph question answering, achieving top-1 ranking performance in the CCKS2023 CKBQA competition.
- We propose an effective SPARQL generation method based on large language models, utilizing mention extraction, entity linking, attribute selection models, and entity combination to provide high-quality inputs for the language models, significantly improving SPARQL generation quality. The model process is shown in Fig. 1.
- Ablation experiments were conducted to assess the importance of each module in SPARQL generation for our approach.

## 2 Related Work

**Retrieval-Based Methods.** Zhang et al. proposed a subgraph retriever (SR) separate from the subsequent reasoner for KBQA. The SR was designed as an efficient dual-encoder capable of updating the question representation when expanding the path, as well as determining when to stop the expansion [3]. He et al. proposed a teacher-student approach for multi-hop KBQA. The teacher network utilized bidirectional reasoning to produce reliable intermediate supervision signals that improved the reasoning of the student network and reduced spurious reasoning [4].

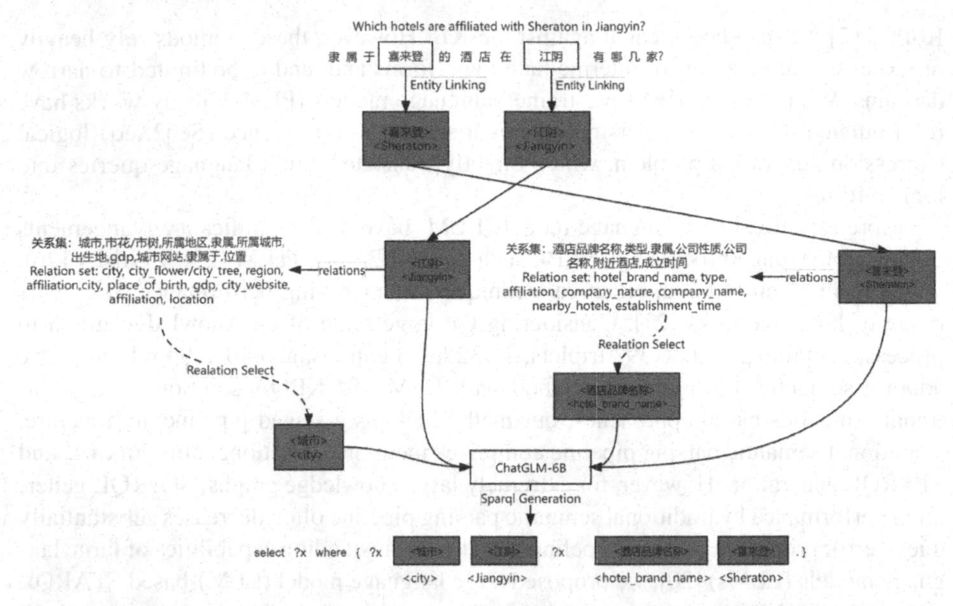


**Fig. 1.** SPAEQL generation with selected entity and relations. Mentions (highlighted in white boxes) need to be linked to entites which are from knowledge base. There are two entities (highlighted in green boxes), we need to obtain all the relations or attributes of each entity, and then use attribute/relation rank model to sort them. The selected entities (in green boxs) and relations (in red boxes) will as input to the SPARQL generation model. The given textual mentions can be utilized to construct focused SPARQL queries, incorporating the most relevant entities and relations.

**Semantic Parsing-Based Methods.** Purkayastha et al. [13] used a Seq2Seq model to generate SPARQL query sketch, and then apply entity and relation linkers to fill in the sketch and produce a complete SPARQL query. Lambda-DCS (lambda dependency-based compositional semantics) [11] is a tree-structured logical Forms, which propose to reduce the complexity in compositionally creating the logical form of a sentence. Cao et al. [12] first parse the original question into the skeleton of KoPL program, a sequence of symbolic functions, and then train an argument parser to retrieve corresponding arguments of these functions.

**Seq2Seq Methods.** Nie et al. proposed a unified intermediate representation (GraphQ IR) that bridges the semantic gap between natural language queries and formal graph query languages. GraphQ IR can produce intermediate representation sequences using composition rules consistent with English to capture natural language semantics while maintaining fundamental graph structures [14]. Cao et al. proposed a Line Graph Enhanced Text-to-SQL (LGESQL) to extract relational features from text without having to construct meta paths. The Line Graph representation allowed messages to propagate more efficiently by considering not just connections between nodes, but also the topology of directed edges [15]. Das et al. first identify different queries with semantically equivalent components, and then construct a new logical form by combining these matching components from the discovered queries [21]. Huang et al. utilize a large model-based algorithm to identify entities and relations within a question, and then generate a query structure with placeholders, which are then populated in a post-processing step [22]. Xiong et al. utilize advanced generative pre-trained language models to generate questions from logical form and then make predictions, the auto-prompter has the ability to paraphrase predicates in a consistent and fluent manner [23].

**LLM-Based Methods.** LLM with billions of parameters have achieved state-of-the-art results on many NLP benchmarks by learning powerful contextual representations of language from large amounts of text data. One key development in LLM is the use of self-attention mechanism [28] and transformer architectures [29]. Another important development is the use of pre-training, where models are first trained on massive datasets and then fine-tuned on downstream tasks. LLM transfers broad linguistic knowledge that significantly improves performance across many language understanding tasks. One remarkable recent development is the launch of ChatGPT [30], a conversational AI system powered by LLMs. ChatGPT has gained widespread public attention for its ability to engage in surprisingly natural conversations, which highlight the substantial progress LLMs have made in language understanding and generation that allows them to partake in coherent human-like dialogue.

## 3 Method

As shown in Fig. 2, The methodology we propose comprises four fundamental components: 1) extracting textual mentions from the input, 2) linking mentions to entities in the knowledge graph, 3) selecting relevant attributes and relations from these entities, and 4) combining these entities to generate SPARQL queries. The specific implementations of each module will be described fully in subsequent sections. The complete descriptions of the individual modules' specific implementations will be provided in subsequent sections.

### 3.1 Mention Extraction

Mention Extraction is the task of identifying the mention span of all entities in the question [31]. Each such span is referred to as an entity mention. The word or sequence of words that refers to an entity is also known as the surface form of the entity. An utterance may contain multiple entity, often also consisting of more than one word. Additionally, a broader classification of entities, such as person, location, and organization, can sometimes be assigned.

Our mention extraction model architecture is composed of a BERT encoder with a token-level classifier on top followed by a Linear-Chain CRF. We first use BERT to encode user question and outputs a sequence of encoded token representations, then a classification model projects each token's encoded representation to the tag space. We also frame mention extraction as a generative task, and attempt to extract mentions using ChatGLM [35].

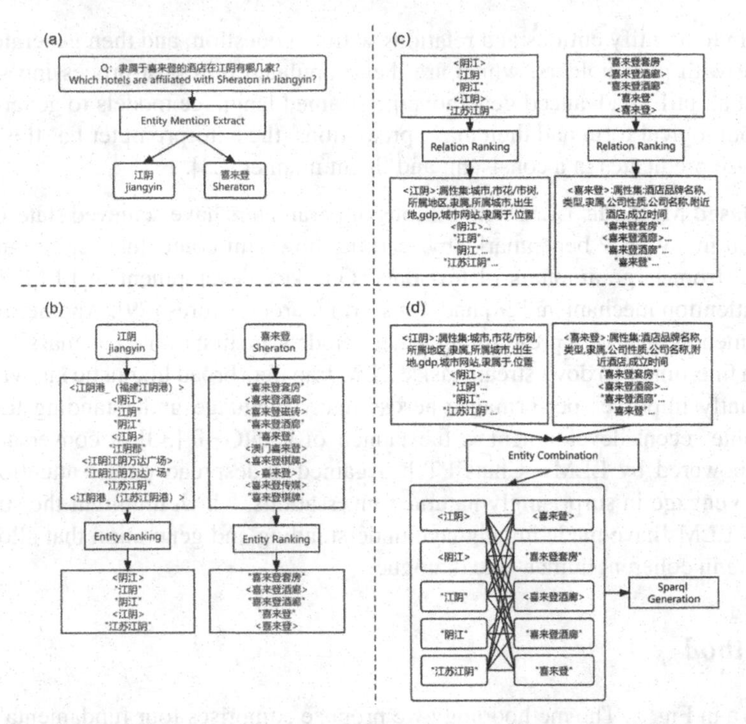


**Fig. 2.** Method Flow: (a) Mention extraction is carried out on the question. (b) For each mention, top-5 candidate entities are selected from the knowledge graph by using Elasticsearch and rules, and then ranked by the entity linking model. (c) Relation selection is applied to choose the most relevant relations for entities from the previous entity linking results. (d) Candidate entities across different mentions are combined and fed into the SPARQL generation model to produce multiple SPARQL queries.

## 3.2 Entity Linking

The task of Entity Linking involves establishing connections between annotated mentions in a given utterance and their corresponding entities within a knowledge base [32–34]. This task was addressed by using popular knowledge bases such as DBpedia, Freebase or Wikipedia. Entity linking serves as a bridge between textual spans and structured entities within a knowledge base, thereby will be beneficial to various downstream tasks like question answering and knowledge extraction. EL aims to link entity mentions in unstructured text to their corresponding entities in a designated knowledge base.

Our entity linking model is trained to assign a score to each candidate entity as shown in (1). Specifically, given the question $q$ and the candidate entity $e\_text$, we use a BERT-based encoder to generate a score indicating the confidence of the link [29].

$$el\_score = sigmoid(AVG(BERT([q : e\_text]))) \tag{1}$$

For every mention, we will select top-5 entities according to their linking confidence scores for the next phase.

### 3.3 Entity Attribute/Relation Select

When given an entity and its relations, Entity Attribute/Relation Select model can select question related relations, thus the model also learns to score each entity relation. Specifically, given the question $q$ and the candidate entity relations $r\_text$, we also use a BERT-based encoder to generate a correlation score between the question and either entity attributes or relations.

$$es\_score = sigmoid(AVG(BERT([q : r\_text]))) \tag{2}$$

The entity and its relation are represented by triples consisting of the entity, relationship, and tail. For each entity, we will select top-5 relationships based on their correlation score for the next phase.

### 3.4 Entity Combination and SPARQL Generation

After the phase of entity linking and entity attribute/relation select, we have obtained top-5 entities for each mention, and each entity has top-5 relations, as key supporting evidence. In the SPARQL generation stage, we attempted different methods.

**Method 1**: The question and all key supporting evidence from different mentions are concatenated as input to the SPARQL generation model, resulting in a single expression.

**Method 2**: Entities from different elements are combined and concatenated within each combination to generate multiple SPARQL queries, which ca be executed against the KB. Unlike Method 1, the approach will produce multiple SPARQL expressions.

Taking the question "What are the hotels affiliated with Sheraton in Jiangyin?" as an example, after mention extraction, entity linking, and entity attribute/relation selection, we obtaine the most relevant knowledge related to this question from the knowledge graph. In Method 1, we filled all relevant information into the prompt, obtaining the complete prompt as shown below.

*请根据问题:\"隶属于喜来登的酒店在江阴有哪几家?\",和候选实体信息:[0]名称: < 喜来登 >,属性集:酒店品牌名称,类型,隶属,公司性质,公司名称,附近酒店,成立时间;[1]名称: < 江阴 >,属性集:城市,市花/市树,所属地区,隶属,所属城市,出生地,gdp,城市网站,隶属于,位置,市长,所在城市,适用地区,分布区域,所属地区,著名景点,位于,行政区域,属于,家乡;[2]名称: < 江阴黄嘉喜来登酒店 >,属性集:实体名称,酒店品牌名称,酒店入住开始时间,是否有鲜花店,是否有酒吧,是否有接机服务,是否有接机服务-营业时间,是否有接送服务-营业时间,是否有温泉,是否有桑拿浴室,是否有允许带宠物,是否有茶室,是否有会议室,是否有桌球室,是否有管家服务,是否有熨衣服务,是否有图书馆,是否有wifi服务,是否有游戏室,是否有礼宾服务;[3]名称: < 镇江富力喜来登酒店 >,属性集:实体名称,酒店品牌名称,房型名称,是否有鲜花店,是否有桑拿浴室,酒店入住开始时间,是否有允许带宠物,是否有温泉,是否有高尔夫球场,是否有保龄球馆,是否有租车服务,是否有大堂吧,是否有多功能厅,是否有网球场,是否有婚宴服务,是否有叫醒服务,是否有礼宾服务,是否有KTV,是否有图书馆,是否有会议室;[4]名称:\"喜来登\",属性集:中文名称,公司名称,对应查询图谱的Sparql的语句为:*

*Please follow the question: \"Which hotels are affiliated to Sheraton in Jiangyin?\", and candidate entity information: [0] Name: < Sheraton >, attribute set: hotel brand* *name, type, affiliation, company nature, company Name, nearby hotels, establishment time;* [1] *Name: < Jiangyin >, attribute set: city, city flower/city tree, region, affiliation, city, place of birth, gdp, city website, affiliation, location, city Length, city, applicable area, distribution area, belonging area, famous scenic spot, location, administrative area, belonging to, hometown;* [2] *name: < Jiangyin Huangjia Sheraton Hotel >, attribute set: entity name, hotel brand name, hotel Check-in start time, whether there is a flower shop, whether there is a bar, whether there is a pick-up service, whether there is a pick-up service-opening hours, whether there is a pick-up service-opening hours, whether there is a hot spring, whether there is a sauna, whether pets are allowed, whether there is a tea room, whether there is a meeting room, whether there is a billiard room, whether there is a butler service, whether there is an ironing service, whether there is a library, whether there is wifi service, whether there is a game room, whether there is a concierge service;* [3] *name: < Sheraton Zhenjiang Hotel >, attribute set: entity name, hotel brand name, room type name, whether there is a flower shop, whether there is a sauna, hotel check-in start time, whether pets are allowed, whether there is a hot spring, whether there is a golf course, Is there a bowling alley, is there a car rental service, is there a lobby bar, is there a multi-function hall, is there a tennis court, is there a wedding banquet service, is there a wake-up call service, is there a concierge service, is there a KTV, is there a book Museum, whether there is a conference room;* [4] *Name: \"Sheraton\", attribute set: Chinese name, company name, the Sparql statement corresponding to the graph is:*

In Method 2, we combined entity information from different mentions, then filled each combination into the prompt, so we could obtain multiple prompts to generate SPARQL statements. For the combination < Sheraton >, < Jiangyin >, the complete prompt is shown below.

*请根据问题:\"隶属于喜来登的酒店在江阴有哪几家?\",和候选实体信息:[0]名称: < 喜来登 >,属性集:酒店品牌名称,类型,隶属,公司性质,公司名称,附近酒店,成立时间;[1]名称: < 江阴 >,属性集:城市,市花/市树,所属地区,隶属,所属城市,出生地,gdp,城市网站,隶属于,位置,市长,所在城市,适用地区,分布区域,所属地区,著名景点,位于,行政区域,属于,家乡,对应查询图谱的Sparql的语句为:*

*Please follow the question: \"Which hotels are affiliated to Sheraton in Jiangyin?\", and candidate entity information: [0] Name: < Sheraton >, attribute set: hotel brand name, type, affiliation, company nature, company Name, nearby hotels, establishment time;* [1] *Name: < Jiangyin >, attribute set: city, city flower/city tree, region, affiliation, city, place of birth, gdp, city website, affiliation, location, city Length, city, applicable area, distribution area, area, famous scenic spot, location, administrative area, belonging, hometown, the Sparql statement corresponding to the graph is:*

After obtaining the complete prompt, we feed it to the LLM to generate SPARQL. We select ChatGLM-6B[1] [35] as the SPARQL generation models. ChatGLM-6B is a pre-trained large language model with 6.2 billion parameters, based on the General Language Model (GLM) architecture. ChatGLM-6B was trained on around 1 trillion tokens of Chinese and English corpus, with additional supervised fine-tuning, feedback

[1] https://github.com/THUDM/ChatGLM-6B.

bootstrap, and reinforcement learning using human feedback. This enables ChatGLM-6B to generate answers that are aligned with human preference, with fluency in both English and Chinese.

We use low-rank adaptation (LoRA) to finetune ChatGLM-6B for SPARQL Generation [36]. The parameter settings used for LoRA fine-tuning are shown in Table 1.

**Table 1.** Knowledge Graph Information.

| Item | Quantity |
|---|---|
| lora_alpha | 32 |
| lora_dropout | 0.1 |
| lora_rank | 8 |
| lora_target | query_key_value |

# 4 Results

The key statistics for the knowledge graph and training data used in this work are presented in Table 2. The knowledge graph contains 66,630,393 triplets, 11,327,935 entities, and 408,794 relations. The training data is comprised of 7,625 examples.

**Table 2.** Knowledge Graph and Data Information.

| Item | Quantity |
|---|---|
| Triplet | 66,630,393 |
| Entity | 11,327,935 |
| Relation | 408,794 |
| Train Case | 7,625 |

## 4.1 Mention Extraction Result

We compared several mention extraction methods on CKBQA dataset, including BERT + CRF, Roberta + CRF, and ChatGLM-6b(LoRA). As shown in Table 3, the ChatGLM-6b(LoRA) model achieved the highest F1 score.

## 4.2 Entity Linking Result

We compared Bert and Roberta on entity linking task. As shown in the Table 4, RoBerta achieved the higher F1 score of 94.48%, compared to 93.64% for Bert. This indicates RoBerta is more effective for this entity linking task, outperforming Bert by 0.84% in terms of F1 score.

**Table 3.** Mention Extraction Result

| Method | F1 Performance (100%) |
|---|---|
| Bert + CRF | 84.5 |
| Roberta + CRF | 85.2 |
| ChatGLM-6b(LoRA) | 89.4 |

**Table 4.** Entity Linking Result

| Method | F1 Performance (100%) |
|---|---|
| Bert | 93.64 |
| RoBerta | 94.48 |

### 4.3 Entity Attribute/Relation Select

F1 scores for Bert and RoBerta models on an entity attribute/relation selection task are presented in Table 5. RoBerta model achieved the higher F1 score of 95.17%, compared to 94.12% for Bert model, indicating RoBerta is more effective for extracting entity attributes and relations, outperforming Bert model by 1.05% based F1 evaluation metric.

**Table 5.** Entity Attribute/Relation Select Result

| Method | Performance (100%) |
|---|---|
| Bert | 94.12 |
| RoBerta | 95.17 |

### 4.4 Entity Combination and SPARQL Generation

At this stage, we compared the impact of different entity combination methods on SPARQL generation. Using the same ChatlGLM-6B model and LoRA fine-tuning parameters, we trained and fine-tuned two SPARQL generation models with different entity combination approaches. Table 6 shows the performance of the two entity combination methods on the training and validation set, whis is evaluated using ROUGE-1, ROUGE-2, ROUGE-L [37]. To evaluate the correctness of SPARQL, we introduced the Pass Rate metric. ChatGLM-6b-Method2 achieved higher scores across all metrics, with notably large improvements in ROUGE-2 (90.11%vs 85.96%) and Pass rate (68.9% vs 61.5%). This suggests that ChatGLM-6b-Method2 is more effective for SPARQL generation.

The pass rate metric measures the ratio of generated SPARQL queries that are syntactically valid and return correct answers on test set.

**Table 6.** Entity Combination and SPARQL Generation Result.

| Method | Acc Performance (100%) | | | |
|---|---|---|---|---|
| | ROUGE-1 | ROUGE -2 | ROUGE -L | Pass Rate |
| ChatGLM-6b-Method1 | 91.72 | 85.96 | 89.09 | 61.5 |
| ChatGLM-6b-Method2 | 94.89 | 90.11 | 91.72 | 68.9 |

## 4.5 End to End Performance

We conducted ablation experiments to evaluate the importance of each module in our pipeline. The results of these experiments are shown in Table 7.

Table 6 shows the incremental impact of on KBQA system from adding different knowledge graph components. We evaluated five system variations (V1-V5) on the CKBQA training dataset.

**System V1** uses only a mention extraction (ME) model and SPARQL generation (SG) module, achieving an F1 score of 45.11%. The lack of entity linking, relation selection, and entity combining modules limits its performance. By analyzing the generated SPARQL, we found that errors often occur due to inconsistent entity formats with the knowledge base, making it impossible to obtain answers through SPARQL.

**System V2** adds an entity linking (EL) module using RoBerta, improving performance to 66.45% F1. Linking mentions to knowledge graph entities provides useful contextual information.

**System V3** further incorporates an entity attribute/relation selection (ERS) module based on Roberta. This model eliminates interference from irrelevant attributes and relationship of entities in the input, increasing F1 to 69.23%.

**System V4** adds an entity combination (EC) module. Through this module, we can assemble entity information from different mentions to generate multiple SPARQL queries. Concurrently, we can determine the relevance of each SPARQL query based on relatedness between entities. The most relevant SPARQL that can retrieve results from the knowledge graph is selected as the final generated query. By utilizing this method, we improved the performance of our system to 73.93% F1 score.

Even after System V4, we still found a limited number of questions for which it was not possible to generate an accurate SPARQL query that could retrieve answers from the knowledge graph. Therefore, we supplemented with an additional KBQA method based on triple retrieval. By integrating this approach, we further improved our system's score to 75.63%.

**Table 7.** Knowledge Graph Information. **ME** means Mention Extract Model; **EL** means Entity Linking Model; **ERS** means Entity attribute/relation Select Model and **EC** means Entity Combination Module and **SG** means SPARQL Generation, and **Retrieval** means Retrieval Method For KBQA.

| System | ME | EL | ERS | EC | SG | Retrieval | Acc (100%) |
|---|---|---|---|---|---|---|---|
| V1 | √ | × | × | × | √ | × | 45.11 |
| V2 | √ | √ | × | × | √ | × | 60.45 |
| V3 | √ | √ | √ | × | √ | × | 69.23 |
| V4 | √ | √ | √ | √ | √ | × | 73.93 |
| V5 | √ | √ | √ | √ | √ | √ | 75.63 |

## 5 Conclusion

In this paper, we proposed large language model (LLM)-based SPARQL generation model, which accepts multiple candidate entities and relations as inputs, reducing the reliance on mention extraction and entity linking performance. And we found an entity combination strategy based on mentions, which can produce multiple SPARQL queries for a single question to boost the chances of finding the correct answer. Finally, we get 1st place in CCKS2023 CKBQA competition with F1 score of 75.63%. In the future, we will delve into research on SPARQL query generation with large language models, especially focus on multiple hops and multi constraints query.

## References

1. Zhang, J., Chen, B., Zhang, L., et al.: Neural, symbolic and neural-symbolic reasoning on knowledge graphs. AI Open **2**, 14–35 (2021). https://doi.org/10.1016/j.aiopen.2021.03.001
2. Ye, X., Yavuz, S., Hashimoto, K., et al.: RnG-KBQA: Generation Augmented Iterative Ranking for Knowledge Base Question Answering. arXiv e-prints https://doi.org/10.48550/arXiv.2109.08678 (2021)
3. Zhang, J., Zhang, X., Yu, J., et al.: Subgraph Retrieval Enhanced Model for Multi-hop Knowledge Base Question Answering (2022). https://doi.org/10.48550/arXiv.2202.13296
4. He, G., Lan, Y., Jiang, J., et al.: Improving multi-hop knowledge base question answering by learning intermediate supervision signals. Proceedings of the 14th ACM International Conference on Web Search and Data Mining, pp. 553–561 (2021)
5. Chen, Y., Wu, L., Zaki, M.J.: Bidirectional Attentive Memory Networks for Question Answering Over Knowledge Bases. arXiv preprint arXiv:1903.02188 (2019)
6. Saxena, A., Tripathi, A., Talukdar, P.: Improving multi-hop question answering over knowledge graphs using knowledge base embeddings. Proceedings of the 58th Annual Meeting of the Association for Computational Linguistics, pp. 4498–4507 (2020)
7. Xu, K., Lai, Y., Feng, Y., et al.: Enhancing key-value memory neural networks for knowledge based question answering. Proceedings of the 2019 Conference of the North American Chapter of the Association for Computational Linguistics: Human Language Technologies, Volume 1 (Long and Short Papers), pp. 2937–2947 (2019)

8. Sun, H., Dhingra, B., Zaheer, M., et al.: Open Domain Question Answering Using Early Fusion of Knowledge Bases and Text. arXiv preprint arXiv:1809.00782 (2018)
9. Sun, H., Bedrax-Weiss, T., Cohen, W.W.: Pullnet: Open Domain Question Answering with Iterative Retrieval on Knowledge Bases and Text. arXiv preprint arXiv:1904.09537 (2019)
10. Pérez, J., Arenas, M., Gutierrez, C.: Semantics and complexity of SPARQL. ACM Trans. Database Syst. **34**(3), 1–45 (2009)
11. Liang, P.: Lambda dependency-based compositional semantics. Computer Science (2013). https://doi.org/10.48550/arXiv.1309.4408.]
12. Cao, S., et al.: KQApro: a dataset with explicit compositional programs for complex question answering over knowledge base. In: Proceedings of the 60th Annual Meeting of the Association for Computational Linguistics (Volume 1: Long Papers). pp. 6101–6119 (2022)
13. Purkayastha, S., Dana, S., Garg, D., Khandelwal, D., Bhargav, G.S.: A deep neural approach to KGQA via SPARQL Silhouette generation. In: 2022 International Joint Conference on Neural Networks. IJCNN, IEEE, pp. 1–8 (2022)
14. Nie, L., et al: GraphQ IR: Unifying the Semantic Parsing of Graph Query Languages with One Intermediate Representation. ArXiv, arXiv:2205.12078 (2022)
15. Cao, R., Chen, L., Chen, Z., Zhao, Y., Zhu, S., Yu, K.: LGESQL: Line graph enhanced text-to-SQL model with mixed local and non-local relations. In: Proceedings of the 59th Annual Meeting of the Association for Computational Linguistics and the 11th International Joint Conference on Natural Language Processing. ACL/IJCNLP 2021, (Volume 1: Long Papers), Virtual Event, August 1–6, 2021, Association for Computational Linguistics, pp. 2541–2555 (2021)
16. Das, R., Zaheer, M., Thai, D., et al.: Case-Based Reasoning for Natural Language Queries Over Knowledge Bases. arXiv preprint arXiv:2104.08762 (2021)
17. Kapanipathi, P., Abdelaziz, I., Ravishankar, S., et al.: Leveraging Abstract Meaning Representation for Knowledge Base Question Answering. arXiv preprint arXiv:2012.01707 (2020)
18. Lan, Y., Jiang, J.: Query Graph Generation for Answering Multi-Hop Complex Questions from Knowledge Bases. Association for Computational Linguistics (2020)
19. Sun, Y., Zhang, L., Cheng, G., et al.: SPARQA: skeleton-based semantic parsing for complex questions over knowledge bases. Proceedings of the AAAI Conference on Artificial Intelligence **34**(05), 8952–8959 (2020)
20. Qiu, Y., Wang, Y., Jin, X., et al.: Stepwise reasoning for multi-relation question answering over knowledge graph with weak supervision. Proceedings of the 13th International Conference on Web Search and Data Mining, pp. 474–482 (2020)
21. Das, R., Zaheer, M., Thai, D., et al.: Case-based reasoning for natural language queries over knowledge bases. Proceedings of the 2021 Conference on Empirical Methods in Natural Language Processing, pp. 9594-9611 (2021)
22. Huang, X., Kim, J.J., Zou, B.: Unseen entity handling in complex question answering over knowledge base via language generation. Findings of the Association for Computational Linguistics: EMNLP 2021, pp. 547–557 (2021)
23. Xiong, G., Bao, J., Zhao, W., et al.: AutoQGS: auto-prompt for low-resource knowledge-based question generation from SPARQL. Proceedings of the 31st ACM International Conference on Information & Knowledge Management, pp. 2250–2259 (2022)
24. Floridi, L., Chiriatti, M.: GPT-3: Its Nature, Scope, Limits, and Consequences. [2023–08–17]. https://doi.org/10.1007/s11023-020-09548-1
25. Chowdhery, A., Narang, S., Devlin, J., et al.: PaLM: Scaling Language Modeling with Pathways (2022). https://doi.org/10.48550/arXiv.2204.02311
26. Touvron, H., Lavril, T., Izacard, G., et al.: Llama: Open and Efficient Foundation Language Models. arXiv preprint arXiv:2302.13971 (2023)

27. Min, B., Ross, H., Sulem, E., et al.: Recent Advances in Natural Language Processing via Large Pre-Trained Language Models: A Survey. arXiv preprint arXiv:2111.01243 (2021)
28. Vaswani, A., Shazeer, N., Parmar, N., et al.: Attention is all you need. Advances in Neural Information Processing Syst. **30** (2017)
29. Kenton, J.D.M.W.C., Toutanova, L.K.: Bert: Pre-training of deep bidirectional transformers for language understanding. Proceedings of NaacL-HLT, **1**, p. 2 (2019)
30. Team O A I. ChatGPT: Optimizing Language Models for Dialogue (2022)
31. Zhao, W.X., Zhou, K., Li, J., et al.: A Survey of Large Language Models. arXiv preprint arXiv:2303.18223 (2023)
32. Lin, Y., Ji, H., Huang, F., et al.: A joint neural model for information extraction with global features. Proceedings of the 58th Annual Meeting of the Association for Computational Linguistics, pp. 7999–8009 (2020)
33. Wu, L., Petroni, F., Josifoski, M., et al.: Scalable Zero-Shot Entity Linking with Dense Entity Retrieval. arXiv preprint arXiv:1911.03814 (2019)
34. Li, B.Z., Min, S., Iyer, S., et al.: Efficient One-Pass End-to-End Entity Linking for Questions. arXiv preprint arXiv:2010.02413 (2020)
35. Du, Z., Qian, Y., Liu, X., et al.: Glm: General Language Model Pretraining with Autoregressive Blank Infilling. arXiv preprint arXiv:2103.10360 (2021)
36. Hu, E.J., Shen, Y., Wallis, P., et al.: Lora: Low-Rank Adaptation of Large Language Models. arXiv preprint arXiv:2106.09685 (2021)
37. Lin, C.Y.: Rouge: a package for automatic evaluation of summaries. Text Summarization Branches Out, pp. 74–81 (2004)

# Robust NL-to-Cypher Translation for KBQA: Harnessing Large Language Model with Chain of Prompts

Guandong Feng[1,2], Guoliang Zhu[1,2], Shengze Shi[1,2], Yue Sun[1], Zhongyi Fan[3], Sulin Gao[4], and Jun Hu[1(✉)]

[1] Institute of Software Chinese Academy of Sciences, Beijing, China
{guandong2022,guoliang2022,,sunyue2021,hujun}@iscas.ac.cn
[2] University of Chinese Academy of Sciences, Beijing, China
shishengze23@mails.ucas.ac.cn
[3] Minzu University of China, Beijing, China
20010662@muc.edu.cn
[4] Harbin Engineering University, Harbin, China
gaosulin@hrbeu.edu.cn

**Abstract.** Knowledge Base Question Answering (KBQA) is a significant task in natural language processing, aiming to retrieve answers from structured knowledge bases in response to natural language questions. NL2Cypher is crucial for accurately querying answers from knowledge bases, but there is limited research in this area or the results are unsatisfactory. Our work explores the convergence of advanced natural language processing techniques with knowledge base question answering (KBQA), focusing on the automated generation of Cypher queries from natural language queries. By leveraging the capabilities of large language model (LLM), our approach bridges the gap between textual questions and structured knowledge representations. The proposed methodology showcases promising results in accurately formulating Cypher queries. We achieved substantial performance in the CCKS2023 Foreign Military Unmanned Systems Knowledge Graph Reasoning Question-Answering Evaluation Task. Our method achieved an F1 score of 0.94269 on the final testing dataset.

**Keywords:** KBQA · LLM · Cypher

## 1 Introduction

Knowledge Base Question Answering (KBQA) has attracted much attention (Barent et al. (2013) [1]) in the field of artificial intelligence and knowledge graph, it aims to extract pertinent entities and relations from a pre-established knowledge graph (KG) to accurately answer natural language questions. KBQA is an effective way of using structured knowledge for information retrieval, but

Supported by China Conference on Knowledge Graph and Semantic Computing.
G. Feng and G. Zhu—Contributed equally.


H. Wang et al. (Eds.): CCKS 2023, CCIS 1923, pp. 317–326, 2023.
https://doi.org/10.1007/978-981-99-7224-1_25

it also faces some challenges, including natural language understanding, entity linking, relation extraction.

Recent progress in deep learning and natural language processing has led to the emergence of large language models (LLM) like GPT-3 [2], BERT [3], etc. These models have demonstrated impressive performance across diverse language tasks, acquiring extensive linguistic knowledge and semantic representations through self-supervised learning on vast quantities of unlabeled text.

This paper introduces an approach for KBQA that utilizes large language models. In particular, the proposed method focuses on employing LLM to extract entities and relations from natural language questions, facilitating semantic comprehension and subsequent transformation into structured Cypher queries, which ultimately enables the retrieval of answers from the knowledge base. This methodology capitalizes on the robust semantic comprehension and generalization capabilities of large language models, effectively enhancing the performance of Knowledge Base Question Answering tasks.

### 1.1 KBQA

Knowledge Base Question Answering (KBQA) systems, designed to fetch relevant entities and relations from a knowledge graph to answer natural language queries, present challenges including understanding natural language, entity linking, and relation extraction. Three predominant KBQA methods exist: template-based, semantic parsing-based, and information retrieval-based (Lan et al. (2021) [4]).

Template-based methods employ pre-defined templates to match queries, thus generating formalized queries. Despite their quick response time and accuracy, they require a substantial library of templates to accommodate diverse user queries, making them labor-intensive.

Semantic parsing-based methods utilize semantic parsing to understand the semantics of natural language queries, transforming them into equivalent logical forms. These forms are then queried and processed by a query engine to obtain answers. This method can accurately return query results if parsing is successful, but errors or ambiguities in parsing can lead to inaccuracies or failure.

Information retrieval-based methods identify the central entity of a query, generate candidate answers, and use scoring and ranking to determine the most suitable answer. This method, whilst effective for simple queries, assumes the query to be simple and the answer to be proximate to the central entity in the knowledge graph.

### 1.2 LLM

Large language models (LLMs) are a type of artificial intelligence that can process and generate natural language texts based on massive amounts of data. LLMs are usually built with deep neural networks, such as transformers, that can learn from large-scale unlabeled or semi-labeled text corpora, such as the

Internet. LLMs can perform various natural language tasks, such as understanding, summarizing, translating, predicting, and creating texts, by taking an input text and repeatedly predicting the next token or word (Zhao et al. (2023) [13]).

LLMs have achieved remarkable results in many natural language processing (NLP) applications, such as question answering, text classification, sentiment analysis, machine translation, text generation, and more. For example, GPT-4 [6] has been tested on various professional and academic benchmarks, and it has shown human-level performance on many of them. Open source model such as LLaMA-2 [10] has shown competitive abilities in common sense reasoning and reading comprehension.

### 1.3 KBQA with LLM

Existing mainstream KBQA methods often require significant resources for fine-tuning on question-answer pairs and understanding complex semantics. Additionally, domain-specific approaches struggle with seamless adaptability to different domains. Leveraging the semantic comprehension capabilities of LLMs offers the advantage of efficiently identifying named entities within questions and grasping the underlying intent. This enables a more effective utilization of semantic cues in question understanding for KBQA. Furthermore, the robust generalization capabilities inherent to LLMs facilitate the straightforward extension of this approach to diverse domains. Consequently, the need for extensive domain-specific fine-tuning can be alleviated, thereby enhancing the versatility and applicability of KBQA techniques.

## 2 Related Work

### 2.1 KBQA

In recent years, KBQA has attracted much attention as a technique that leverages the structured information in knowledge graphs to answer natural language questions. KBQA involves question analysis, candidate generation, ranking, and answer generation. KBQA finds applications like Meituan's query services and medical knowledge graph-based health assistance. It's a vital research area in natural language processing and knowledge graphs.

Recent methods mainly focus on utilizing knowledge from the KG itself. Ye et al. (2022) [12] proposes a ranking-and-generation approach that uses a contrastive ranker to rank candidate logical forms, and then introduces a customized generation model to combine the final logical form based on the question and the top-ranked candidate logical forms. Saxena et al. (2022) [8] demonstrate the Transformer model can serve as a scalable and generic knowledge graph embedding (KGE) model for knowledge graph completion (KGC) and knowledge graph question answering.

### 2.2 In-Context Learning

Since the emergence of LLMs, many studies were conducted to show the power of in-context learning. Min et al. (2022) [5] explored in-context learning with large language models like GPT-3 for new tasks using limited input-label pairs. They discovered that successful in-context learning hinges on input characteristics rather than real demonstrations. Wang et al. (2023) [11] compared in-context learning with supervised learning using identical pre-trained models and demonstrations. The research demonstrated that gold labels significantly affect in-context performance, particularly for larger models.

## 3 Our Method

Our method comprises two integral modules: Named Entity Recognition (NER) and NL2Cypher. The NER Module serves to identify and link entities mentioned in questions to corresponding entities in the knowledge graph, forming a foundational step for query generation. The NL2Cypher Module serves as the core of our approach, converting natural language questions into structured Cypher queries. The large language model used in this module is ChatGPT. This conversion process draws on the enriched information provided by the NER module. To accomplish this, our methodology employs a series of prompts, including Instruction, NER, Scheme, Question, Few-Shot, and Check prompts, collectively guiding the transformation process. The full process is shown in the following Fig. 1

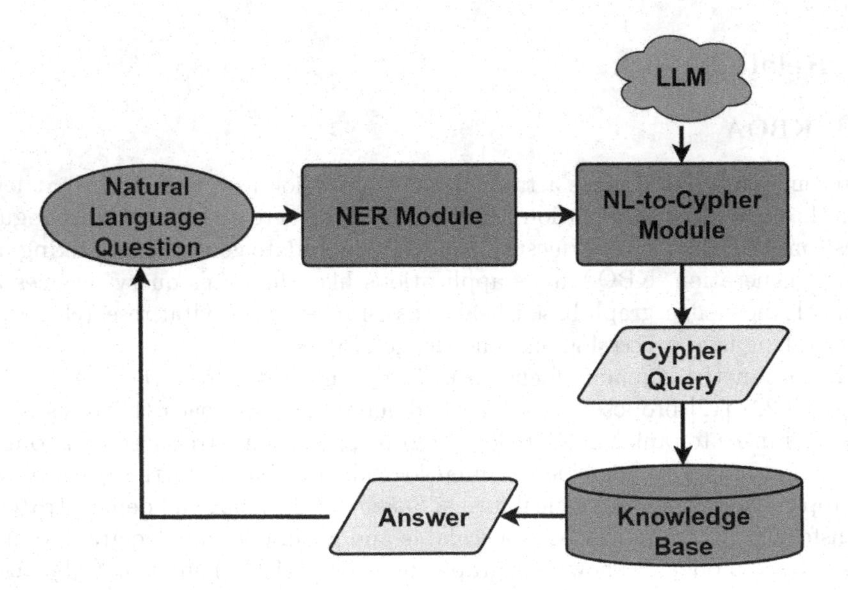


**Fig. 1.** Full process of KBQA, NL2Cypher is the core module

### 3.1 Name Entity Recognition

**Substring Match.** The core objective of Substring Matching is to identify potential mentions from the questions and accurately link them to the knowledge graph. By applying this method, we have accomplished a precision level of 0.9 in correctly identifying the mentioned terms along with their respective labels within the training dataset. This method involves five types of mentions, namely entity name, entity label, relationship, attribute label, and attribute label value. Initially, we adopt a simple approach, to perform exact matches between the mentioned terms in the question and the entries in three files: entity.csv, relation.csv, and attribute.csv. Substring matching is deemed successful only when every character aligns accurately. In instances where direct matches cannot be established, the task is passed on to the subsequent entity linking module for further processing.

**Entity Linking.** Entity linking is the task of identifying and linking the entities mentioned in a natural language question to the corresponding entities in the knowledge graph. Entity linking is an essential step for natural language to Cypher conversion, as it enables the system to map the natural language expressions to the graph elements.

To perform entity linking, we first use OpenAI's API of ChatGPT to extract possible entities from the given question. ChatGPT is a large language model that captures context and semantics. The question is combined with a command prompt to generate relevant entities. For example, given the question "Who are the actors who played in movies directed by Steven Spielberg?", ChatGPT may generate a list of entities like "Tom Hanks, Harrison Ford, Jurassic Park, Schindler's List, etc."

Next, we match the extracted entities to the entity list in the knowledge graph using Sentence-BERT [7], which computes semantic similarity between phrases. Sentence-BERT embeddings of knowledge graph entities are pre-computed. For each extracted entity, we compute the embedding and find the closest match by cosine similarity.

Finally, we check matches using the matched length ratio between extracted and candidate entities. This filters out false positives or ambiguous matches. By extracting, matching, and filtering entities, we can effectively link entities from natural language to the knowledge graph.

**Label Fixes.** Due to the significant reliance of our generated Cypher statements on the labels of elements within the graph database, the preceding NER phase holds vital importance. Errors in this phase can propagate to subsequent stages, including downstream llm models. The objective here is not only to accurately identify mentions but also to assign accurate labels. This part aims to rectify label inaccuracies for certain multi-labeled mentions.

Specifically, for certain mentioned terms, such as "ability", "chinese name", "alias", "English full name", etc., they play distinct roles as both relationship and attribute labels in different questions. Therefore, we need to perform label correction for the mentioned terms identified during entity linking. The specific approach involves starting from the mentioned entity names and entity labels in the question and conducting a subgraph query to narrow down the scope. This subgraph query aims to retrieve all relationships and attribute labels within a one-hop or two-hop range. These labels are then stored in a list. Subsequently, the list is traversed, and if the label associated with the mentioned term differs from the matching result obtained during entity linking, the new result overrides the original one.

### 3.2 NL2Cypher

As demonstrated by Zhao et al. (2021) [14], tailoring prompts to specific domains enhances the performance of large language models in generating reasonable answers.

In order to fully explore the potential of large language model while ensuring their compliance with our task requirements, we have devised a series of prompts to provide guidance and instructions to the model. These prompts are carefully designed to elicit specific responses aligned with the objectives of our task. The overall structure is depicted in the following Fig. 2.

**Instruction Prompt.** Due to the inherent diversity and variability in outcomes produced by large language models, the application of prompts becomes essential to constrain and guide the model's outputs towards the desired objectives of the task at hand. The objective of this section is to teach the large language model the specific goals and output requirements of our task, in order to obtain compliant and standardized output results.

**NER Prompt.** The purpose of this section is to furnish our Named Entity Recognition (NER) results to the large language model. The generation of Cypher statements that correspond precisely to entries within the knowledge graph is imperative for accurate result retrieval. Therefore the NER results will aid the Language Model in generating Cypher query statements with greater precision.

**Schema Prompt.** The goal of this part is to query the surrounding information of related entities to provide LLM for reasoning. Specifically, extract the subgraph Schema involved in the entity, and obtain the list of label, name, property, and relation. Through this schema, the large language model can better understand the context of the entity to achieve better generation results.

**Question Prompt.** This part aims to judge the type of the query problem through the method of keyword matching, including single-hop/multi-hop relationship query, quantity query, attribute query, maximum value query, etc. The

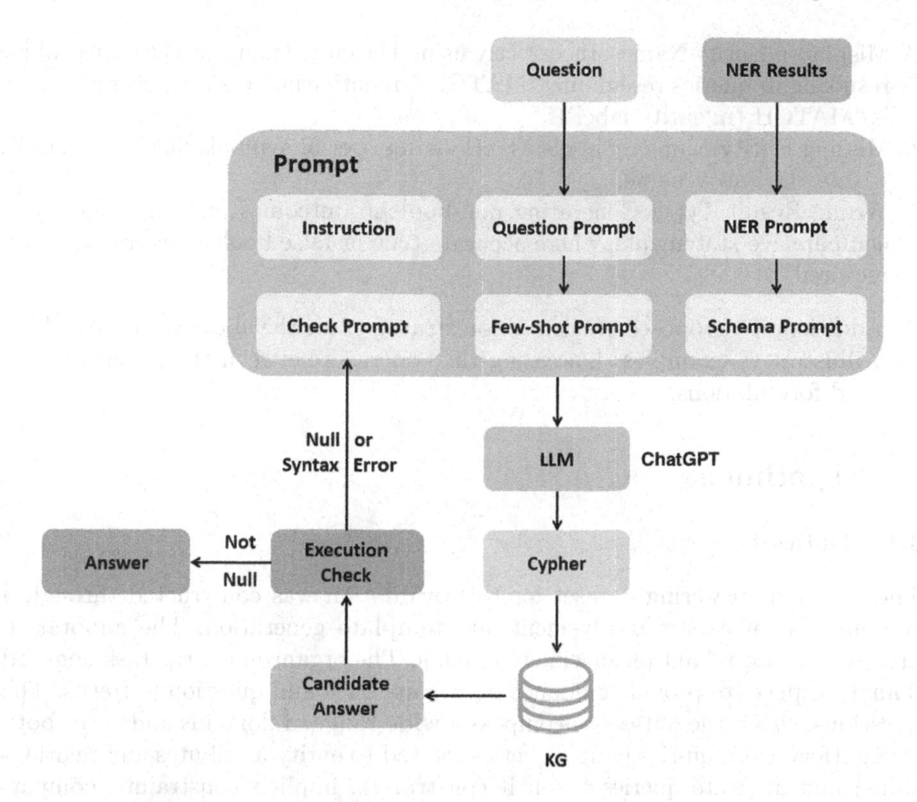


**Fig. 2.** Framework of NL2Cypher process: Takes natural language question and its NER results as input, combined with different prompts to generate a Cypher statement to get answer from Knowledge Graph

Cypher statements corresponding to identical query types exhibit remarkable similarity. Thus, furnishing the specific query type to the large-scale model can substantially enhance the efficiency of Cypher statement generation.

**Few-Shot Prompt.** Su et al. (2022) [9] shows that large language models can perform in-context learning, where they learn a new task from a few task demonstrations, without any parameter updates. In the realm of specialized domain tasks, the utilization of few-shot learning frequently results in noteworthy enhancements in performance, surpassing the outcomes achieved through zero-shot methodologies. To facilitate in-context learning, a collection of 3–4 representative samples per question type is curated for input into the large language model. This approach further refines the model's grasp of distinct query types and yields commendable improvements.

**Check Prompt.** For the result that the query result is empty, provide error correction ideas for LLM. The following scenarios are encompassed:

1. Misplaced Entity Names: Incorrectly using the entity name as the entity label, resulting in queries resembling "MATCH (n:'entity name')", the correct result is "MATCH (n:'entity label')".
2. Missing Entity Name Symbols: Overlooking special symbols like "/", " ", or "·" within the entity name.
3. Wrong Result Type: Generating non-boolean outcomes (e.g., n, n.name) in comparative statements, where accurate true or false boolean expressions are required.

Additionally, proposed are correction strategies for the above scenarios, along with illustrative examples showcasing the contrasts between the erroneous and rectified formulations.

# 4 Experiment

## 4.1 Dataset

The question-answering dataset for this evaluation was constructed through a combination of expert involvement and template generation. The annotation process was not reliant on specific templates. The organizers of the task engaged domain experts to provide guidance on corpus style and question patterns. The questions within the dataset encompass a wide range of domains and cover both straightforward inquiries (single queries related to entity attributes and relationships) and intricate queries (explicit constraints, implicit constraints, comparisons, Boolean operations, multi-hop queries, etc.).

The primary knowledge base utilized is derived from the fusion of knowledge graphs formed through the construction and evaluation tasks of the CCKS2021 and CCKS2022 Foreign Military Unmanned Systems Knowledge Graphs. This task includes a training dataset of 4,000 instances, a validation dataset of 1,000 instances, and a testing dataset of 1,000 instances. The annotated data comprises questions, Cypher queries, and corresponding answers.

## 4.2 Results

In this section, we present the comprehensive results of our proposed approach on both the validation and test sets, each composed of 1000 question-answer pair. Our model achieved promising performance, as indicated by the F1-scores on the validation and test sets respectively as presented in Table 1.

**Table 1.** Results on Validation dataset and Test dataset

| Metrics | Validation | Test |
|---|---|---|
| Accuracy | 0.91137 | 0.94437 |
| Recall | 0.90978 | 0.94278 |
| F1-score | 0.90969 | 0.94269 |

These results underscore the effectiveness of our approach in addressing the challenges posed by the dataset. Notice that there is a gap between results on two datasets, that is because we improved prompt quality in the process.

To further assess the contribution of individual components within our model, we conducted ablation experiments on the training dataset, demonstrating the effectiveness of different type of prompts. The results of these experiments reaffirm the significance of each prompt in enhancing the model's performance. Detailed summaries of the ablation experiments are presented in Table 2.

**Table 2.** Ablation Study

| Prompt Removed | F1-score |
|---|---|
| w/o NER Result Prompt | 0.867 |
| w/o Schema Prompt | 0.886 |
| w/o Question Type Prompt | 0.870 |
| w/o Few-Shot Prompt | 0.895 |
| w/o Check Prompt | 0.886 |
| All Prompts | 0.907 |

It can be concluded that all types of prompt contribute to the final result, with NER prompt and Question Type prompt being the most impactful, which is consistent with intuitive expectations.

## 5 Conclusion

In conclusion, our work has effectively showcased the potential of integrating large language model with knowledge base question answering, as evidenced by our top-ranking achievement in the CCKS2023 Knowledge Graph Reasoning Question-Answering Evaluation Task. A significant contribution of our work lies in the innovative design of prompts for large language models. We tailored our prompts to capture domain-specific intricacies, thereby enhancing the model's ability to generate relevant and accurate Cypher queries. This innovative prompt design played a pivotal role in achieving our outstanding performance, underscoring the potential for optimizing large language models for task-specific applications in knowledge graph reasoning and question answering.

**Acknowledgements.** We extend our sincere gratitude to the organizers of the competition for providing us with the invaluable platform and comprehensive dataset that facilitated our research efforts. Furthermore, we are deeply thankful for the guidance and insights provided by our mentors. This collaborative effort and the valuable resources made available to us were essential in achieving the commendable results presented in this paper.

## References

1. Berant, J., Chou, A., Frostig, R., Liang, P.: Semantic parsing on freebase from question-answer pairs. In: Proceedings of the 2013 Conference on Empirical Methods in Natural Language Processing, pp. 1533–1544 (2013)
2. Brown, T., et al.: Language models are few-shot learners. Adv. Neural. Inf. Process. Syst. **33**, 1877–1901 (2020)
3. Devlin, J., Chang, M.W., Lee, K., Toutanova, K.: BERT: pre-training of deep bidirectional transformers for language understanding. In: Proceedings of the 2019 Conference of the North American Chapter of the Association for Computational Linguistics: Human Language Technologies, Volume 1 (Long and Short Papers), Minneapolis, Minnesota, June 2019, pp. 4171–4186. Association for Computational Linguistics (2019). https://doi.org/10.18653/v1/N19-1423. https://aclanthology.org/N19-1423
4. Lan, Y., He, G., Jiang, J., Jiang, J., Zhao, W.X., Wen, J.R.: A survey on complex knowledge base question answering: methods, challenges and solutions. arXiv preprint arXiv:2105.11644 (2021)
5. Min, S., et al.: Rethinking the role of demonstrations: what makes in-context learning work? arXiv preprint arXiv:2202.12837 (2022)
6. OpenAI: GPT-4 technical report. arXiv preprint arXiv:2303.08774 (2023)
7. Reimers, N., Gurevych, I.: Sentence-BERT: sentence embeddings using Siamese BERT-networks. In: Proceedings of the 2019 Conference on Empirical Methods in Natural Language Processing and the 9th International Joint Conference on Natural Language Processing (EMNLP-IJCNLP), January 2019 (2019). https://doi.org/10.18653/v1/d19-1410
8. Saxena, A., Kochsiek, A., Gemulla, R.: Sequence-to-sequence knowledge graph completion and question answering. arXiv preprint arXiv:2203.10321 (2022)
9. Su, H., et al.: Selective annotation makes language models better few-shot learners. arXiv preprint arXiv:2209.01975 (2022)
10. Touvron, H., et al.: Llama 2: open foundation and fine-tuned chat models. arXiv preprint arXiv:2307.09288 (2023)
11. Wang, X., et al.: Investigating the learning behaviour of in-context learning: a comparison with supervised learning, July 2023
12. Ye, X., Yavuz, S., Hashimoto, K., Zhou, Y., Xiong, C.: RNG-KBQA: generation augmented iterative ranking for knowledge base question answering. In: Proceedings of the 60th Annual Meeting of the Association for Computational Linguistics (Volume 1: Long Papers), January 2022 (2022). https://doi.org/10.18653/v1/2022.acl-long.417
13. Zhao, W.X., et al.: A survey of large language models. arXiv preprint arXiv:2303.18223 (2023)
14. Zhao, Z., Wallace, E., Feng, S., Klein, D., Singh, S.: Calibrate before use: improving few-shot performance of language models. In: International Conference on Machine Learning, pp. 12697–12706. PMLR (2021)

# In-Context Learning for Knowledge Base Question Answering for Unmanned Systems Based on Large Language Models

Yunlong Chen[1], Yaming Zhang[1], Jianfei Yu[1(✉)], Li Yang[2], and Rui Xia[1]

[1] School of Computer Science and Engineering, Nanjing University of Science and Technology, Nanjing, China
{ylchen,ymzhang,jfyu,rxia}@njust.edu.cn

[2] Wee Kim Wee School of Communication and Information, Nanyang Technological University, Singapore, Singapore

**Abstract.** Knowledge Base Question Answering (KBQA) aims to answer factoid questions based on knowledge bases. However, generating the most appropriate knowledge base query code based on Natural Language Questions (NLQ) poses a significant challenge in KBQA. In this work, we focus on the CCKS2023 Competition of Question Answering with Knowledge Graph Inference for Unmanned Systems. Inspired by the recent success of large language models (LLMs) like ChatGPT and GPT-3 in many QA tasks, we propose a ChatGPT-based Cypher Query Language (CQL) generation framework to generate the most appropriate CQL based on the given NLQ. Our generative framework contains six parts: an auxiliary model predicting the syntax-related information of CQL based on the given NLQ, a proper noun matcher extracting proper nouns from the given NLQ, a demonstration example selector retrieving similar examples of the input sample, a prompt constructor designing the input template of ChatGPT, a ChatGPT-based generation model generating the CQL, and an ensemble model to obtain the final answers from diversified outputs. With our ChatGPT-based CQL generation framework, we achieved the second place in the CCKS 2023 Question Answering with Knowledge Graph Inference for Unmanned Systems competition, achieving an F1-score of 0.92676.

**Keywords:** ChatGPT · Chain-of-Thought · In-Context Learning

## 1 Introduction

As an important task in Natural Language Processing (NLP), Knowledge Base Question Answering (KBQA) aims to generate accurate and complete query statements from user-provided natural language questions (NLQs), and these query statements are then used to retrieve relevant information from the knowledge base and provide accurate answers. In this work, we focus on the CCKS


H. Wang et al. (Eds.): CCKS 2023, CCIS 1923, pp. 327–339, 2023.
https://doi.org/10.1007/978-981-99-7224-1_26

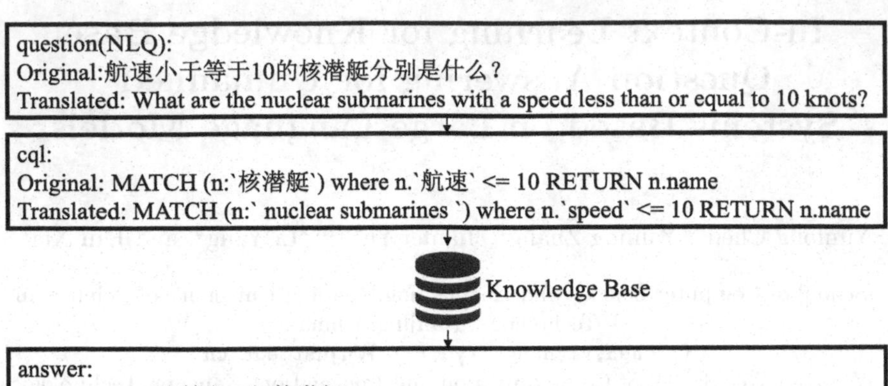


**Fig. 1.** Illustration of an example in the evaluation task.

2023 Question Answering with Knowledge Graph Inference for Unmanned Systems competition, which is a KBQA evaluation task where cypher query language (CQL) serves as the query statements. Figure 1 gives an example of the CCKS2023 competition.

In the literature, most existing studies on KBQA can be categorized into two types: information retrieval-based (IR-based) approaches and semantic parsing-based (SP-based) approaches. Both of them require first identifying the subject within the NLQ and linking it to an entity in the knowledge base (KB). The former line of work aims to derive answers by reasoning within a question-specific graph extracted from the KB with the assistance of those linked entities [1,2], whereas the latter line of work aims to obtain answers by executing a parsed logic form based on the linked entities [3,4]. Since the annotation in the dataset of the CCKS2023 competition contains manually annotated CQLs, we follow the latter line of approaches in this work.

However, the majority of existing SP-based approaches are built upon LSTM or pre-trained models like BERT, which are constrained by their scale or their pre-training data and may encounter challenges in effectively generating suitable knowledge base query codes based on NLQs. With the recent advancements of pre-trained language models, many Large Language Models (LLMs) have been shown to achieve surprisingly good performance on many question answering datasets under zero-shot or few-shot settings. These LLMs have also showcased an impressive capacity to deeply comprehend sentence semantics and accurately translate them into multiple languages, and even generate code when required. Therefore, we aim to explore the potential of LLMs on the Chinese KBQA task for unmanned systems.

Specifically, we propose a ChatGPT-based CQL generation framework, consisting of six parts. The first part involves an auxiliary model that takes the given NLQ as input and predicts structural information for each clause separately. The second part comprises a proper noun matcher, which identifies explicitly mentioned proper nouns existing in the KB from the given NLQ. The third part consists of a demonstration example selector that employs a key-information-based similarity calculation criterion to retrieve demonstration samples for the given NLQ based on the aforementioned results. The fourth part encompasses a prompt constructor that constructs input text by integrating demonstration samples, NLQ, and task-specific prior knowledge. The fifth part incorporates a ChatGPT-based generation model, which inputs the constructed text into ChatGPT to generate CQL. Subsequently, post-processing is applied to the generated CQL. Lastly, the sixth part introduces an ensemble model, in which multiple answers retrieved from the knowledge base by the post-processed CQL are combined through a voting mechanism to obtain the final result.

We conduct experiments on the dataset provided by the competition, and the results show the high efficiency of our generative framework. Therefore, we achieved the second place in the CCKS 2023 Question Answering with Knowledge Graph Inference for Unmanned Systems competition with an F1-score of 0.92676.

## 2 Related Work

### 2.1 Large Language Model

Large Language Models (LLMs) typically possess a vast number of learnable parameters and undergo extensive training on enormous text datasets, examples of which include ChatGPT [5], LLaMA [6], OPT [7], PaLM [8], CodeX [9], and so on. With the advancement of LLMs, traditional pre-trained models like BERT [10], RoBERTa [11], BART [12], T5 [13], have faced great challenges. The ability of LLMs to adapt to downstream tasks without the need for retraining, but task-specific instructions, has greatly reduced the cost of solving downstream tasks.

### 2.2 In-Context Learning

As mentioned in Sect. 2.1, LLMs typically demonstrate emergent abilities [14,15] with increasing model and corpus size, i.e., the ability to learn from the given examples present in the context, known as In-Context Learning (ICL). This ability helps LLMs in better adapting to downstream tasks. While solely relying on task-specific instructions may not lead to superior performance compared to fine-tuned models in some downstream tasks, introducing ICL can often result in considerable improvements in LLMs' performance on downstream tasks.

### 2.3 Chain-of-Thought

Chain-of-Thought (CoT) [16] is an extremely efficient and easy prompting strategy that endows LLMs with reasoning capabilities, enabling LLMs to decompose and comprehend complex tasks. Specifically, CoT leverages several given examples with inferred answers to assist LLMs in comprehending the reasoning process of complex tasks, thus performing reasoning on the target problem and obtaining results. In general, CoT leverages several pre-given exemplars with inferred answers to help LLMs understand the reasoning process of intricate tasks.

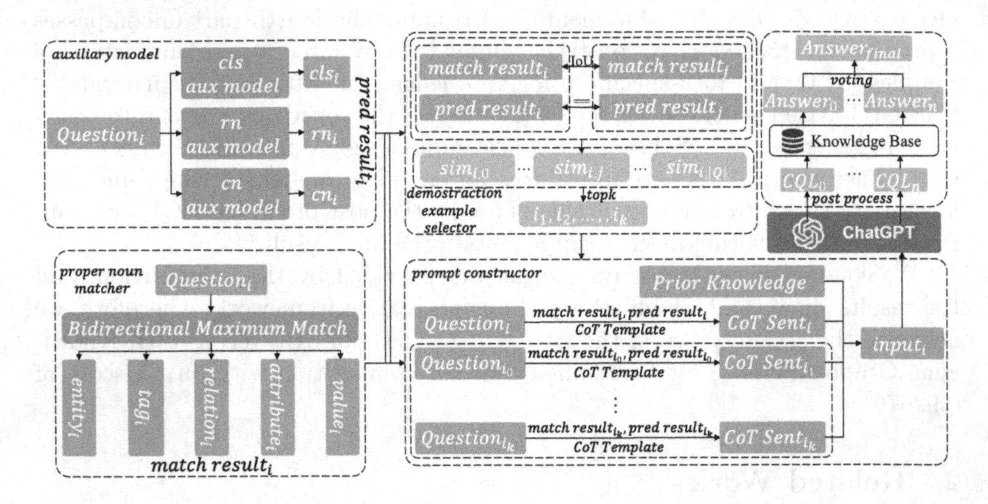


**Fig. 2.** The overall architecture of our ChatGPT-based KBQA framework.

## 3 Methodology

Recently, LLMs have showcased robust generalizability across a diverse spectrum of tasks by leveraging few-shot in-context learning. Notably, LLMs possess the capability to transform unstructured sentences into structured and executable code, rendering them valuable assets in KBQA [17].

However, the CQL solely generated by ChatGPT falls short of our expectations. Therefore, we observe CQL's overall structure and summarize empirical knowledge to design processing techniques and auxiliary tasks. These aid ChatGPT in capturing key information and parsing CQL structures from NLQs, enabling it to adapt to downstream tasks and generate high-quality CQL.

In general, our ChatGPT-based CQL generation framework consists of six steps (excluding KB construction and answer retrieval), as illustrated in Fig. 2 and Fig. 3 provides a visualized example of using our generative framework to generate CQL from NLQ.

1. Three auxiliary tasks to predict structural information for CQL clauses.
2. Bidirectional maximum matching-based proper noun matching for NLQ.
3. Selecting demonstration examples based on the aforementioned results.
4. Combining NLQ, demonstration examples, and prior knowledge into CoT format as input text.

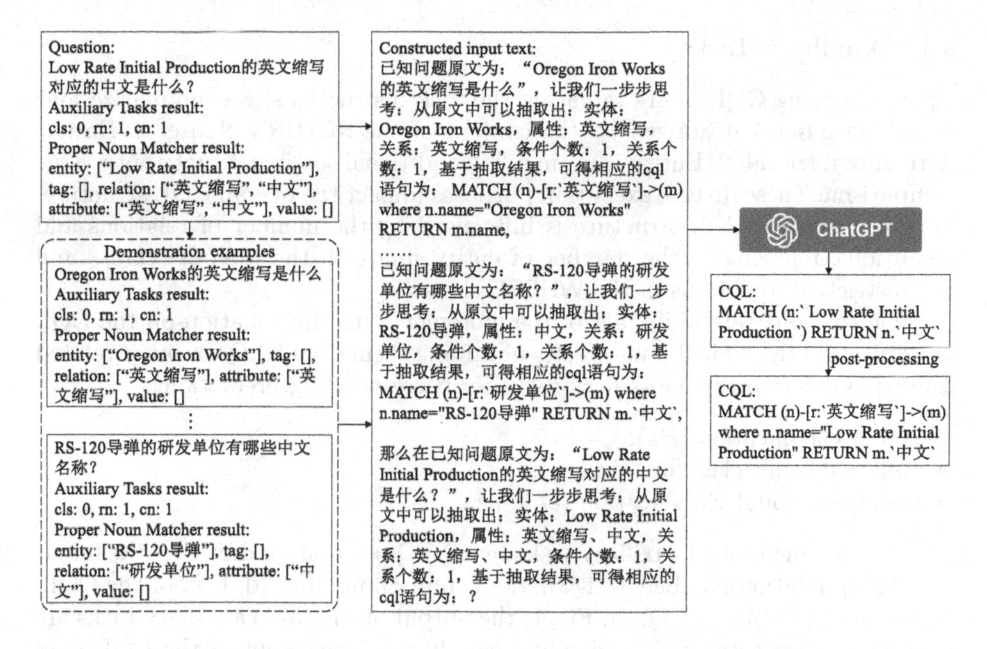


**Fig. 3.** A visualized example of generating CQL from NLQ.

5. ChatGPT generates CQL based on the constructed input text, followed by post-processing of the generated CQL.
6. Voting for the answers retrieved from the given KB by CQLs.

The overall workflow is as follow:

Firstly, within the **auxiliary model**, the structural information of different clauses in CQL ($pred\ result_i$) is predicted based on NLQ ($Question_i$). Meanwhile, the **proper noun matcher** identifies the proper nouns ($match\ result_i$) explicitly mentioned in $Question_i$ and existing in the KB.

Subsequently, both $pred\ result_i$ and $match\ result_i$ are fed into the **demonstration example selector** to compute the similarity between different samples, thereby selecting demonstration examples $(i_1, i_2 \ldots, i_k)$ for each sample.

Following that, within the **prompt constructor**, the $Question_i$, $pred\ result_i$, and $match\ result_i$ are firstly combined into CoT format to obtain $CoT\ Sent_i$. Next, $Prior\ Knowledge$, $CoT\ Sent_i$, and corresponding demonstration examples $(CoT\ Sent_{i_1}, \ldots, CoT\ Sent_{i_k})$ are combined to form $input_i$. Then, $input_i$ is fed into **ChatGPT** to generate CQL, followed by post-processing.

By repeating the aforementioned steps, multiple CQLs $(CQL_0, \ldots, CQL_n)$ are acquired. Executing these CQLs in the given KB yields multiple answers $(Answer_0, \ldots, Answer_n)$. Employing a **voting** mechanism on these answers yields the most reliable response as the final outcome $(Answer_{final})$.

### 3.1 Auxiliary Tasks

Upon analyzing CQLs and evaluating the task instructions, we summarize four execution intent categorized by the content of the RETURN clause: 1. Entity-Attribute retrieval; 2. Entity counting; 3. Conditional sorting; 4. Attribute-value comparison. These diverse execution purposes impact the overall CQL structure. Furthermore, the CQL structure is influenced by the number of relations and conditions, referring to the number of entity jumps in the MATCH clause and the restrictive conditions in the WHERE clause.

Due to the potential impact of the aforementioned information on the CQL structure and the difficulty in directly obtaining them from NLQs or other related sources, we design three auxiliary tasks for corresponding predictions:

- Intent classification ($cls$).
- Relation count classification ($rn$).
- Condition count classification ($cn$).

The aforementioned auxiliary tasks take NLQs as input and predict the intent (0/1/2/3), relation number (0/1/2), and condition number (0/1/2) of the CQL. For example, given the CQL in Fig. 1, the output of the three auxiliary tasks are 0, 0, and 1, respectively. These values signify that the intent of the CQL is to retrieve an entity's attribute, there are no relations mentioned in the CQL, and there is one condition specified.

Notably, these three auxiliary tasks share NLQs as input and yield similar outputs, all treated as simple classification results, thus allowing them to utilize a consistent model architecture. Any pre-trained model like BERT-chinese and mT5 can be used to extract global features from the text, followed by a feed-forward network for output prediction of the auxiliary tasks.

### 3.2 Proper Noun Matcher

In NLQs, some proper nouns are either explicitly or implicitly mentioned, which are likely to appear in the CQLs. Therefore, it is essential to extract the mentioned proper nouns from NLQs. To this end, we employ the bidirectional maximum matching based on the given proper noun vocabulary to all NLQs, obtaining the proper nouns that appear in them.

During bidirectional maximum matching, we observed that unconstrained execution could introduce noise into the matching process for relation and attribute proper nouns. Although designing a universal set of constraints is challenging, creating NLQ-specific constraints is more feasible. Therefore, our approach gives precedence to extracting entities and tags from the NLQ. Subsequently, by leveraging relevant knowledge from KB, we integrate the entities corresponding to the tags into the matched entities. Taking the entities as the starting point and considering two-hop relations as the scope, all the involved entities are referred to as entity set, and all the involved relations are referred to as *candidate* relation set. Using the entity set, we retrieve associated attributes from the given knowledge base, creating a *candidate* attribute set.

The NLQ's bidirectional maximum matching results in the *matched* attribute and relation sets. The final matching results for attribute and relation sets are obtained by intersecting the *candidate* sets with the *matched* sets.

### 3.3 Demonstration Example Selector

For NLQ-specific demonstration example selection, an appropriate similarity calculation criterion is vital. The conventional method, such as using BERT or similar models for cosine similarity computation between global features (GF-Sim), may not yield accurate similarity results due to substantial noise present in NLQs. This noise considerably affects the results of GF-Sim. With approximately 35% of the text containing valid information (proper nouns) while the rest being noisy, the impact on GF-Sim is significant.

Based on the above findings, we propose a key-information-based similarity criterion (KI-Sim) for NLQs. It focuses on key components in NLQ like proper nouns and other influential details impacting CQLs, which is computed as follows:

$$Similarity(i,j) = \sum_{k}^{e,t,r,a,v} w_k * \mathrm{IoU}(i_k, j_k) + \sum_{k}^{cls,rn,cn} w_k * (i_k == j_k) \quad (1)$$

where, $i, j$ represent NLQs' id, $e, t, r, a, v$ stand for entity, tag, relation, attribute, and value, $cls, rn, cn$ indicate the predicted intent, relation number, and condition number, $i_k, j_k$ refer to proper nouns or auxiliary task predictions from NLQs with id $i, j$, and $w_k$ signifies the corresponding similarity weight.

### 3.4 Prompt Constructor

**Prior Knowledge.** To bolster ChatGPT's grasp and alignment with the downstream task, we propose to incorporate task-specific prior knowledge and integrate it into text (see Appendix 1) to feed into ChatGPT. These prior knowledge are derived from observations of CQLs and hold general applicability for this downstream task, rather than being specific to any particular NLQ.

**In-Context Learning.** To enhance ChatGPT's CQL generation, we utilize KI-Sim (Sect. 3.3) to select demonstration examples for ICL. This aids ChatGPT in CQL generation, thereby improving the quality of the generated CQLs.

**Chain-of-Thought.** Intuitively, directly generating CQL from NLQ is challenging. Yet, analyzing the composition and syntax of CQLs and NLQs reveals a high likelihood of shared proper nouns. Drawing inspiration from the CoT method, we split the CQL generation task into two sub-tasks:

1. Matching relevant proper nouns from NLQ.
2. Generating CQL based on NLQ and the matched proper nouns.

As elucidated in Sect. 3.2, with the completion of sub-task 1, ChatGPT now focuses on addressing sub-task 2. To achieve this, a template was devised (see Appendix 2) that integrates matched proper nouns and NLQ following the CoT approach, presenting them jointly to ChatGPT.

### 3.5 ChatGPT-Based Generation Model

After inputting the constructed text into ChatGPT, it generates the corresponding CQL for the given NLQ. Concerning the CQL, three common situations arise:

- Entities matched are mis-classified as tags.
- The number of relation does not match the results from auxiliary tasks.
- Matching with unprovided proper nouns.

Among these situations, the first two can result in inaccurate CQL execution and should be prevented or rectified. In contrast, the third situation is favorable. As described in Sect. 3.2, only explicit proper nouns can be matched, leaving implicit ones unmatched. This implies ChatGPT's successful identification of implicit proper nouns in the NLQ. In this case, post-processing methods can be used to map non-proper nouns to the provided vocabulary for correction.

Consequently, we present three post-processing methods to address these situations to get the final ChatGPT-generated CQL:

- Reclassify the mis-classified tags as entities and position them correctly.
- Use condition truncation and filling for isolated condition clause correction.
- Utilize fuzzy matching to map implicit proper nouns to vocabulary and make CQL modifications correspondingly.

### 3.6 Ensemble Model

Executing CQL in the KB doesn't always ensure correct answers, and sometimes no answers are found. However, generating new CQLs could enhance the retrieval success. Thus, for each NLQ, we generate multiple CQLs, retrieve corresponding answers, and apply a voting mechanism to ascertain the final answer.

## 4 Experiment

### 4.1 Dataset

We conduct experiments based on the competition's dataset, which encompasses knowledge base construction data, as well as training, validation (preliminary round), and test (final round) datasets. The training set includes annotations, while validation's annotations are released with the un-annotated test set in the final round. Annotations include answers and the CQL used for retrieval from KB. We will validate our generative framework on this dataset.

### 4.2 Evaluation

In the experiments, the main evaluation concern is precise answer retrieval for NLQs. The evaluation metrics include Macro Precision (Eq. (2)), Macro Recall (Eq. (3)), and Averaged F1 (Eq. (4)), which are defined below:

$$P = \frac{1}{|Q|}\sum_{i=1}^{|Q|} P_i, \quad P_i = \frac{|A_i \cap G_i|}{|A_i|} \tag{2}$$

$$R = \frac{1}{|Q|}\sum_{i=1}^{|Q|} R_i, \quad R_i = \frac{|A_i \cap G_i|}{|G_i|} \tag{3}$$

$$F1 = \frac{1}{|Q|}\sum_{i=1}^{|Q|} \frac{2P_iR_i}{P_i + R_i} \tag{4}$$

where $|Q|$ denotes the number of NLQs in the dataset, $A_i, G_i$ denotes the player's and ground-truth answer sets to the question whose id is $i$, respectively.

### 4.3 Implementation

**Similarity.** During similarity computation, entity weights are set to 5, tag weights to 3, relation weights to 3, attribute weights to 1, value weights to 0.5, *cls* weights to 0.5, *rn* weights to 0.3, and *cn* weights to 0.3.

**ChatGPT.** The ChatGPT we used in this paper is gpt-3.5-turbo-0613. It should be noted that we set the temperature parameter to 1 (default) to ensure the diversity of CQLs when ChatGPT generates responses multiple times.

**Auxiliary Task.** We use the mT5-large as the pre-trained model. For the auxiliary tasks, the global random seed is 33. The batch size is 32, trained for 100 epochs. Initial learning rates for the backbone and non-backbone part are set at 1e−6 and 1e−4, respectively. Cross-entropy loss is employed for loss calculation.

### 4.4 Main Results

In Table 1, the performance of ChatGPT with different processing techniques is presented, where the last row shows the performance of our proposed ChatGPT-based CQL generation framework.

In the preliminary round, with only prior knowledge and voting mechanism, our F1 score on the validation set is 0.83865, obtaining the second place. In the final round, our ChatGPT-based CQL generation framework achieves an F1 score of 0.92676 on the test set, obtaining the second place.

**Table 1.** Main Results. Note the Prior indicate the prior knowledge, the Ensemble indicate the ensemble model, the Post indicate the post-processing in ChatGPT-based Generation Model.

| +Prior | +Ensemble | +ICL+CoT | +Post | Averaged F1 (Validation) | Averaged F1 (Test) |
|---|---|---|---|---|---|
| ✓ | × | × | × | 0.72539 | \ |
| ✓ | ✓ | × | × | 0.83865 | 0.86204 |
| ✓ | ✓ | ✓ | × | \ | 0.91561 |
| ✓ | ✓ | ✓ | ✓ | \ | **0.92676** |

### 4.5 Ablation Study

As shown in Table 1, all different processing techniques can improve the final performance, but their effects are different:

**Ensemble Model.** The essence of this technique is to allow ChatGPT to generate multiple CQLs and vote on the answers. The multiple generations can help ChatGPT re-understand NLQ and increase the diversity of generated CQLs.

**ICL+CoT.** The essence of this technique is to enable ChatGPT to capture and learn implicit relations that may exist in downstream tasks based on given demonstration examples. By using the decomposed sub-tasks, ChatGPT can achieve a deeper understanding of the downstream task and adapt to it, generating higher quality and more robust CQLs.

**Post-processing.** The essence of this technique is to manually correct the generation errors of ChatGPT without interfering with its process of generating CQLs. Instead, it intervenes in the results generated by ChatGPT, ensuring that the results do not contain factual errors.

### 4.6 Auxiliary Task Results

**Table 2.** The performance on three auxiliary tasks

| Auxiliary Tasks | Accuracy (%) |
|---|---|
| Intent classification | 99.0 |
| Relation count classification | 97.0 |
| Condition count classification | 98.2 |

Based on the performance of the auxiliary tasks in Table 2, we can find that the proposed model performs well on the three auxiliary tasks. Therefore, it is generally useful to incorporate the auxiliary task-related information into our generative framework.

### 4.7 Similarity Comparison

To verify KI-Sim's effectiveness (Sect. 3.3), we present the demonstration examples in Table 3. The global features are extracted from bert-base-chinese.

**Table 3.** Demonstration examples based on different similarity calculation criterion

| NLQ | | Original | Translated |
|---|---|---|---|
| | | **最大飞行速度**小于等于**460**的实体有几个? | How many entities have a **maximum flying speed** less than or equal to **460**? |
| GF-Sim | top1 | **阿姆德-500M/2M沉底水雷**的**产国**是哪个? | What is the **origin country** of the **AMD-500M/2M Submarine Mine**? |
| | top2 | **94式90毫米轻迫击炮**的**口径**是多少? | What is the **caliber** of the **Type-94 90mm Light Mortar**? |
| | top3 | **弹径**为**1.37**的**舰地(潜地)导弹**有哪些? | Which **Ship-to-Ground (Submarine-to-Ground) Missile** has a **caliber** equal to **1.37**? |
| KI-Sim | top1 | **最大飞行速度**大于**252**的实体有几个? | How many entities have a **maximum flying speed** greater than **252**? |
| | top2 | **最大飞行速度**等于**850**的实体有几个? | How many entities have a **maximum flying speed** equal to **850**? |
| | top3 | **最大飞行速度**等于**745**的实体有几个? | How many entities have a **maximum flying speed** equal to **745**? |

Key information in the NLQs is highlighted using bold, with scolid and dashed underlines denoting their presence and absence in the top NLQ, respectively. It is evident that the demonstration examples selected by KI-Sim are more similar to the top NLQ. This underscores the effectiveness of KI-Sim.

## 5 Conclusion

In this paper, we proposed a ChatGPT-based CQL generation framework, which consists of six components: an auxiliary model that predicted structural information for CQLs based on given NLQs, a proper noun matcher that extracted explicit proper nouns, a demonstration example selector that used KI-Sim to select demonstration examples, a prompt constructor that concatenated the NLQ, demonstration examples, and prior knowledge in the form of a Chain-of-Thought, a ChatGPT-based generation model that generated CQLs using the concatenated text, and an ensemble model that produced more reliable results by voting on diversified answers. Experimental results validate the effectiveness of our generative framework, achieving a remarkable second-place rank in the CCKS 2023 Question Answering with Knowledge Graph Inference for Unmanned Systems competition.

**Acknowledgements.** This work as supported by the Natural Science Foundation of China (62076133 and 62006117), and the Natural Science Foundation of Jiangsu Province for Young Scholars (BK20200463) and Distinguished Young Scholars (BK20200018).

## Appendix 1

Given a Chinese question for querying a knowledge graph, the question includes simple queries about entity attributes and relationships, as well as complex queries involving explicit and implicit constraints, comparisons, boolean logic, and multi-hop scenarios. Emulating the provided code style and syntax, the analysis encompasses entities, attributes, labels, and relationships to generate Cypher Query Language (CQL) for knowledge graph answers. CQL consists of several sections: MATCH, identifying nodes and relationships with specified types, properties, and directions; WHERE, filtering results with logical and comparison operators; RETURN, providing query outcomes using aggregation, sorting, and limiting functions; and WITH, forwarding results to subsequent clauses. You need to identify entities, attributes, labels, relationships, and constraint conditions in the question. In CQL syntax, use parentheses () to represent nodes, square brackets [] to represent relationships, colons : to represent tags, periods . to represent attributes, and arrows -> to indicate relationship directions. For example: [r:Technology Used], [r:Manufacturer Capability], n.Weight, (n)-[r:Origin Country]->(m)-[r1:Military Branch Involved Project]->(l). Operators such as =, <, >, AND, OR, NOT are used to compare values and filter results. Functions such as count(), min(), max(), avg(), sum() are used to perform calculations on results. Note that if the question is boolean in nature, the CQL's return value is either True or False. For example, question: "Was MQ-4C first flown in 2013?" Return value: True. If the question involves comparing entities and inquiring about an entity, the CQL's return value is the entity name. For example, question: "In comparison to the caliber of the M1 anti-aircraft gun, is the caliber of the M29 81mm mortar smaller?" Corresponding CQL: MATCH (n) where n.name = "M29 81mm Mortar" or n.name = "M1 Anti-Aircraft Gun" RETURN n.name ORDER BY n.'caliber' asc limit 1 Return value: M29 81mm Mortar. If the question involves comparing entity attribute values, the CQL's return value is True or False. For example, question: "Is the delivery quantity of the ScanEagle UAV greater than that of the MQ-1 Predator UAV?" Corresponding CQL: MATCH (n), (m) where n.name = "ScanEagle UAV" and m.name = "MQ-1 Predator UAV" RETURN n.'delivery quantity' > m.'delivery quantity' Return value: False. Please independently determine the question type, analyze logical relationships in the question, and generate CQL accordingly.

## Appendix 2

Given the original question text: "{question}", let's think step by step: From the original text, we can extract: Entity: {entity}, Tag: {tag}, Attribute: {attribute},

Value: {value}, Relation: {relation}, Condition Count: {cn}, Relation Count: {rn}. Based on the extracted results, the corresponding CQL can be obtained as: {cql}.

## References


1. Yan, Y., et al.: Large-scale relation learning for question answering over knowledge bases with pre-trained language models. In: Proceedings of the 2021 Conference on Empirical Methods in Natural Language Processing, pp. 3653–3660 (2021)
2. Zhang, J., et al.: Subgraph retrieval enhanced model for multi-hop knowledge base question answering. arXiv preprint arXiv:2202.13296 (2022)
3. Gu, Y., Pahuja, V., Cheng, G., Su, Y.: Knowledge base question answering: a semantic parsing perspective. arXiv preprint arXiv:2209.04994 (2022)
4. Yu, G., et al.: Beyond I.I.D.: three levels of generalization for question answering on knowledge bases. In: 2021 Proceedings of the Web Conference, pp. 3477–3488 (2021)
5. OpenAI. ChatGPT: optimizing language models for dialogue (2022)
6. Touvron, H., et al.: LLaMA: open and efficient foundation language models. arXiv preprint arXiv:2302.13971 (2023)
7. Zhang, S., et al.: OPT: open pre-trained transformer language models. arXiv preprint arXiv:2205.01068 (2022)
8. Chowdhery, A., et al.: PaLM: scaling language modeling with pathways. arXiv preprint arXiv:2204.02311 (2022)
9. Chen, M., et al.: Evaluating large language models trained on code. arXiv preprint arXiv:2107.03374 (2021)
10. Devlin, J., Chang, M.-W., Lee, K., Toutanova, K.: BERT: pre-training of deep bidirectional transformers for language understanding. arXiv preprint arXiv:1810.04805 (2018)
11. Liu, Y., et al.: RoBERTa: a robustly optimized BERT pretraining approach. arXiv preprint arXiv:1907.11692 (2019)
12. Lewis, M.: BART: denoising sequence-to-sequence pre-training for natural language generation, translation, and comprehension. arXiv preprint arXiv:1910.13461 (2019)
13. Raffel, C., et al.: Exploring the limits of transfer learning with a unified text-to-text transformer. J. Mach. Learn. Res. **21**(1), 5485–5551 (2020)
14. Wei, J., et al.: Emergent abilities of large language models. arXiv preprint arXiv:2206.07682 (2022)
15. Brown, T., et al.: Language models are few-shot learners. In: Advances in Neural Information Processing Systems, vol. 33, pp. 1877–1901 (2020)
16. Wei, J., et al.: Chain-of-thought prompting elicits reasoning in large language models. In: Advances in Neural Information Processing Systems, vol. 35, pp. 24824–24837 (2022)
17. Li, T., Ma, X., Zhuang, A., Gu, Y., Su, Y., Chen, W.: Few-shot in-context learning for knowledge base question answering. arXiv preprint arXiv:2305.01750 (2023)

# A Military Domain Knowledge-Based Question Answering Method Based on Large Language Model Enhancement

Yibo Liu[1(✉)], Jian Zhang[2], Fanghuai Hu[2], Taowei Li[1], and Zhaolei Wang[2]

[1] National University of Defense Technology, Wuhan 430014, China
liuyibo@nudt.edu.cn

[2] Haiyizhi Information Technology (Nanjing) Co., Ltd, Nanjing 210000, China

**Abstract.** With the rise of big language model technology, the application of big models in the military field is increasingly being valued. How to combine big language models with knowledge graph technology to improve the effectiveness of knowledge Q&A is currently a key research direction for improving military knowledge services. Based on the big language model technology, this paper implements knowledge Q&A in the military field by using template learning and template matching methods. For the key steps of knowledge linking and template matching, this paper uses the knowledge linking and semantic matching technology enhanced by the big language model. Finally, experimental verification was conducted in the test set provided by the CCKS (China Conference on Knowledge Graph and Semantic Computing) conference, and F1 reached 0.869. In summary, this paper provides a new solution for natural language Q&A in the military field using a large language model. This method achieves high accuracy while reducing dependence on training corpus data.

**Keywords:** Big language model · military knowledge Q&A · knowledge linking · template matching

## 1 Introduction

As a structured knowledge representation method, a knowledge graph can accurately capture the correlation and semantic information between entities, providing strong support for knowledge integration, query, and inference. Knowledge Q&A technology is a key technology in the field of knowledge graphs, which can match and fuse structured knowledge in the knowledge base with natural language to obtain question answers. Since the emergence of the ChatGPT large model [1], pre-training on large-scale text data has enabled the model to learn richer semantic information. Large model technology has demonstrated outstanding capabilities in semantic understanding and generation tasks. How to combine these two and utilize their respective advantages to build a powerful knowledge Q&A system, achieving intelligent management and application of massive knowledge, is a research hotspot in the current field.


H. Wang et al. (Eds.): CCKS 2023, CCIS 1923, pp. 340–352, 2023.
https://doi.org/10.1007/978-981-99-7224-1_27

The military field is a typical professional business field, with a vast and complex knowledge system and a large number of professional terms in relevant texts. The massive amount of military knowledge is difficult to simply meet diverse military needs through traditional manual retrieval and analysis, especially in the rapidly changing battlefield. There is an urgent need for more intelligent and precise knowledge acquisition and application methods. This paper conducts military domain Q&A research based on large models and knowledge graphs to address this challenge.

The main contributions of this paper are as follows:

(1) In the knowledge linking process, this paper uses small sample prompt learning technology based on a large language model to achieve automatic recognition of knowledge elements in the question, based on precise knowledge element matching. Similar calculations are embedded in the large language model to link the identified knowledge elements to the knowledge in the knowledge base.
(2) In the process of identifying knowledge elements and linking them to knowledge in the knowledge base, to improve the accuracy of knowledge linking, this paper uses the LoRa fine-tuning method to fine-tune the knowledge linking task of large language models.
(3) In the template matching process, this paper first replaces the identified knowledge elements to obtain templates with slots or user questions and then uses similar calculations based on large language model embedding to obtain the optimal template corresponding to the user question.

The remaining part of this paper is organized as follows: Sect. 2 introduces the work related to knowledge Q&A and large model technology, Sect. 3 introduces the method and model proposed in this paper, Sect. 4 introduces the experiments and results, and Sect. 5 summarizes the work of this paper.

## 2 Related Work

### 2.1 Knowledge Q and A Technology

Knowledge graph question answering technology, as one of the important research directions in the field of natural language processing, aims to organically integrate structured knowledge in the knowledge graph with natural language questions, thereby achieving intelligent question answering [2]. Knowledge Q&A connects the entities, relationships, and attributes in the knowledge graph with natural language, enabling intelligent query and inference of structured knowledge. Ultimately, complex questions raised by users are transformed into accurate answers through graph databases and query languages. In terms of technical implementation, there are currently three main types of mainstream methods [3]: rule-based question-answering methods [4], information retrieval-based question-answering methods [5], and semantic analysis-based question-answering methods [6]. In recent years, knowledge graph-based question-answering systems have achieved significant results in fields such as healthcare and finance, providing new ideas for the military question answering questions.

### 2.2 Combining Big Models with Knowledge Q and A

Large model technologies such as BERT and GPT pre-train models on large-scale text data through deep learning, enabling them to learn richer semantic representations and demonstrate strong capabilities in question-and-answer tasks. These large models can understand context, make inferences, and generate fluent natural language responses. In the field of question and answer, big model technology has made significant progress in machine reading comprehension, dialogue generation, and other areas.

By combining a large model with a knowledge graph, the accuracy and intelligence level of a question-answering system can be further improved. By deeply analyzing the semantics and context of the question, the large model can accurately identify the entity information in the user's question and link it to the corresponding entity in the knowledge base, thus achieving accurate question answering. At the same time, the large model can also identify the implicit intentions of users, help determine the response strategy of the system, and provide information that is more in line with user needs. This application not only improves the intelligence level of the question-answering system but also provides users with a more efficient and accurate interactive experience.

Although instruction fine-tuning is more efficient than pre-training (only requiring the processing of fine-tuning datasets), full fine-tuning of all parameters still requires a significant amount of computational power. There are currently multiple efficient parameter tuning schemes that can significantly reduce tuning costs while achieving the same performance as full parameter tuning. Parameter Efficient Fine Tuning (PEFT) can only fine-tune small or additional model parameters and fix most of the pre-training parameters, greatly reducing training costs [7]. Current research shows that compared to full parameter tuning, PEFT performs slightly worse on large language models that have not been tuned, but its performance is close to models that have already been tuned.

PEFT methods are mainly divided into three categories: Adapter Tuning, Prefix Tuning [8], and P-Tuning [9, 10].

## 3 Method

To construct an efficient and accurate military domain knowledge Q&A method based on large language model technology, this paper will provide a detailed introduction to the overall architecture of the method, mainly including building the Template and fine-tuning LLM, template matching for the question, and CQL generation, as shown in Fig. 1, Fig. 2, and Fig. 3.

### 3.1 Building the Template and Fine-Tuning LLM

The goal of template construction is to extract common question templates from training data to cover diverse questions in military Q&A. These question templates cover various question types, entities, relationships, and attributes. Through in-depth analysis and careful organization of training data, we can extract a series of question templates, providing strong support for the subsequent template-matching stage. Template learning mainly includes the following tasks:

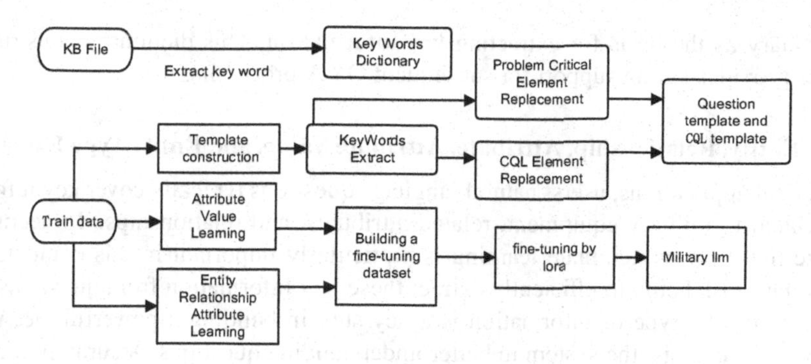


**Fig. 1.** Building the Template and fine-tuning LLM

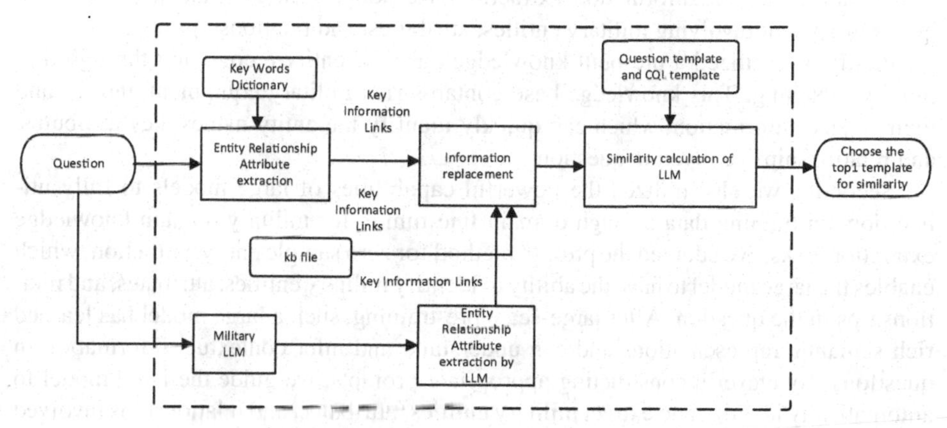


**Fig. 2.** Template matching for the question

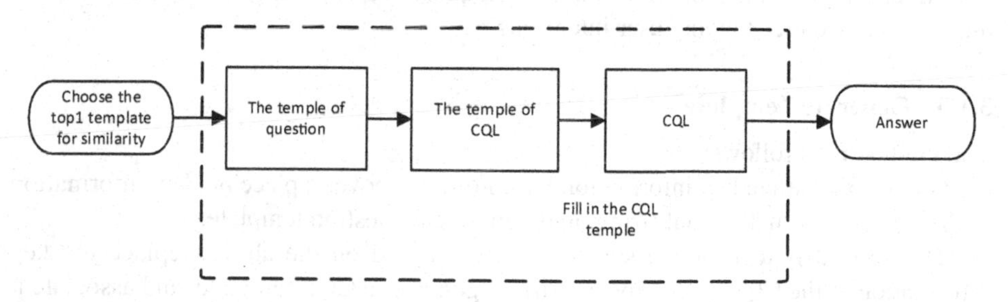


**Fig. 3.** CQL generation

### 3.1.1 Building a Key Information Dictionary Based on Graph Files

Establishing a dictionary of key elements helps to capture key information in the question more accurately. By parsing the graph file, we can extract entity names, entity types, relationships, attributes, attribute values, and corresponding association relationships from the graph. Organize this key information into key-value pairs and store them in a dictionary as the basis for extracting key information. This dictionary construction process provides strong support for subsequent Q&A processing.

### 3.1.2 Entity, Relationship, Attribute, Attribute Value, and Entity Type Extraction

In practical applications, users' natural language questions typically cover key information related to military equipment, related attributes, and relationships. To accurately analyze these issues, template learning is particularly important in this context, with the primary task being to efficiently extract these key information from questions. The extraction of this type of information is a key step in building a powerful Q&A system, which can assist the system in better understanding questions, accurately locating entities, and ultimately generating accurate and accurate answers.

In the process of information extraction, we adopt various strategies to capture question content involving military entities, attributes, and relationships.

Firstly, we utilized a pre-built knowledge base for entity recognition through dictionary matching. This knowledge base contains rich military equipment entities and their related information, which can quickly identify the entity names, key attributes, and relationships involved in questions.

Secondly, we also utilized the powerful capabilities of large models to fully utilize domain training data through domain fine-tuning for military domain knowledge extraction tasks. We adopted the prompt method for zero sample entity extraction, which enables the large model to have the ability to identify military entities, attributes, and relationships in the question. After large-scale pre-training, such a large model has learned rich semantic representations and can understand and infer contextual information in questions. By cleverly constructing appropriate prompts, we guide the large model to automatically identify and extract military entities, attributes, and relationships involved in the question based on the context of the question. This innovative method enables our system to extract useful information from questions without clear training samples, thereby better-understanding user intentions.

### 3.1.3 Generate Template

The method is as follows:

One is to generate key information based on the above, replace the key information in the question with a special string, and generate a question template.

The second is to generate key information based on the above, replace the key information in the CQL with a special string, generate a CQL template, and associate it with the question template to save.

### 3.1.4 Fine Tuning of Military Large Models

Despite the excellent processing and learning capabilities of the universal large model, it may not necessarily be the best choice in certain specific, vertical industry application scenarios. Therefore, in response to this situation, industry-customized large models are gradually emerging. These industry-customized large models have been optimized and adjusted specifically for specific business scenarios and applications to provide better performance in specific contexts.

Furthermore, the limitations of large models are not always fundamental issues. Although large models exhibit excellent performance on widely used data and tasks, their effectiveness may not be as expected in specific and highly personalized scenarios.

Based on the above considerations, this study conducted a fine-tuning operation on the general large model to create an industry-specific large model suitable for the military industry. Through this step, we aim to better meet the needs of specific fields and provide more precise and efficient solutions for Q&A tasks in the military industry.

(1) Building a fine-tuning dataset

When constructing a large-scale military domain model based on graph Q&A, we adopted a series of key steps aimed at fully utilizing large-scale model technology. Firstly, we constructed a fine-tuning dataset for text extraction and text linking based on rich training data to ensure that the model performs well in military applications.

The dataset we constructed mainly includes two aspects, namely key element extraction data and entity link data. In the process of constructing key element extraction data, we borrowed Chatie's technical route and constructed effective prompts, and extracted key information by training the question text and key elements in the data. This step helps guide the model to better understand the question and accurately extract the required information from the text.

On the other hand, the construction of entity link data is carried out through training files. We compared the text in the question with the text in CQL to construct a dataset of entity links. This step aims to establish an association between the semantic information in the question and the entities in the graph, providing support for subsequent knowledge queries.

(2) Large model fine-tuning

The fine-tuning method is a strategy for fine-tuning a pre-trained large language model on a natural language format instance set. This method is closely related to supervised fine-tuning and multi-task prompt training methods. In the fine-tuning process, we first need to collect or construct instances that are suitable for the instruction format. Next, we use these formatted instances to fine-tune the large language model in a supervised manner (for example, using sequence-to-sequence loss for training). Through this instruction fine-tuning method, the large language model exhibits excellent generalization ability on tasks that have not been touched before, even in multilingual environments.

Although fine-tuning is a more efficient method compared to pre-training (as only the fine-tuning dataset needs to be processed), full parameter instruction fine-tuning still requires significant computational resources [11]. However, there are currently multiple efficient parameter tuning schemes that can significantly reduce the cost of tuning while maintaining the same performance as full parameter tuning.

In this study, we used the LoRa fine-tuning method [12]. The basic idea is to add a channel next to the original pre-training weights, which is processed through dimensionality reduction and dimensionality enhancement operations [13]. During the training process, we maintain the pre-training weights unchanged and only adjust the dimensionality reduction matrix A and the dimensionality increase matrix B. However, the input and output dimensions of the model remain unchanged, and when outputting, the dimensionality reduction matrix B and dimensionality increase matrix A are overlaid with the parameters of the pre-trained weights. In the actual fine-tuning process, the dimensionality reduction matrix A and dimensionality increase matrix B are usually at the MB level, while the weight of large language models is usually at the GB level. By using the LoRa fine-tuning method, we can effectively reduce the computational burden during the fine-tuning process while maintaining high performance. The fine-tuning process of this paper is shown in Fig. 4.

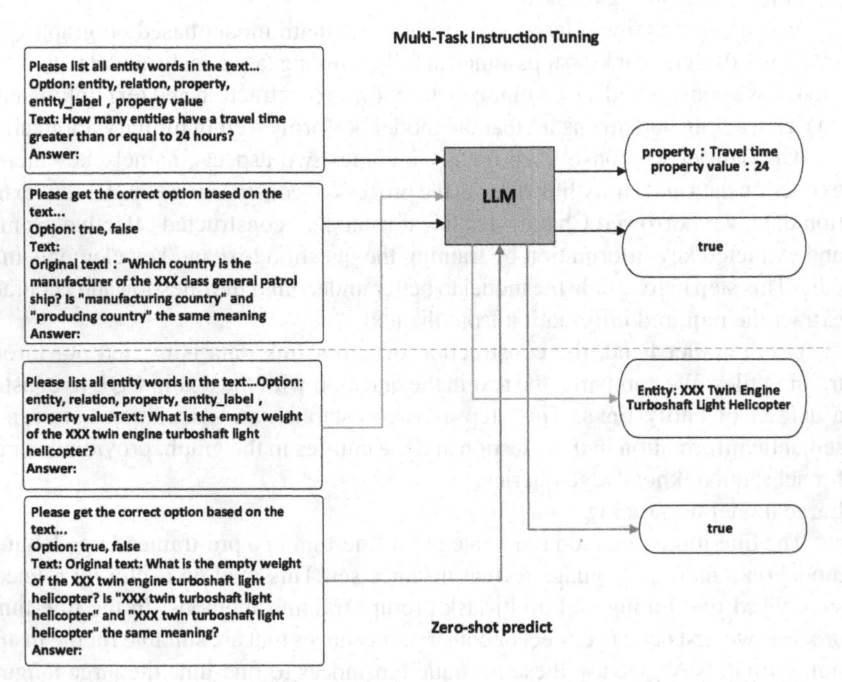


**Fig. 4.** Fine-tuning process of information extraction and knowledge linking in the military field for a large model

## 3.2 Template Matching for the Question

In the template matching stage, we matched the user's natural language questions with the constructed question template. By fully utilizing the semantic understanding ability of large models, our system can accurately identify entities, relationships, and attributes in question sentences, and map them to corresponding templates. Once the matching is successful, the system will generate corresponding CQL (Cypher Query Language) query statements for information retrieval in the knowledge graph.

This template-matching process is a meaningful connection between the user's natural language question and the existing question template. By mapping the elements in the question to the template structure, we can more accurately grasp the user's intentions, thus laying a solid foundation for subsequent knowledge graph queries. The semantic understanding ability of large models plays a crucial role here, enabling our system to efficiently analyze questions and retrieve information, providing users with accurate and satisfactory answers.

### 3.2.1 Dictionary-Based Information Extraction and Linking

We used dictionaries to extract key information in response to user input issues. This key information is matched with the information in the knowledge graph, resulting in meaningful linking results.

The extraction of key information plays a crucial role in this process. By extracting key elements from user questions, we can more accurately capture users' query intentions. Subsequently, comparing this key information with the entities, relationships, and attributes in the knowledge graph established a valuable connection with the knowledge graph. This connection process helps to provide users with relevant and in-depth answers, while also injecting higher intelligence and accuracy into our graph-based question-answering system.

Therefore, dictionary-driven key information extraction plays a crucial role in the framework of the entire question-and-answer system, providing an important bridge for the effective connection between knowledge graphs and user questions.

### 3.2.2 Extraction and Linking Based on Large Models

We use fine-tuned specialized military domain large-scale models to extract key information for the issues provided by users. The extracted key information will then be matched and associated with the information in the knowledge graph (KB) file. If no corresponding matching terms are found during the matching process, we will vectorize the extracted key information and calculate the similarity with the information vectors in the knowledge graph. In this process, we will select the key information with the highest score, construct a prompt with the extracted key information, and then input it into a large model to obtain the link results of the key information.

In this process, fine-tuning the military field's large model plays a crucial role. By fine-tuning the model to address issues in the military field, we have made it more professional and accurate. This provides strong support for the extraction of key information, enabling us to better understand the user's query intent.

Meanwhile, the vectorization and similarity calculation of key information provides us with an effective way to handle situations where direct matches cannot be found in the knowledge graph. By calculating the similarity between key information and knowledge graph information in vector space, we can find the most relevant information and construct more guiding prompts, further guiding the large model to generate accurate link results.

Overall, this method integrates fine-tuning techniques, vectorization calculations, and large-scale model processing, providing a powerful solution for key information extraction and linking based on graph question-answering systems.

### 3.2.3 Key Information Fusion and Replacement

Merge the final results in Sects. 3.2.1 and 3.2.2, and if there is a conflict, first consider trusting the results obtained in Sect. 3.2.1. Subsequently, we use special characters to replace the key information that appears in the merged question, to obtain the question template.

The main purpose of this consolidation process is to summarize the results of two independent steps to form a more comprehensive and consistent question template. In situations where there may be conflicts, we prefer to use the results of Sect. 3.2.1 as they are given a higher level of trust throughout the entire process.

We have achieved the construction of a question template by embedding the merged key information into the question text in the form of special characters. This process provides a strong foundation for subsequent template matching and question analysis, helping the system to accurately understand the user's query intention and generate more accurate answers.

### 3.2.4 Template Similarity Matching

The templates obtained in Sect. 3.2.3 will be calculated for similarity with each template in the pre-established template library. In this process, we will evaluate the similarity between each template and the resulting template, and ultimately select the template with the highest similarity score as our extraction template.

This step aims to find the template that best matches the template we generated from the existing template library. By calculating similarity, we can quantitatively compare the similarity between various templates to determine the optimal extraction template. This method ensures that we have selected the template with the most adaptability and matching degree, which helps to provide more accurate guidance for subsequent question matching and information extraction.

## 3.3 CQL Generation

In the knowledge graph question answering system, CQL (Cypher Query Language) is a query language used to query information in graph databases. To achieve effective knowledge graph Q&A, we first map the natural language questions raised by users to the corresponding CQL templates through question template matching. Next, we replace the key information in the question with fixed symbols in the CQL template to generate the final CQL query statement.

### 3.3.1 Matching Question Templates to CQL Templates

A question template is a predefined series of question types, each of which is associated with a corresponding CQL template. The question template captures the different ways and semantics of user questioning, while the CQL template defines how to construct query statements to obtain the required knowledge graph information. By matching question templates, the system can quickly select suitable CQL templates based on the type of question entered by the user, thereby reducing query space and improving query efficiency.

For example, the question template: "What are the concepts A with attribute A less than or equal to the numerical value A?" can be matched to the corresponding CQL template: "MATCH (n: 'Concept A') where n. 'Attribute A' < = numerical value A RETURN n. name".

### 3.3.2 Key Information Replacement and CQL Generation

Once the question template is successfully matched, the next key step is to fill in the key information from the question into the CQL template, which includes replacing the entities, attributes, relationships, and other information involved in the question with the corresponding positions in the CQL template. For example, the key information in the example is: {"Attribute A": "Speed", "Concept A": "Nuclear Submarine", "Value A": "10"}.

By replacing key information, we obtained a complete CQL query statement: "MATCH (n: 'Nuclear Submarine') where n. 'Speed' < = 10 RETURN n. name". It can be directly executed in the knowledge graph database to obtain information that matches user issues.

Finally, execute the CQL query statement and return the final result.

# 4 Experiments and Results

## 4.1 Experimental Data

The Experimental dataset used in this paper is an open dataset provided by the CCKS2023 Conference's Foreign Military Unmanned System Knowledge Graph Reasoning Question and Answer Evaluation Task. The training set of this dataset contains a total of 4000 question-and-answer pairs, and the test set contains a total of 1000 questions. The sample data for Q&A is shown in Table 1.

**Table 1.** The sample data for Q&A.

| No | Question sentence | CQL | Answer |
|---|---|---|---|
| 1 | Is the crew of the xxx armored vehicle 3 people? | MATCH (n) where n.name = \ "XXX armored vehicle \" RETURN n.' passengers ' = 3 | True |
| 2 | How many entities have a maximum range greater than 3900? | MATCH (n) where n.' Range' > 3900 RETURN count(n) | 200 |

## 4.2 Evaluation Indicators and Results

The CCKS conference provided 1000 questions as test data and released evaluation methods, with evaluation indicators including Macro Precision, Macro Recall, and Average

F1 value. The final ranking is based on the Average F1 value. Set it as the set of questions, the set of answers given by the contestant to the question, and the standard set of answers to the question. The relevant calculation formula is as follows:

$$\text{MacroPrecision} = \frac{1}{|Q|}\sum\nolimits_{i=1}^{|Q|} P_i, P_i = \frac{|A_i \bigcap G_i|}{|A_i|} \tag{1}$$

$$\text{MacroRecall} = \frac{1}{|Q|}\sum\nolimits_{i=1}^{|Q|} R_i, R_i = \frac{|A_i \bigcap G_i|}{|G_i|} \tag{2}$$

$$\text{AveragedF1} = \frac{1}{|Q|}\sum\nolimits_{i=1}^{|Q|} \frac{2P_iR_i}{P_i + R_i} \tag{3}$$

According to the official final evaluation results, the final rating of this experiment is shown in Table 2.

**Table 2.** The results of this experiment.

| Model | P | R | F1 |
|---|---|---|---|
| llm-kbqa | 0.86919 | 0.8722 | 0.86902 |

From the results, it can be seen that by introducing key technologies such as template construction, template matching, CQL generation, and fine-tuning entity linking and knowledge extraction using large models, our system can efficiently meet users' Q&A needs, providing strong support for knowledge acquisition and decision support in the military field.

In the template construction phase, we systematically organized the problem templates to cover a variety of problem types and scenarios. Next, in the template matching process, we intelligently match the user's natural language questions with the pre-constructed template to capture the core elements of the problem.

In the process of generating CQL, we use Cypher Query Language, a specialized query language, to construct knowledge graph query statements. This helps to accurately locate relevant information in the knowledge graph.

In addition, we have also introduced large model technology to fine-tune the system's professionalism and accuracy in the military field through entity linking and knowledge extraction. This enables our system to better understand the meaning behind the problem, effectively link to key information in the knowledge graph, and provide users with more accurate answers and decision support.

In summary, the military domain question and answer method based on large models and knowledge graphs proposed in this paper achieves efficient knowledge query and information extraction in complex fields through the fusion of multiple technical means.

## 5 Conclusion

This study delves into the implementation of a military domain knowledge base question answering system, fully utilizing the powerful capabilities of large model technology, and providing innovative solutions for solving knowledge acquisition and decision support questions in complex fields. Through the integration of key technologies such as template construction, template matching, CQL generation, entity linking of large models, and fine-tuning of knowledge extraction, we have successfully implemented an efficient Q&A system. Experiments have shown that the system performs well in handling complex questions in the military field, providing users with high-quality and accurate answers and support.

However, there are still some issues worth further research and exploration in this study. Firstly, although large model technology has achieved significant results in question-answering systems, there are still certain challenges in terms of domain adaptability and data scarcity. Future research can further explore how to further improve the performance of large models in military Q&A through more targeted data and training methods.

In addition, this study can also consider how to further optimize the construction and maintenance of knowledge graphs to adapt to the constantly changing needs of the military field. At the same time, we can explore how to introduce multimodal information, such as image and video data, to further enrich the context and content of the question, and enhance the diversity and practicality of the system.

Finally, with the continuous development of artificial intelligence technology, we can integrate graph-based question-answering systems with other advanced technologies such as natural language generation and emotion analysis to create a more intelligent and user-friendly military question-answering system, providing users with more comprehensive and personalized services.

## References

1. Antaki, F., Touma, S., Milad, D., et al.: Evaluating the performance of chatgpt in ophthalmology: an analysis of its successes and shortcomings. Ophthalmology Science 100324 (2023)
2. Bordes, A., et al.: Translating embeddings for modeling multi-relational data. Advances in Neural Information Processing Syst. **26** (2013)
3. Woods, W.A.: Progress in natural language understanding: an application to lunar geology. Proceedings of the June 4–8, 1973, national computer conference and exposition, pp. 441–450. ACM Digital Library (1973)
4. Dong, L., et al.: Question answering over freebase with multi-column convolutional neural networks. Proceedings of the 53rd Annual Meeting of the Association for Computational Linguistics and the 7th International Joint Conference on Natural Language Processing Vol. 1, pp. 260–269. Long Papers (2015)
5. Zhang, Y., et al.: Question Answering Over Knowledge Base with Neural Attention Combining Global Knowledge Information. arXiv preprint arXiv: 1606.00979 (2016)
6. Duan, N.: Overview of the nlpcc-iccpol 2016 shared task: open domain chinese question answering. Natural Language Understanding and Intelligent Applications: 5th CCF Conference on Natural Language Processing and Chinese Computing, NLPCC 2016, and 24th International Conference on Computer Processing of Oriental Languages, ICCPOL 2016, Kunming, China, December 2–6, 2016, Proceedings 24. Springer International Publishing (2016)
7. Hu, Z., et al.: LLM-Adapters: An Adapter Family for Parameter-Efficient Fine-Tuning of Large Language Models. arXiv preprint arXiv:2304.01933 (2023)

8. Li, X.L., Liang, P.: Prefix-Tuning: Optimizing Continuous Prompts for Generation. arXiv preprint arXiv:2101.00190 (2021)
9. Lester, B., Al-Rfou, R., Constant, N.: The Power of Scale for Parameter-Efficient Prompt Tuning. arXiv preprint arXiv:2104.08691 (2021)
10. Liu, X., et al.: P-Tuning v2: Prompt Tuning can be Comparable to Fine-Tuning Universally Across Scales and Tasks. arXiv preprint arXiv:2110.07602 (2021)
11. He, J., et al.: Towards a Unified View of Parameter-Efficient Transfer Learning. arXiv preprint arXiv:2110.04366 (2021)
12. Hu, E.J., et al.: Lora: Low-Rank Adaptation of Large Language Models. arXiv preprint arXiv: 2106.09685 (2021)
13. Aghajanyan, A., Zettlemoyer, L., Gupta, S.: Intrinsic Dimensionality Explains the Effectiveness of Language Model Fine-Tuning. arXiv preprint arXiv:2012.13255 (2020)

# Advanced PromptCBLUE Performance: A Novel Approach Leveraging Large Language Models

Hongshun Ling[1], Chengze Ge[1,2], Jie Wang[1], Fuliang Quan[1(✉)], and Jianping Zeng[2]

[1] Huimei Technology, Hangzhou, China
{linghongshun,gechengze,wangjie-sf,quanfuliang}@huimei.com
[2] School of Computer Science, Fudan University, Shanghai, China
zjp@fudan.edu.cn

**Abstract.** PromptCBLUE is a Chinese medical NLP benchmark that converts sixteen different medical NLP tasks into text generation task through prompt learning. It requires completing all tasks with just one large language model backbone and keeping the fine-tuning parameters under 1% of the model size, posing huge challenges. To address this issue, we have proposed an innovative approach. Initially a two-stage fine-tuning process is employed. For intricate tasks, a task-decomposition prompting construction method is introduced, enhancing algorithmic efficacy. Moreover, the integration of focus loss is explored to further amplify algorithm performance. Using this approach, a score of 72.28 was achieved in the CCKS2023 PromptCBLUE general competition track. This research contributes new insights to the advancement of large language models in the medical field.

**Keywords:** LLM · PromptCBLUE · AIGC · prompt learning

## 1 Introduction

With the introduction of ChatGPT and GPT-4, a new wave of the large language model revolution is sweeping across the globe. The launch of the PromptCBLUE benchmark has further expanded the scope of CBLUE [12] (Chinese Biomedical Language Understanding Evaluation) by encompassing 16 sub-tasks, including information extraction, medical concept normalization, medical text classification, medical dialogue understanding and generation. This compilation establishes the first large language model (LLM) benchmark tailored to Chinese medical scenarios.

The prevailing approach typically involves employing pre-trained language models as encoder or decoder components and developing distinct downstream task models or separate fine-tuning for each sub-task. However, this method

H. Ling and C. Ge—Contributed equally to this work.


H. Wang et al. (Eds.): CCKS 2023, CCIS 1923, pp. 353–361, 2023.
https://doi.org/10.1007/978-981-99-7224-1_28

necessitates constructing an individual model for each downstream task, leading to complexity, and poses challenges in enabling multi-task training across multiple models. CCKS2023 PromptCBLUE introduces a framework for instructing large language models, accomplishing the unification of Chinese medical natural language processing (NLP) tasks, and paving the way for optimizing large language models for medical applications.

In alignment with this trajectory, we have also introduced a medical-oriented large language model named HuimeiGPT. In response to the PromptCBLUE evaluation task within the context of the CCKS competition, this paper presents a diverse array of refined fine-tuning strategies tailored to the HuimeiGPT model. Notably, our model achieved a score of 72.28 on the final leaderboard of the general competition track, underscoring its pronounced efficacy.

The main contributions of this article are as follows:

- To captures shared representations from diverse subtasks and enhance performance across three sub-categories, we introduce a hierarchical two-stage fine-tuning strategy.
- To address complex tasks, we introduce a task-decomposition-based prompt learning strategy, which enhances the model's performance on these tasks.
- For information extraction tasks, we employ a multi-loss function approach to enhance the model's focus on specific entities.

## 2 Related Work

### 2.1 Medical Information Extraction

The field of medical information extraction has gained traction in enhancing clinical decision-making and research. Approaches include Named Entity Recognition (NER) for medical terms, Relation Extraction for knowledge graphs, Event Extraction for clinical events, and Deep Learning for context-rich understanding. Challenges persist in handling unstructured texts and domain variations. In recent years, medical information extraction has predominantly centered around transformer-based pre-trained language models [4,9].

### 2.2 Prompt Learning and Parameter-Efficient Tuning

With the development of large language models, many open-source models have been swiftly applied across diverse domains, including ChatGLM [2,11], LLaMA [8], etc. As the parameter scale of large language models reaches several billions, the cost of fine-tuning all parameters becomes increasingly prohibitive. This has promoted research into prompt engineering, where task instructions are formulated as natural language prompts to elicit desired model behaviors without updating parameters. For instance, the 175B-parameter GPT-3 [1] leverages prompt engineering to achieve strong performance on many NLP tasks through few-shot prompting, without any gradient update. Meanwhile, recent research shows that efficient parameter fine-tuning techniques like LoRA [3] or

P-Tuning [6,7] can yield considerable results on specific datasets. These methods often require only a small number of fine-tuned parameters to achieve significant performance gains. However, directly applying prompt engineering or efficient fine-tuning methods like P-Tuning-V2 and LoRA often leads to mediocre performance.

## 3 Methodology

CCKS released the PromptCBLUE benchmark by re-developing the CBLUE benchmark and converting 16 different medical NLP tasks into prompt-based language generation tasks, forming the first Chinese medical LLM benchmark. 94 instructional fine-tuning templates were adopted with the input field as string input to the LLM and the target field as the string output that the LLM needs to generate. The difficulty lies in completing the 16 different tasks using only one LLM backbone and keeping the total parameters of the efficient fine-tuning module under 1% of the LLM backbone.

Our methods propose three optimization approaches to enhance the performance of large language models. First, a hierarchical two-stage fine-tuning is utilized to improve the overall model effectiveness. Second, building on the optimized overall performance, two customized optimization techniques are designed for certain task characteristics: For complex tasks that can be decomposed into several simpler sub-tasks, a task-decomposition prompting method is employed to increase model scores. For tasks with critical information, a focus loss is integrated into the loss calculation.

### 3.1 Hierarchical Two-Stage Fine-Tuning

As illustrated in Fig. 1, the entire process is divided into two stages. In the first stage, we utilize HuimeiGPT as the base model, employing the LoRA [3] technique to fine-tune the model across all 16 subtasks. This initial integration captures shared representations and promotes cross-task knowledge transfer. We use the model fine-tuned by LoRA as a new base model. Subsequently, in the second stage, we capitalize on identified subtask categories. We categorize all tasks into three main classes: Information Extraction, Classification or Textual Entailment, and Others. Then we perform efficient parameter fine-tuning (P-Tuning [6,7]) separately for these three major categories. This stratified approach allows us to optimize model performance for each task type effectively. This process enables the model to capture nuanced task characteristics for each subcategory, enhancing overall performance. In the end, three P-Tuning fine-tuned models were generated, with the overall fine-tuning parameters being less than 1% of the LLM backbone parameters, meeting the competition requirements.

Experimental results underscore the effectiveness of our hierarchical two-stage fine-tuning method, showcasing superior performance compared to conventional single-stage fine-tuning and other multi-task learning approaches. This method not only advances multi-task learning but also offers insights into hierarchical modeling for diverse task sets.

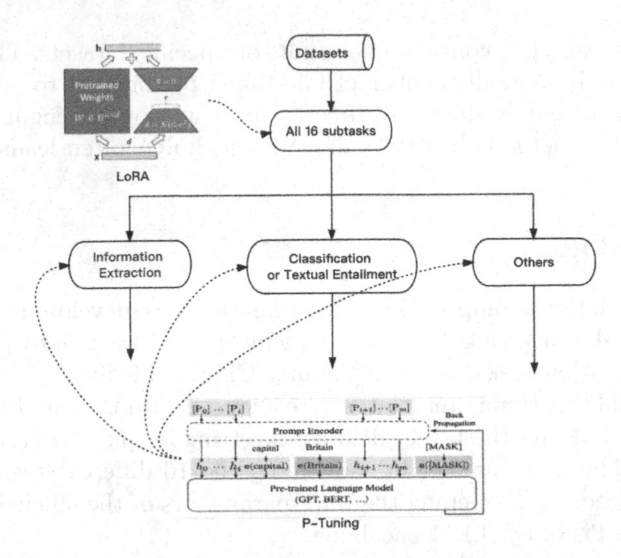


**Fig. 1.** A Two-Stage Fine-Tuning Approach

## 3.2 Task-Decomposition Prompting

Drawing insights from the work of CoT [10] and Zero-Shot CoT [5]. In the realm of complex tasks within CBLUE, like CMeIE, CHIP-CDEE, and IMCS-V2-SR, a fruitful approach involves breaking them down into two distinct sub-tasks. For instance, let's consider the IMCS-V2-SR task, which entails identifying entities like disease symptoms from conversational content and determining whether it is positive or negative at the same time.

As illustrated in Fig. 2, in the original training data, the input consists of medical dialogues between patient and doctor, while the output involves diverse symptom-related information and polarity classifications. The model's predictions may occasionally deviate from accuracy, particularly in cases where it fails to identify all relevant symptoms. To address this, we applied a task-decomposition prompting approach. Initially, the model is provided with information about four distinct symptoms, following which it is prompted to perform polarity classifications for each symptom. Remarkably, we observed that through this structured methodology, the model's performance was enhanced, even when specific symptoms were not explicitly provided within the validation data.

## 3.3 Objective Function

Token loss is a baseline loss calculation method for large language models. It involves tokenizing the label sentences into individual tokens. The cross-entropy loss is then computed between the tokens generated by the model and the tokens from the labels. This loss derived from comparing tokens one-by-one is referred to as token loss. It provides a basic training signal for language models by evaluating the token-level probability distribution. The token loss solely relies on all tokens

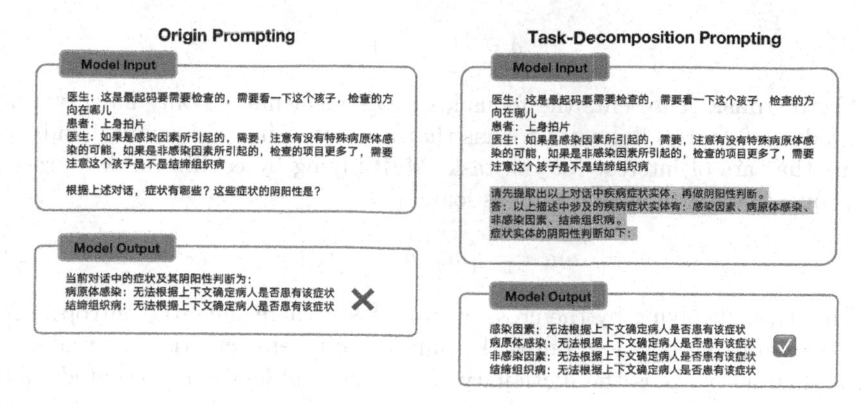


**Fig. 2.** Task-Decomposition Prompting

cross-entropy, while the enhancement stems from the assimilation of losses tied to the positions of entity and relation types. This incorporation enables the model to hone its focus on pivotal information. For instance, in the sentence: <Head and tail entity pairs with laboratory test relations are as follows: the head entity is myeloma, and the tail entity is serum calcitonin. Head and tail entity pairs with surgical treatment relations are as follows:>. The focus information is: <laboratory test> <myeloma> <serum calcitonin> <surgical treatment> (as depicted in Fig. 3).

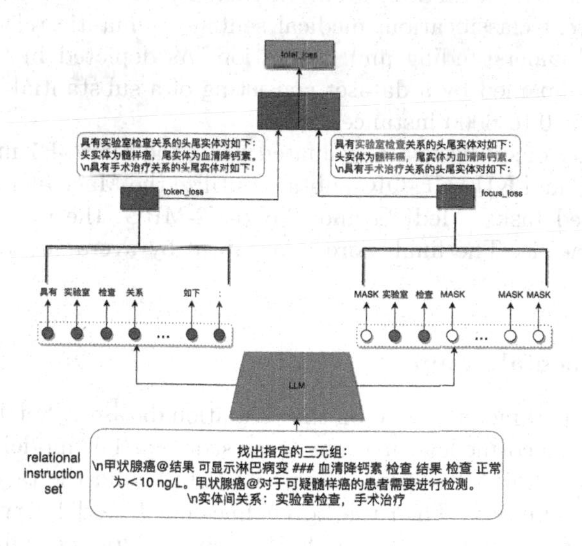


**Fig. 3.** The loss function of the model we have designed is obtained by weighted summing the token loss and the focus loss.

Vector **a** is used to represent the token loss part. Since cross entropy loss needs to be calculated for each token position, **a** is a vector of length number of tokens with all values equal to 1.

$$a = \begin{bmatrix}1\ 1 \cdots 1\end{bmatrix} \tag{1}$$

Vector **mask** represents the focus loss. The elements $m_i$ only have two possible values, 0 or 1. 0 indicates tokens that are not of interest, while 1 indicates tokens that are of interest for the task. Multiplying by coefficient $\lambda$ represents the proportional weight of the focus loss.

$$mask = \begin{bmatrix}m_0\ m_1 \cdots m_n\end{bmatrix} \tag{2}$$

Finally, multiplying by the cross entropy loss, where the cross entropy $i$ represents each token position, $len$ is the number of tokens, $c$ is the index of words in the dictionary, $M$ is the dictionary size. The total loss is represented by the following formula:

$$(a + \lambda \cdot mask) \times \sum_{i}^{len} \sum_{c=1}^{M} y_{ic} log(P_{ic}) \tag{3}$$

# 4 Experiments

## 4.1 Datasets and Evaluation Metrics

The datasets consists 16 subtasks in medical text information extraction (entity recognition, relation extraction, event extraction), medical concept normalization, medical text classification, medical sentence semantic relation judgment, medical dialog understanding and generation. As depicted in the Fig. 4, each subtask is accompanied by a dataset consisting of a substantial volume of data ranging from 3000 to 6000 instances.

The majority of subtasks are evaluated using the Micro-F1 metric, however, the CHIP-CTC and KUAKE-QIC subtasks utilize the Macro-F1 metric. For the generation-based tasks, MedDG and IMCS-V2-MRG, the evaluation employs the ROUGE metric. The final score is obtained by averaging the scores of all individual subtasks.

## 4.2 Experimental Setup

We tune hyper-parameters based on the validation datasets. Setting the learning rate to 5e-5 with a cosine learning rate decay strategy. The model was trained for 5 epochs. For the LoRA fine-tuning, rank is set to 8, and the target module was configured as <Q,K,V>. Then the LoRA fine-tuned model serves as the backbone model. For certain tasks, we apply P-Tuning-V2 for specialized fine-tuning, allowing further improvement on those task performances. P-Tuning-V2 has trainable prefix encodings that are concatenated to the key and value layers of each transformer layer. There is a hyperparameter called $prefix\ sequence\ length$ that controls the length of the prefix encodings. Through extensive experiments,

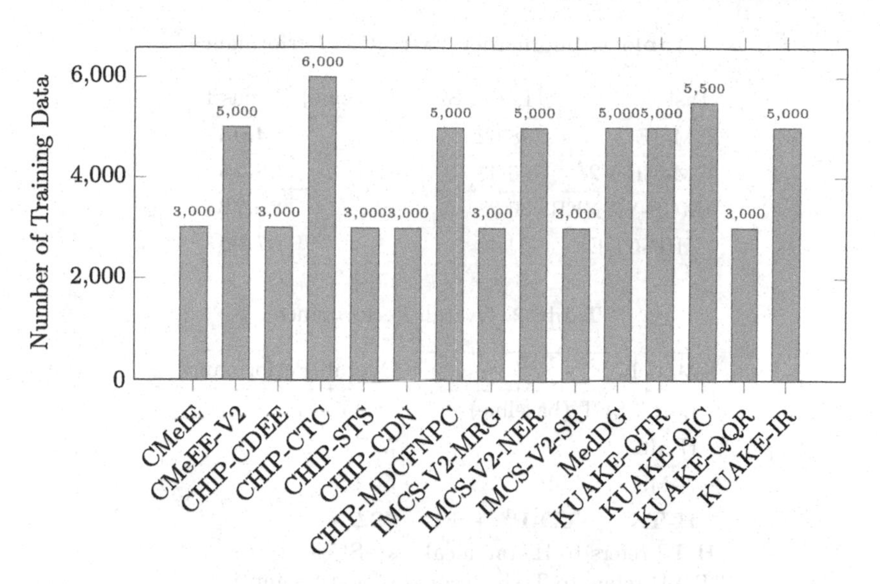


**Fig. 4.** Number of Training Data for Each Subtask

we find prefix sequence length of 128 works the best. Setting it too small makes the prefix encodings not effective enough, while setting it too large makes the attention mechanism focus too much on the prefix encodings and too little on the original key and value layers, both leading to inferior performance.

### 4.3 Medical Information Extraction

For the medical information extraction task, we optimized the model objective function and proposed adding two losses, as Eq. 3 shown. How to set the value of the $\lambda$ is the key issue. Through experimentation, we found the optimal performance with $\lambda$ of Focus Loss set to 0.25. Setting $\lambda$ too high causes the model to overfocus on the salient information, inadvertently disregarding the instruction template in the output. On the other hand, setting $\lambda$ too low is equivalent to solely computing the token loss. Experimental results demonstrate that our focus loss strategy significantly enhances the performance of such tasks, as illustrated in Table 1.

### 4.4 Overall Experiments Result

Experimental overall performance are presented in Table 2. In the baseline, only LoRA is used for supervised fine-tuning, with an average score of 68.25 on 16 tasks. First, using hierarchical two-stage fine-tuning improves the average score to 70.57 (+2.32). Second, based on the task-decomposition prompting, further improving the average score to 71.46 (+0.89). Finally, incorporating focus loss, the average score is fixed at 72.28 (+0.82).

**Table 1.** Information Extraction Performance

| Task | LoRA SFT (baseline) | Focus Loss |
|---|---|---|
| CMeIE | 48.62 | **54.96** |
| CMeEE-V2 | 66.43 | **68.25** |
| IMCS-V2-NER | 86.28 | **89.78** |
| CHIP-CDEE | 64.43 | **67.05** |

**Table 2.** Overall Performance

| Model | Overall Performance |
|---|---|
| LoRA SFT (baseline) | 68.25 |
| H-MS | 70.57 |
| H-MS + TD-P | 71.46 |
| **H-TS[a] + TD-P[b] + FL[c]** | **72.28** |

[a]H-TS refers to Hierarchical Two-Stage.
[b]TD-P refers to Task-Decomposition Prompting.
[c]FL refers to Focus Loss.

## 5 Conclusion

This paper proposes a novel approach for the PromptCBLUE benchmark, utilizing the large language model (LLM) developed by Huimei Technology. Firstly, model performance is enhanced through the construction of a hierarchical two-stage fine-tuning process. Secondly, for intricate tasks, a task-decomposition prompting construction method is introduced. Finally, the incorporation of an important word loss mechanism is integrated into the model's loss calculation. This research introduces innovative ideas for the advancement of large models in the medical field. Ultimately, our model achieved a score of 72.28 on the final leaderboard of the general competition track.

## References


1. Brown, T.B., et al.: Language models are few-shot learners. In: Advances in Neural Information Processing Systems 33: Annual Conference on Neural Information Processing Systems 2020, NeurIPS 2020, 6–12 December 2020, virtual (2020). https://proceedings.neurips.cc/paper/2020/hash/1457c0d6bfcb4967418bfb8ac142f64a-Abstract.html
2. Du, Z., et al.: GLM: general language model pretraining with autoregressive blank infilling. In: Proceedings of the 60th Annual Meeting of the Association for Computational Linguistics (Volume 1: Long Papers), Dublin, Ireland, pp. 320–335. Association for Computational Linguistics (2022). https://doi.org/10.18653/v1/2022.acl-long.26. https://aclanthology.org/2022.acl-long.26
3. Hu, E.J., et al.: Lora: low-rank adaptation of large language models. In: The Tenth International Conference on Learning Representations, ICLR 2022, Virtual

Event, 25–29 April 2022. OpenReview.net (2022). https://openreview.net/forum?id=nZeVKeeFYf9
4. Kalyan, K.S., Rajasekharan, A., Sangeetha, S.: AMMU: a survey of transformer-based biomedical pretrained language models. J. Biomed. Informatics **126**, 103982 (2022). https://doi.org/10.1016/j.jbi.2021.103982
5. Kojima, T., Gu, S.S., Reid, M., Matsuo, Y., Iwasawa, Y.: Large language models are zero-shot reasoners. In: NeurIPS (2022). http://papers.nips.cc/paper_files/paper/2022/hash/8bb0d291acd4acf06ef112099c16f326-Abstract-Conference.html
6. Liu, X., et al.: P-tuning: prompt tuning can be comparable to fine-tuning across scales and tasks. In: Proceedings of the 60th Annual Meeting of the Association for Computational Linguistics (Volume 2: Short Papers), ACL 2022, Dublin, Ireland, 22–27 May 2022, pp. 61–68. Association for Computational Linguistics (2022). https://doi.org/10.18653/v1/2022.acl-short.8
7. Liu, X., et al.: GPT understands, too. CoRR abs/2103.10385 (2021). https://arxiv.org/abs/2103.10385
8. Touvron, H., et al.: Llama: open and efficient foundation language models (2023)
9. Wang, B., Xie, Q., Pei, J., Tiwari, P., Li, Z., Fu, J.: Pre-trained language models in biomedical domain: a systematic survey. CoRR abs/2110.05006 (2021). https://arxiv.org/abs/2110.05006
10. Wei, J., et al.: Chain-of-thought prompting elicits reasoning in large language models. In: NeurIPS (2022). http://papers.nips.cc/paper_files/paper/2022/hash/9d5609613524ecf4f15af0f7b31abca4-Abstract-Conference.html
11. Zeng, A., et al.: GLM-130B: an open bilingual pre-trained model. In: The Eleventh International Conference on Learning Representations (ICLR) (2023)
12. Zhang, N., et al.: CBLUE: a Chinese biomedical language understanding evaluation benchmark. In: Proceedings of the 60th Annual Meeting of the Association for Computational Linguistics (Volume 1: Long Papers), Dublin, Ireland, pp. 7888–7915. Association for Computational Linguistics (2022). https://aclanthology.org/2022.acl-long.544

# Author Index





H. Wang et al. (Eds.): CCKS 2023, CCIS 1923, pp. 363–364, 2023.
https://doi.org/10.1007/978-981-99-7224-1

Printed in the United States
by Baker & Taylor Publisher Services

Printed in the United States
by Baker & Taylor Publisher Services